Physics of Thin Films

Advances in Research and Development

VOLUME 11

CONTRIBUTORS TO THIS VOLUME

G. Hass

H. Holloway

W. R. Hunter

A. K. Jonscher

C. E. C. Wood

Physics of Thin Films

Advances in Research and Development

Edited by

GEORG HASS

*U.S. Army Electronics and Development Command
Night Vision and Electro-Optics Laboratory
Fort Belvoir, Virginia*

MAURICE H. FRANCOMBE

*Research and Development Center
Westinghouse Electric Corporation
Pittsburgh, Pennsylvania*

VOLUME 11

1980

ACADEMIC PRESS

A Subsidiary of Harcourt Brace Jovanovich, Publishers

New York   London   Toronto   Sydney   San Francisco

COPYRIGHT © 1980, BY ACADEMIC PRESS, INC.
ALL RIGHTS RESERVED.
NO PART OF THIS PUBLICATION MAY BE REPRODUCED OR
TRANSMITTED IN ANY FORM OR BY ANY MEANS, ELECTRONIC
OR MECHANICAL, INCLUDING PHOTOCOPY, RECORDING, OR ANY
INFORMATION STORAGE AND RETRIEVAL SYSTEM, WITHOUT
PERMISSION IN WRITING FROM THE PUBLISHER.

ACADEMIC PRESS, INC.
111 Fifth Avenue, New York, New York 10003

United Kingdom Edition published by
ACADEMIC PRESS, INC. (LONDON) LTD.
24/28 Oval Road, London NW1 7DX

LIBRARY OF CONGRESS CATALOG CARD NUMBER: 63-16561

ISBN 0-12-533011-1

PRINTED IN THE UNITED STATES OF AMERICA

80 81 82 83   9 8 7 6 5 4 3 2 1

Contents

CONTRIBUTORS TO VOLUME 11	vii
PREFACE	ix
ARTICLES PLANNED FOR FUTURE VOLUMES	xi
CONTENTS OF PREVIOUS VOLUMES	xiii

Preparation and Testing of Reflectance Coatings for Diffraction Gratings in the Extreme Ultraviolet

W. R. Hunter and G. Hass

I. Introduction	1
II. Coatings for Grating Ruling	2
III. Techniques for Coating Gratings	9
IV. Measuring Grating Efficiencies	17
References	33

Progress, Problems, and Applications of Molecular-Beam Epitaxy

Colin E. C. Wood

I. Introduction	36
II. Some Considerations in System Design	36
III. *In Situ* Assessment Techniques	40
IV. Initial Stages of Growth	48
V. Doping	52
VI. III–V Binaries Other Than GaAs	66
VII. Ternary III–V Alloys	69
VIII. Device Applications	75
IX. Technological Developments	86
X. Projections	92
XI. Conclusions	95
References	96

Thin-Film IV–VI Semiconductor Photodiodes

H. Holloway

I. Introduction	106
II. Techniques for Vacuum Deposition of IV–VI Layers	108

III.	Properties of IV–VI Layers on Insulating Substrates	113
IV.	Some Properties of IR Photodiodes	122
V.	Thin-Film IV–VI Photodiodes	128
VI.	Thin-Film Photodiodes for 3–5 μm Operation	157
VII.	Thin-Film Photodiodes for 8–12 μm Operation	166
VIII.	Unconventional Thin-Film Devices	168
IX.	Thin-Film Photodiode Arrays	191
X.	Conclusions	198
	References	199

The Universal Dielectric Response: A Review of Data and Their New Interpretation

A. K. Jonscher

I.	Introduction	206
II.	Definitions and Basic Relations	209
III.	Methods of Presentation of Dietectric Data	217
IV.	Basic Mechanisms of Polarization	223
V.	Empirical Classification of Loss Characteristics	226
VI.	Survey of Experimental Data	235
VII.	Currently Accepted Interpretations	287
VIII.	The Many-Body Interpretation	296
IX.	Concluding Comments	312
	References	314
	Note Added in Proof	317

AUTHOR INDEX	319
SUBJECT INDEX	329

Contributors to Volume 11

Numbers in parentheses indicate the pages on which the authors' contributions begin.

G. HASS (1), U.S. Army Electronics Research and Development Command, Night Vision and Electro-Optics Laboratory, Fort Belvoir, Virginia 22060

H. HOLLOWAY (105), Research Staff, Ford Motor Company, Dearborn, Michigan 48121

W. R. HUNTER (1), E. O. Hulburt Center for Space Research, U.S. Naval Research Laboratory, Washington, D.C. 20375

A. K. JONSCHER (205), Chelsea College, University of London, London SW6, England

C. E. C. WOOD (35), Department of Electrical Engineering and National Research and Resource Facility for Submicron Structures, Cornell University, Ithaca, New York 14853

Preface

The review articles contained in this eleventh volume of *Physics of Thin Films* cover four diverse topics.

The first review by W. R. Hunter and G. Hass discusses the preparation and testing of reflectance coatings for diffraction gratings in the extreme ultraviolet. These gratings are used extensively for spectroscopy in space research where, due to the high cost of experiments, high efficiency, stability, and uniformity are at a premium. Areas such as variation in blaze and chemical changes in coatings which degrade optical properties, and effective experimental methods used for optimizing grating performance are described.

The second article by C. E. C. Wood provides a critical evaluation of the now extensive bibliography on molecular beam epitaxy studies of III–V compounds. The main emphasis in this field thus far has been on GaAs and its solid solutions, and impressive results already have been achieved on microwave device structures made by this method. Considerable progress also has been demonstrated on novel electrooptic device structures such as light-emitting diodes, heterojunction lasers, and high-efficiency solar cells.

Important developments have occurred in vacuum-deposited IV–VI compound films, as described by H. Holloway in the third review. Here, the emphasis is entirely on the development of infrared photodetectors and the subject matter complements the earlier review in Vol. 3 of this publication by D. E. Bode on lead salt detectors. Holloway describes the preparation of thin-film photodiodes, in particular for the 3–5 μm and 8–12 μm spectral bands, and discusses some novel approaches peculiar to thin-film geometries which can be used for optimizing detector performance. These approaches include the application of interference effects (through thickness control) to increase quantum efficiency and the use of pinched-off and lateral-collection photodiode structures in order to achieve low-capacitance devices without increasing noise or degrading detectivity.

The fourth article by A. K. Jonscher presents a new theoretical treatment of a broad selection of experimental dielectric data. The treatment departs radically from the usual interpretations, which are based on the Debye model of dielectric relaxation, and outlines a unified presentation based on the concept of many-body interactions as the dominant mechanism in the dielectric response. Despite its general relevance to both bulk and thin film ac measurements, the Editors felt that this review represented a valuable extension of the description of dc response behavior of thin films published by Jonscher and Hill in Volume 8 of this publication.

G. Hass
M. H. Francombe

Articles Planned for Future Volumes

Preparation and Reflectance at Various Angles of Incidence of Evaporated Films for Front Surface Mirrors
  G. Hass and W. R. Hunter

Laser Coatings
  H. E. Bennett and J. M. Bennett

Superconducting Thin Films
  M. Ashkin, J. A. Gavaler, M. A. Jonocka, and J. H. Parker

Ferroelectric Films
  M. H. Francombe, S. Y. Wu, and W. J. Takei

Photoemissive Films
  G. Ghoch

Preparation and Properties of Thin-Film Solar Cells by Spray Pyrolysis
  K. L. Chopra

Contents of Previous Volumes

Volume 1

Ultra-High Vacuum Evaporators and Residual Gas Analysis
Hollis L. Caswell

Theory and Calculations of Optical Thin Films
Peter H. Berning

Preparation and Measurement of Reflecting Coatings for the Vacuum Ultraviolet
Robert P. Madden

Structure of Thin Films
Rudolf E. Thun

Low Temperature Films
William B. Ittner, III

Magnetic Films of Nickel-Iron
Emerson W. Pugh

AUTHOR INDEX · SUBJECT INDEX

Volume 2

Structural Disorder Phenomena in Thin Metal Films
C. A. Neugebauer

Interaction of Electron Beams with Thin Films
C. J. Calbick

The Insulated-Gate Thin-Film Transistor
Paul K. Weimer

Measurement of Optical Constants of Thin Films
O. S. Heavens

Antireflection Coatings for Optical and Infrared Optical Materials
J. Thomas Cox and Georg Hass

Solar Absorptance and Thermal Emittance of Evaporated Coatings
Louis F. Drummeter, Jr. and Georg Hass

Thin Film Components and Circuits
N. Schwartz and R. W. Berry

AUTHOR INDEX · SUBJECT INDEX

Volume 3

Film-Thickness and Deposition-Rate Monitoring Devices and Techniques for Producing Films of Uniform Thickness
Klaus H. Behrndt

The Deposition of Thin Films by Cathode Sputtering
Leon I. Maissel

Gas-Phase Deposition of Insulating Films
L. V. Gregor

Methods of Activating and Recrystallizing Thin Films of II–VI Compounds
A. Vecht

The Mechanical Properties of Thin Condensed Films
R. W. Hoffman

Lead Salt Detectors
D. E. Bode

AUTHOR INDEX · SUBJECT INDEX

Volume 4

Precision Measurements in Thin Film Optics
H. E. Bennett and Jean M. Bennett

Nucleation Processes in Thin Film Formation
J. P. Hirth and K. L. Moazed

Evaporated Single-Crystal Films
J. W. Matthews

The Growth and Structure of Electrodeposits
Kenneth R. Lawless

Thin Glass Films
W. A. Pliskin, D. R. Kerr, and J. A. Perri

Hot-Electron Transport and Electron Tunneling in Thin Film Structures
C. R. Crowell and S. M. Sze

AUTHOR INDEX · SUBJECT INDEX

Volume 5

Interference Photocathodes
D. Kossel, K. Deutscher, and K. Hirschberg

Design of Multilayer Interference Filters
Alfred Thelen

Oxide Layers Deposited from Organic Solutions
H. Schroeder

The Preparation and Properties of Semiconductor Films
M. H. Francombe and J. E. Johnson

The Preparation of Films by Chemical Vapor Deposition
W. M. Feist, S. R. Steele, and D. W. Readey

AUTHOR INDEX · SUBJECT INDEX

Volume 6

Anodic Oxide Films
C. J. Dell'Oca, D. L. Pulfrey, and L. Young

Size-Dependent Electrical Conduction in Thin Metal Films and Wires
D. C. Larson

Optical Properties of Metallic Films
F. Abelès

Interactions in Multilayer Magnetic Films
Arthur Yelon

Diffusion in Metallic Films
C. Weaver

AUTHOR INDEX · SUBJECT INDEX

Volume 7

Electron Diffraction Analysis of the Local Atomic Order in Amorphous Films
D. B. Dove

The Preparation and Use of Unbacked Metal Films as Filters in the Extreme Ultraviolet
W. R. Hunter

Properties and Applications of III–V Compound Films Deposited by Liquid Phase Epitaxy
H. Kressel and H. Nelson

Electromigration in Thin Films
F. M. d'Heurle and R. Rosenberg

Built-Up Molecular Films and Their Applications
V. K. Srivastava

AUTHOR INDEX · SUBJECT INDEX

Volume 8

Dielectric Film Materials for Optical Applications
Elmar Ritter

Inhomogeneous and Coevaporated Homogeneous Films for Optical Applications
R. Jacobsson

Discontinuous and Cermet Films
Z. H. Meiksin

Electrical Conduction in Disordered Nonmetallic Films
A. K. Jonscher and R. M. Hill

Topologically Structured Thin Films in Semiconductor Device Operation
H. C. Nathanson and J. Guldberg

SUBJECT INDEX

Volume 9

Transparent Conducting Films
J. L. Vossen

Metal-Dielectric Interference Filters

Surface Plasma Oscillations and Their Applications
H. Raether

Magnetic Bubble Films
P. Chaudhari, J. J. Cuomo, R. J. Gambino, and E. A. Giess

AUTHOR INDEX · SUBJECT INDEX

Volume 10

Spectrally Selective Surfaces for Photothermal Solar Energy Conversion
R. E. Hahn and B. O. Seraphin

The Use of Evaporated Films for Space Applications—Extreme Ultraviolet Astronomy and Temperature Control of Satellites
G. Hass and W. R. Hunter

Scattering by All-Dielectric Multilayer Bandpass Filters and Mirrors for Lasers
  Jay M. Eastman

Thin Films for Integrated Optics
  D. B. Ostrowsky and C. Vanneste

Correction of Optical Elements by the Addition of Evaporated Films
  J. R. Kurdock and R. R. Austin

AUTHOR INDEX · SUBJECT INDEX

Preparation and Testing of Reflectance Coatings for Diffraction Gratings in the Extreme Ultraviolet

W. R. Hunter

E. O. Hulburt Center for Space Research
U.S. Naval Research Laboratory
Washington, D.C.

AND

G. Hass

U.S. Army Electronics Research and Development Command
Night Vision and Electro-Optics Laboratory
Fort Belvoir, Virginia

I. Introduction	1
II. Coatings for Grating Ruling	2
III. Techniques for Coating Gratings	9
1. Cleaning	9
2. Overcoating	12
3. Effect of Coatings on Grating Performance	12
4. Intermetallic Diffusion	14
IV. Measuring Grating Efficiencies	17
1. Efficiency Maps	18
2. Efficiency vs. Wavelength	25
3. Efficiency in Conical Diffraction	28
4. Measurement of Stray Light from Gratings	30
References	33

I. Introduction

Reflecting diffraction gratings are extensively used in optical devices over a spectral region extending from the vacuum ultraviolet (VUV) to the far infrared. Their use in space research, infrared applications, and as tuning elements for lasers, among other things, has stimulated a great amount of research on coatings for gratings, and important progress toward the development of reflecting coatings with improved efficiency has been made in recent years.

By far the most widely used technique for depositing coatings is evaporation in high vacuum. With no other method can films of highest reflectance and of any desired thickness be prepared with so complete a measure of control. However, coatings for gratings also pose special problems. For example, the performance of a grating is governed, in part, by the accuracy with which the shape, or figure, of the grating surface approximates the ideal shape. Even though the blank on which the grating is to be ruled may have a sufficiently accurate figure, this figure can be altered by a nonuniform deposition on the blank, which can degrade the diffracted wavefront. The same is true in the overcoating of completed gratings. In addition, the materials used in making the grating will determine how it will be treated during coating. For example, replica and holographic gratings have plastic or photoresist, respectively, as part of the substrate. Consequently, they cannot be heated as the coating is applied and, in addition, must be cleaned quite carefully so as not to damage their delicate surface before a highly reflecting coating is applied.

Two other factors affecting the performance of gratings are the efficiency and stray, or scattered, light. If the grating is to be used in the VUV, these two factors are extremely important, especially for space experiments. Therefore, grating efficiencies must be measured so that the most efficient grating can be selected for the experiment, and suitable tests made to determine stray light levels. Both types of information are important to manufacturers of gratings so that they can improve their product.

Because this chapter is concerned with the preparation of vacuum-deposited coatings for gratings, a brief discussion of coatings in which gratings are ruled is presented, followed by an account of the problems and results encountered in over-coating gratings. Finally, techniques used for measuring grating efficiencies and stray light are described.

II. Coatings for Grating Ruling

Conventional gratings are ruled by drawing a diamond stylus across a metal film that has been vacuum-deposited on a substrate. The ruling edge of the diamond is shaped to impart *blaze* to the grooves (see Section IV,1) and forms the grooves by displacing some of the metal at and beneath the surface because of the pressure it exerts. The distance at which the bottom of the groove lies below the undisturbed surface depends on the pressure applied to the diamond, its shape, and the softness of the metal film. The film must be thick enough to prevent the diamond from penetrating to the substrate because generally the substrates are hard enough to cause rapid wear of the diamond.

There is no general rule relating the thickness of the coating to the grating characteristics. A practical rule has the coating thickness twice the groove depth, and because the groove depth depends on blaze angle and groove density (number of grooves/mm), the coating thickness will be a function of both these factors. For shallow blaze angles, 1–10°, which are usually associated with medium to large groove densities, 300–3600 grooves/mm, coating thicknesses on the order of 5000 Å are usually sufficient. Some echelle gratings, however, have only tens of grooves/mm, with much steeper blaze angles, 60–70°, and may require coating thicknesses of some micrometers. Most coatings for grating ruling are much thicker than those intended as reflecting coatings, which range in thickness from a few hundred angstroms to 1000–2000 Å, and are subject to all the problems associated with thick coatings. For example, as the thickness of a coating increases, its roughness also increases. If there are strains in the film, they can increase with thickness and may cause adherence problems. Finally there is the problem of adherence itself, even though the thickness may not be excessive. Sometimes gratings are ruled with a liquid lubricant on the surface to prevent sticking of the metal to the diamond. If adherence is poor, surface tension of the lubricant may curl the extremities of the coating, permitting the lubricant to work its way beneath the coating, destroying adherence completely. Usually adherence can be assured by using a binder layer, such as chromium or nichrome.

Because the ruling of gratings is a mechanical procedure, the most important properties of the metal film are its mechanical properties with respect to deformation and burnishing. Generally soft or malleable metals are preferred. For many years aluminum was the principal metal for grating ruling; it is still used but has been replaced to a great extent by gold. One of the drawbacks of aluminum coatings is the hard oxide film that forms on exposure to air. Although the hardness of the oxide is less than that of diamond, the oxide can cause appreciable wear of the diamond stylus during the course of a ruling. Silver has been used for ruling coarse gratings for infrared work. Another soft metal, indium, has a tendency to stick to the diamond, causing streaks on the grating surface that contribute to the stray light.

In addition to being soft, the metal must burnish well so that a smooth groove surface is produced. Figure 1 (*1*) shows the difference between the surface of an evaporated film of gold approximately 3 μm thick and the

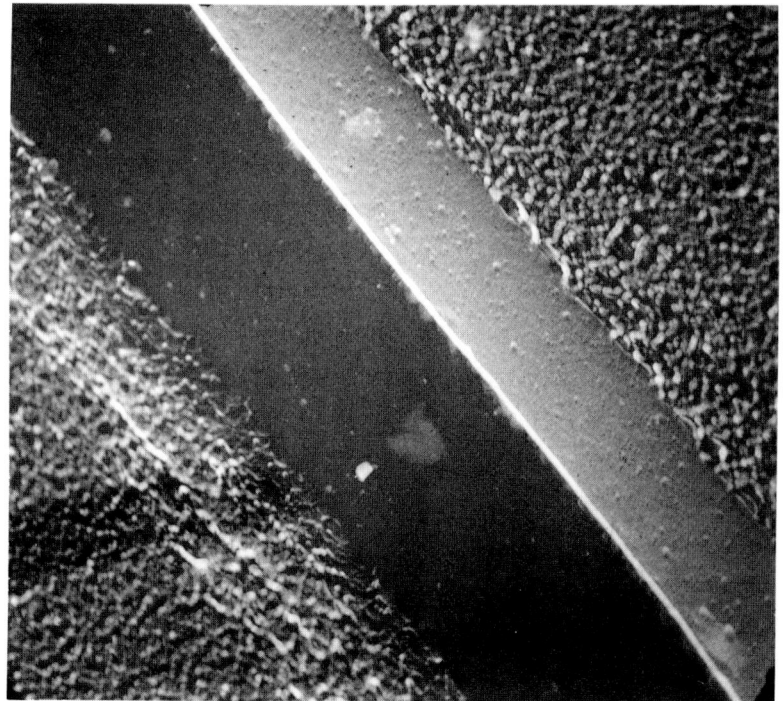

Fig. 1. A single groove ruled by a diamond tool in a gold film 3 μm thick. The groove walls are much smoother than the surface of the vacuum-deposited gold film (1).

burnished surface left by the diamond. This groove was ruled by a stylus with a 90° angle at its point that produced a groove with each side sloping 45° with respect to the unruled surface. The apparent asymmetry is caused by the shadowing technique used for electron microscopy. The photograph shows that the burnished surfaces are much smoother than the surface of the evaporated film.

The factors governing the rulability of an evaporated metal coating are not known, and the evaporation conditions required for producing good rulable films have been obtained empirically. Even so, the uncertainty is such that two coatings of the same metal, produced under apparently identical coating conditions, may not rule in exactly the same manner. Furthermore, some experiences have shown that coatings with good optical properties do not always rule well and that fast evaporations may not be as useful as slower evaporations. For example, fast evaporations of aluminum, with deposition rates of 500 Å/sec or more, result in coatings with optimum optical properties and that have a specular appearance even though some micrometers in thickness. However, fast evaporations of aluminum have also been known to produce films containing voids, sometimes 1500 Å or more deep, that are exposed on ruling. Aluminum has been deposited at about 40 Å/sec at a pressure of approximately 10^{-6} torr for some grating rulings (1). At this comparatively low deposition rate, films a few thousand angstroms thick have a specular appearance but as the thickness increases to micrometers, light scattering due to surface roughness causes the surface to have a milky appearance.

Deposition rates for gold are usually kept below 100 Å/sec when the source is a boat, and between 10 and 15 Å/sec when filaments are used as the source. Fast evaporations are more likely to eject small metal globules (*spit*) from the metal with a velocity large enough to reach the substrate and adhere to the coating. These small metallic spheres are sometimes smeared out by the diamond, as is shown in Fig. 2 (*1*), which results in a local *error of ruling*. They can also lift the diamond so that small areas of the grating surface remain unruled as shown in Fig. 3 (*2*). The effect of these imperfections on the stray light will depend on the spit density. Spit in gold films appears to have a higher density if the gold is evaporated from a boat, rather than a filament. It is also present in aluminum films but to a lesser degree.

Figure 4 (*2*) shows what appear to be holes in a gold coating in which a grating has been ruled. One explanation is that the substrate was not completely polished; however, it is more probable that they are similar to the voids encountered with fast aluminum evaporations. Holes of this type do not appear to interfere with the groove formation but may contribute to stray light.

It is well known that as the film thickness increases, the surface roughness also increases. This is not altogether a disadvantage for grating ruling because textured surfaces seem to have less of a tendency to stick to the diamond than very smooth surfaces and usually produce better rulings. Unfortunately, some of the surface roughness may be preserved in the grooves and contribute to stray light. Figure 5 (*1*) shows some grooves of a ruling in gold with a groove density of 313 grooves/mm, and a blaze angle of 1°20′. Generally rulings proceed toward the steep side of the grooves so that in the photograph the individual groove, separated from the others by one spacing, was ruled last. Remnants of the surface roughness can be seen at one side of the grooves, suggesting that the diamond had not touched the surface in that area. If this were true, however, the roughness should extend to the edge of the preceding

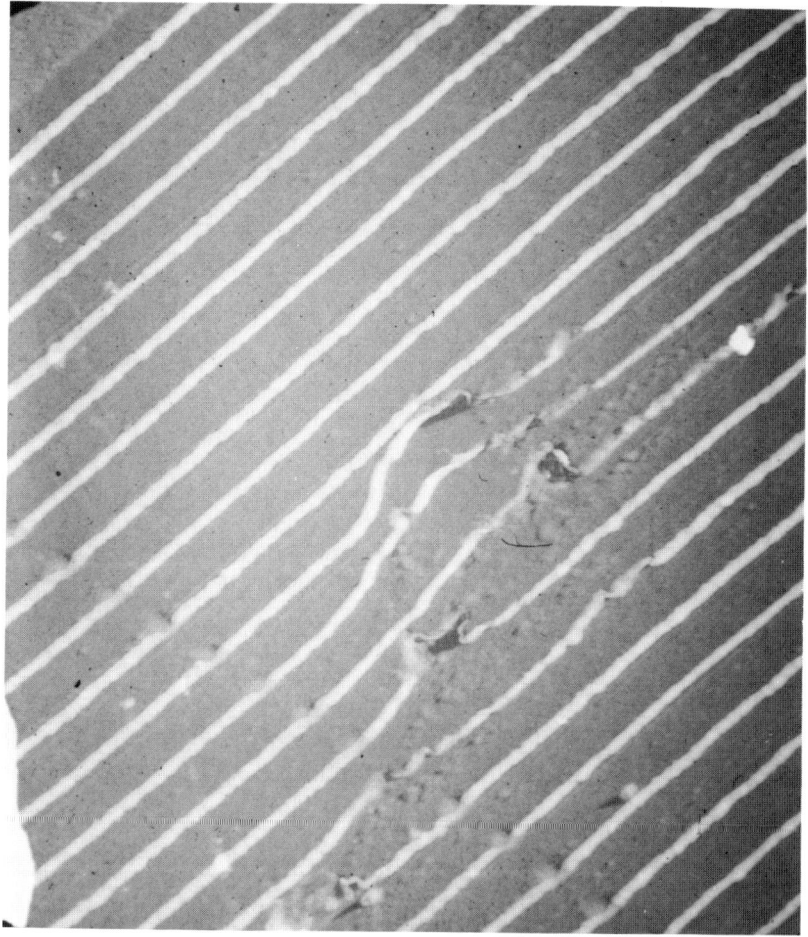

Fig. 2. A spit globule spread out by the ruling diamond has produced a local *error of ruling* (1).

groove, which it does not. Furthermore, the roughness, as judged by the number of small protruberances, is less on the shallow side of the groove than on the nonruled surface. One explanation is that the diamond displaces the gold at the steep side of the groove so that it forms a ridge somewhat above the level of the nonruled surface. This ridge is then burnished at the proper angle during the subsequent stroke of the

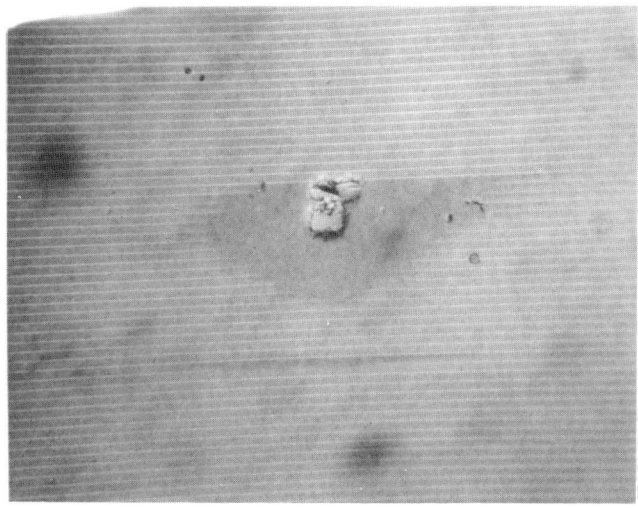

FIG. 3. A spit globule too large to be spread out by the ruling diamond lifts the diamond, leaving a nonruled area (2).

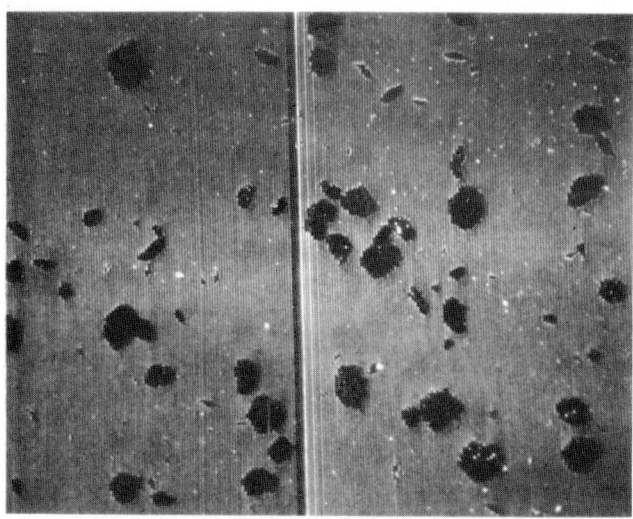

FIG. 4. Holes, or voids, in a gold coating that have not produced local error of ruling but have left nonruled areas (2).

FIG. 5. Rulings in a gold coating showing remnants of roughness at the shallow side of the groove (*1*).

diamond, but the diamond pressure, although correct for ruling the grating, is insufficient to burnish the entire groove facet. While this explanation is conjectural, it gains credence from the fact that the burnished width of the individual groove is approximately the same as the distance from the steep sides of the other grooves to the beginnings of the vestigial roughness.

Figure 5 clearly shows that some of the original surface roughness can persist in the ruled area of a grating, where it will contribute to stray light.

There have been some investigations made of metal alloys, for example

Al–Si and Au–Ge, as suitable coatings in which to rule gratings. However, the inability to reproduce films that rule in the same manner, despite the use of identical evaporation conditions, indicates the desirability of a statistical approach to the problem. Thus there must be enough ruling attempts with different coatings to produce statistically significant results. Unfortunately the economic forces governing the production of gratings conspire to keep the number of ruling attempts below the level of statistical significance.

III. Techniques for Coating Gratings

Because of the large expense in ruling a grating, manufacturers usually supply a replica, rather than an original grating. The techniques of replication do not necessarily provide a replica grating surface with highest efficiency. For example, if the replica surface is aluminum, the natural oxide will reduce the grating efficiency at wavelengths longer than 1000 Å so that an overcoating of Al + MgF_2 may be required. For shorter wavelengths a different metal coating will be required, perhaps platinum or osmium. Thus replica gratings, and also original gratings may require overcoating to ensure highest efficiency. This section outlines the techniques required to produce such overcoatings.

1. CLEANING

To assure good adherence of a vacuum-deposited coating, the substrate must be thoroughly cleaned. Cleaning glass blanks for mirror coatings is relatively simple and can be done manually by scrubbing, or chemically by rinsing with solvents, or both. Gratings, however, have a quite delicate surface. Scrubbing cannot be used and a great deal of caution must be exercised in selecting a solvent that will not damage replica or holographic gratings. The techniques for cleaning gratings have been arrived at by trial and error, and a number of these procedures are described in this section. We begin with physical descriptions of gratings, which will help to make clear the reasoning behind the selection of cleaning procedures.

a. Holographic Gratings (3). Holographic gratings are made by spinning photoresist onto the surface of a suitable shaped supporting blank, usually of glass or silica. Photoresist is a light-sensitive, liquid organic material that polymerizes or depolymerizes, depending on whether or not it is a negative or positive resist, on exposure to light of the proper wavelength. Spinning is a process wherein the supporting blank is rotated at about 4000 rpm, while a given amount of photoresist is dropped onto its center. Centrifugal force causes the resist to spread out over the blank and the result is a thin (1 μm), uniform coating of photoresist that is baked for about 1 h at 60°C to drive off all solvents and harden the resist. After these preparations, the coated blank is exposed to recombining coherent beams of light and the coating etched to remove the nonpolymerized resist. The final result is a grating in photoresist on the blank. Manufacturers usually coat the gratings with aluminum before they are delivered to a customer.

The uncoated photoresist diffraction grating is a rather delicate object. If it must be cleaned no acetone or alcohol may be used and any other solvent must be tested first to see if it dissolves the resist. The best procedure is to store the grating under clean conditions so as to avoid cleaning procedures.

Even after coating, the holographic grating must be cleaned with care and again neither acetone nor alcohol should be used. The authors have cleaned them by rinsing, or by a vapor degreasing process, with "precision grade" trichlorotrifluoroethane (TF). Holographic gratings can also be cleaned with detergent and water by submerging the grating completely and passing absorbent cotton very gently over the surface parallel to the grooves.

b. Holographic Gratings Etched into the Supporting Blank (3). The procedure for making these gratings begins as above but changes after the etching step that produces the photoresist grating. The photoresist grating is ion-etched into the underlying blank by using an ion beam that etches through the photoresist and into the blank. Thus the photoresist is completely removed. Because the etching rates in photoresist and the blank are usually different, the groove depth in the blank will differ from that in the photoresist, and so allowances for this difference are made when exposing and etching the photoresist.

Usually the supporting blank for the holographic grating is glass or fused silica; however, Choyke *et al.* (4) have reported the etching of a holographic grating with sinusoidal groove profiles into a silicon carbide substrate. SiC is material even more resistant to mechanical and chemical damage than fused silica.

A holographic grating etched into its supporting blank is very sturdy and can, in principle, survive any cleaning process that the blank can endure.

c. Master Conventional Gratings. Blanks for ruled gratings are coated as outlined in Section II. Once the blank has been ruled, the grating must be treated with care although it is probably not as vulnerable to damage as a holographic grating. Any organic solvent can be used for cleaning, as well as some acids and bases, depending on the metal in which the grating is ruled. Perhaps the only catastrophe that can befall a master grating during cleaning is the separation of the ruled coating from the blank. If, however, two metals are used for the coating, one as a binding layer for the other (for example, chromium is used as a binding layer for gold or silver to glass or fused silica), the possibility of separation is reduced considerably. Contact with ionic solutions, however, may produce electrolytic action between the two metals that will damage the grating.

d. Replica Conventional Gratings. Master conventional gratings are expensive because of the time consumed in ruling and, to a lesser extent, by the stringent requirements put on the figure of the substrate. In order to avoid this expense, master gratings are relicated, a process that places a layer of epoxy between the replica grating surface and its substrate. Because of the epoxy layer, the figure of the replica substrate need not be as accurate as that of the master substrate, a factor that reduces the cost of a replica grating.

The basic replication process is as follows. The master grating is overcoated (*in vacuo*) with a parting agent such as evaporated silicone diffusion pump oil, which is in turn overcoated with another metal layer. After removing the coated master from the vacuum coating system, its coated surface is cemented with a layer of epoxy to the replica substrate. After the epoxy has cured, the parting agent allows the master and replica to be separated, usually with no damage to either. When completed, a replica grating consists of the substrate, a thin layer of epoxy, and the metal layer that was originally evaporated onto the parting agent on the master grating.

Replica gratings can be cleaned with organic solvents if they will not attack the epoxy layer. They can also be cleaned with detergent and water, just as the holographic gratings described above.

It is also possible to replicate holographic gratings. Such replicas are cleaned in the manner described under replica conventional gratings.

For almost all cleaning procedures, the final step is a short glow discharge cleaning of the grating to be overcoated. This type of cleaning is actually a bombardment of the grating surface by electrons and ions generated at pressures of about 10^{-3} Torr with potentials of a few thousand volts. Exposure of a grating surface to such a discharge will raise its temperature. Therefore, glow discharge cleaning should be used very cautiously for holographic and replica gratings. In practice the grating is exposed to the discharge for only a minute or less. Discharges are less likely to damage master gratings and should not damage gratings etched into glass, fused silica, or SiC.

2. Overcoating

Generally coating techniques for gratings are identical with those used for coating glass blanks for mirrors. In some cases, however, the procedure must be altered slightly, in particular if the grating is likely to be heated excessively during the coating process. As mentioned in the preceding section, holographic and replica gratings are delicate and can be easily damaged by incorrect cleaning procedures. Heat can have the same effect. No systematic studies have been done to ascertain just at what temperature these two type of gratings fail to survive. The authors suggest that they should not be heated above 50°C. Such a limit precludes depositions onto heated holographic and replica gratings. Consequently, if such gratings are to be coated with platinum or iridium, they would not have maximum efficiency because these coating materials must be deposited on a hot substrate for highest reflectance (5). There are, however, metals that have highest reflectance when deposited on room temperature surfaces (40°C), for example, Os, Re, Ru, Au, and Al (5).

One should bear in mind that when high-powered electron guns are used to evaporate metals such as Os, the melt is carried to a rather high temperature during evaporation. Sometimes exposing a delicate grating surface for a time sufficient to obtain the necessary coating thickness may allow it to absorb enough heat to damage the surface.

Master gratings can survive higher temperatures without damage, as long as the temperature is kept below that at which recrystallization of the coating materials occur. There is, however, the problem of intermetallic diffusion, which is discussed in Section III,4.

Gratings etched into glass, fused silica, or SiC can withstand any evaporation conditions that can be applied to the substrate.

3. Effect of Coatings on Grating Performance

Most of the reflecting coatings for the VUV can be applied to gratings if the proper precautions are taken in cleaning and preparation. Such coatings are usually necessary if the gratings are to be used in the VUV because, unless specified differently, gratings are supplied with aluminum surfaces that have become oxidized through exposure to air. Reflectance studies of aluminum (6) have shown that the natural oxide that forms on aluminum when it is exposed to air causes a loss in reflectance in the VUV that becomes larger as the wavelength decreases. For example, the loss at 2000 Å is negligible, but at 1216 Å the reflectance decreases from about 90% for unoxidized aluminum to about 35% during 24 h. On further exposure the reflectance decreases even more.

Because of the poor reflecting properties of oxidized aluminum, most aluminum-surfaced gratings are coated to improve their efficiencies before they are put in use. Figure 6 (7) shows the measured efficiency vs. wavelength of a replica grating with 600 grooves/mm, radius of curvature of 40 cm, and blazed for 1200 Å. The measurements were made at near normal incidence at the center of the grating. The curve labeled "unflashed Al replica grating" is typical of the efficiencies to be expected from Al-surfaced replicas as received from suppliers. The grating was flash-coated (rapid evaporation) with aluminum and its efficiency measured at 2 h and again at 11 days after the coating was deposited. A fresh aluminum coating raises the efficiency considerably, especially at about 1600 Å. However, the increase is not permanent as the efficiency measurement after 11 days shows, because as the oxide layer thickens the reflectance drops.

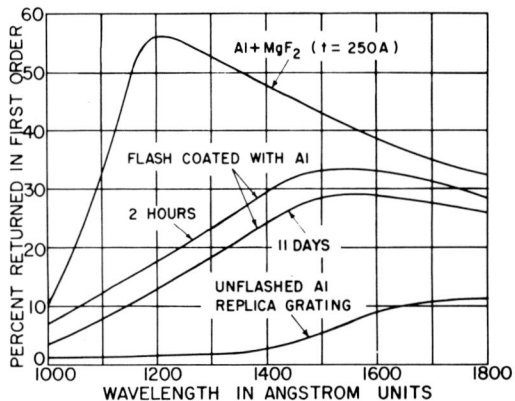

FIG. 6. Improvement in grating efficiency produced by application of a fresh aluminum coating (flash coating) and an Al + MgF$_2$ coating to an unflashed Al replica grating. The grating had a 40 cm radius of curvature, 600 grooves/mm, was blazed at 1200 Å, and was measured at its center only (7).

A permanent increase in efficiency can be obtained by overcoating the grating surface with an Al + MgF$_2$ film combination. Mirrors and gratings with this type of coating were found to retain their reflectances and efficiencies for years. The measured efficiency at 1200 Å, approximately 56%, is the maximum efficiency in the wavelength region shown. The maximum at 1200 Å is due to two causes: (1) the coating reflectance drops sharply to wavelengths less than 1200 Å, and (2) since the blaze is at 1200 Å, the efficiency drops toward longer wavelengths even though the reflectance of the coating is approximately constant at these wavelengths.

The characteristics of the preceding grating are typical of many used in the VUV. The groove spacing is large compared to the wavelength, the depth of the grooves is small compared to the groove spacing, and the angle of incidence is small. Under these conditions, scalar diffraction theory as used by Loewen and Neviere (8) can be applied to estimate the grating performance, which means that polarization effects are either small or negligible. This simplification can be used if the ratio of wavelength to groove spacing (λ/d) is less than about 0.2. If, in addition, the groove depth is shallow, which is the case for blaze angles of 15° or smaller, scalar theory can be used for values of λ/d up to 0.4. Loewen and Nevier have investigated the effects of coatings on plane gratings in the scalar domain and found that the efficiency of the grating can be predicted with good accuracy by calculating the efficiency, assuming a perfectly conducting coating, which simplifies the calculation considerably, and taking the product of this calculated efficiency and the reflectance of the overcoating. This simplified theory is also applicable to holographic gratings, subject to the same conditions.

Although the diffraction theory is only applicable to plane gratings, estimates of the performance of concave gratings can be obtained by taking into account the change of blaze across the grating surface. This is done by averaging a number of "plane" grating calculations over the range of blaze angles encountered on the concave grating surface (9).

4. Intermetallic Diffusion

One must not assume that any grating can be coated with any overcoating material. This is especially true in the case of metallic overcoatings because of the tendency of some metals to interdiffuse. Figure 7 (10) shows how interdiffusion of aluminum and gold can affect the efficiency of a grating in the VUV. This grating was a replica with a gold surface. It is coated with Al + MgF$_2$ for highest efficiency at 1216 Å

PREPARATION AND TESTING OF REFLECTANCE COATINGS 15

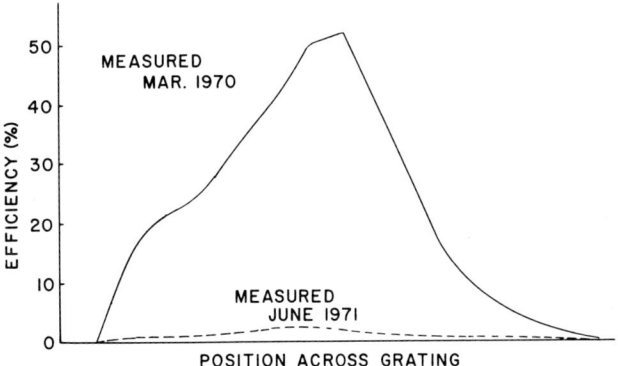

FIG. 7. Effect of intermetallic diffusion on the efficiency of a grating in the second order at 1216 Å (*10*).

and the efficiency map in the second order, labeled "Mar 1970," was obtained from measurements at 1216 Å shortly thereafter. Fifteen months later the efficiency had decreased drastically, as shown by the dashed curve labeled "June 1971." Subsequent investigation confirmed the practically complete interdiffusion of aluminum and gold in the intervening 15 months even though the grating had been kept at room temperature. At temperatures of 100°C, intermetallic diffusion of 1000 Å thick aluminum and gold films can be complete within 1 h.

Figure 8 (*10*) shows the ratio of efficiencies before to the efficiencies

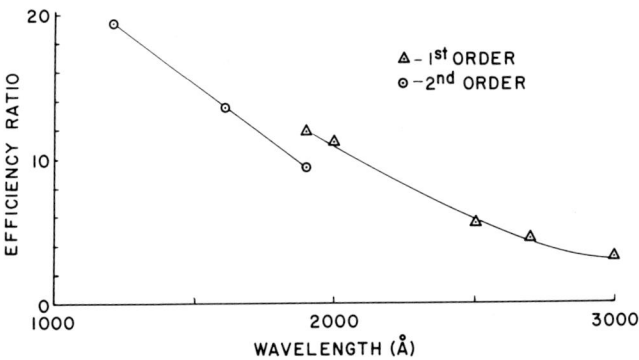

FIG. 8. Ratio of grating efficiencies before and after intermetallic diffusion of the grating of Fig. 7 as a function of wavelength (*10*).

after intermetallic diffusion as a function of wavelength. The effect was greatest at 1216 Å and decreased with increasing wavelength. The loss in efficiency is caused by the change in optical properties due to the interdiffusion as well as a roughening of the surface. Hunter et al. (10) found that a coating of $Au/Al/MgF_2$ on a plane mirror had an increase in scattering at 1216 Å of a factor of 600 after intermetallic diffusion was complete.

Intermetallic diffusion can be prevented by using a barrier layer between metals that tend to interdiffuse (10). The natural oxide that forms on an aluminum film after it is exposed to air for a few days prevents intermetallic diffusion of aluminum and gold even up to temperatures of 250°C. If Au is to be coated with Al, there is no protective oxide layer, but an evaporated coating of SiO of only 25 Å thickness can be used to prevent intermetallic diffusion.

Intermetallic diffusion between other metals can also occur, but it is usually unimportant unless the rate of diffusion is high at room temperature. For example, Pt and Al interdiffuse but the rate at room temperature is too low to cause a significant loss in reflectance over a number of years. However, after 2–3 h exposure at a temperature of 100°C, the coating appears rough to the unaided eye.

Unfortunately, very few of the materials used as coatings in the VUV or otherwise have been studied with reference to their tendency to interdiffuse, either at room temperature or elevated temperatures. Such an investigation could be very important in both space research and solar heating. There is a tendency in VUV solar spectroscopy to use compound telescopes, cassegrain or gregorian, to image the sun on the spectrograph slit. During the rocket or satellite flight, the secondary mirrors may have the equivalent of tens of solar constants or more impinging on them. The rise in temperature at the surface of these mirrors will depend on how efficiently they can be cooled, but they must be kept cool enough to prevent intermetallic diffusion if different metals are in direct contact. Cooling in space, however, is not an easy task. It can be accomplished by reradiation from the mirror and its support, or by expensive circulating liquid cooling systems, which ultimately rely on radiation cooling. Therefore, a knowledge of the interdiffusion properties of materials at different temperatures may enable scientists to design coatings that will better withstand higher temperatures so that cooling is not quite such a difficult problem.

The same problem, intermaterial diffusion, will occur with solar heat collectors used with concentrating optics although in this application the temperatures may be much higher than in space experiments because high temperatures are required for efficient operation of some solar heating systems. At high temperatures, some of the coating materials used for barrier layers may interdiffuse with metals or other dielectrics. Hence the study of interdiffusion would appear to be a fertile field of research.

IV. Measuring Grating Efficiencies

Spectroscopy in the VUV is conducted extensively in space research from rockets and satellites. The very large cost of such experiments means that every effort must be made to ensure their success. Therefore the efficiency of all spectroscope components must be measured to select those with highest efficiency. Measuring the efficiency of gratings can be a complex task. The measuring techniques have been described in the literature (*11, 12*) and are not repeated here except in a very brief outline form. Also, the results of a few measurements are given to illustrate the characteristics of gratings.

The efficiency of a grating at a given wavelength in a given order is defined as the ratio of the energy diffracted into that order to the incident energy. In principle, such a measurement is made by illuminating the grating with monochromatic radiation of measured intensity and measuring the intensity of the diffracted order. In practice, rather specialized equipment is necessary for accurate measurements. Generally a VUV monochromator with a concave grating supplies the illumination and an external device is used to hold and manipulate the test grating and radiation detector. The detector must be movable so that, for a fixed angle of incidence, it can be placed in the beam of the diffracted order to be measured, and so that it can be used to measure the intensity of the incident beam.

The efficiencies of concave gratings are seldom uniform across the grating surfaces, and so the intensity of the beam emerging from the monochromator is almost always nonuniform. If efficiency measurements are made by illuminating the entire test grating surface, the intensity distribution in the incident beam must be known and a complicated unfolding procedure used to obtain the efficiency of the test grating. However, it is sometimes desirable to know the variation in efficiency over the surface of the test grating, especially if the application may require masking of some portions of the ruled area. The problem of nonuniform illumination has been solved by illuminating only a small portion of the test grating at a time and making a number of measurements to obtain the efficiency as a function of position on the test grating surface. The result is an *efficiency map* of the grating surface.

Although most of the gratings used in the VUV are concave, plane gratings are used in Fastie–Ebert instruments and quite often test rulings are made on plan surfaces. Therefore, a grating measuring device should be capable of measuring both kinds of gratings. Since concave gratings have focusing properties and plan gratings do not, the two may require different treatment during measurement. A thorough discussion of the problems of measurements of gratings has been published by Michels *et al.* (*11*) and Hunter and Prinz (*12*).

1. Efficiency Maps

One of the most striking characteristics of concave gratings is their nonuniformity, which is usually caused by the change in blaze angle at different positions on the grating surface (*13*). A grating is said to be blazed for a given wavelength in a given order if the direction of specular reflectance of the incident ray from the groove facet coincides with the direction of the diffracted ray.

Because of the change in blaze, it is possible for the efficiency to vary from practically zero to some tens of percent within a distance of a few centimeters across the grating surface. As an example, Fig. 9 shows an efficiency map across the center of a grating that has a radius of curvature

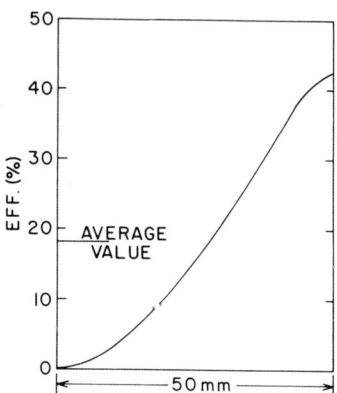

FIG. 9. Efficiency map of second order at 1608 Å across the center of a grating with a 1 m radius of curvature, 600 grooves/mm, an Al + MgF$_2$ coating, and with a blaze angle of 4°45'. Angle of incidence is 0°.

of 1 m, with 600 grooves/mm at a wavelength of 1608 Å in the second order. The angle of incidence is 0°. This grating has a ruled width of 50 mm and a blaze angle of 4°45'. It also has a coating of Al + MgF$_2$ adjusted for maximum specular reflectance at 1216 Å. The abscissa represents the width of the ruled area. At the left-hand side, the grating efficiency is 0.16% and at the right-hand side the efficiency has increased to 42.4%, an increase of a factor of 265. The average value across the surface is 18.1%.

Large changes in blaze across a grating surface may result in a grating with high efficiency in only a very small area, and the average efficiency over the grating surface can be quite small. In order to avoid large blaze changes and the corresponding low average efficiency, gratings can be ruled in three separate rulings, with the diamond set for the proper blaze at the center of each ruling. Such a ruling is called a *tripartite* ruling and Fig. 10 (*13*) shows efficiency maps at 584, 736, and 1216 Å, all in the first order. The maps show clearly (1) the tripartite nature of the ruling, (2) that the grating is blazed almost exactly at 736 Å, and (3) how the blaze shifts as the wavelength departs from the blaze wavelength.

The efficiency map shown in Fig. 9 is for a so-called monopartite ruling, one for which the orientation of the diamond is fixed during the entire ruling.

Generally as the grating radius, or the blaze angle, is decreased, the

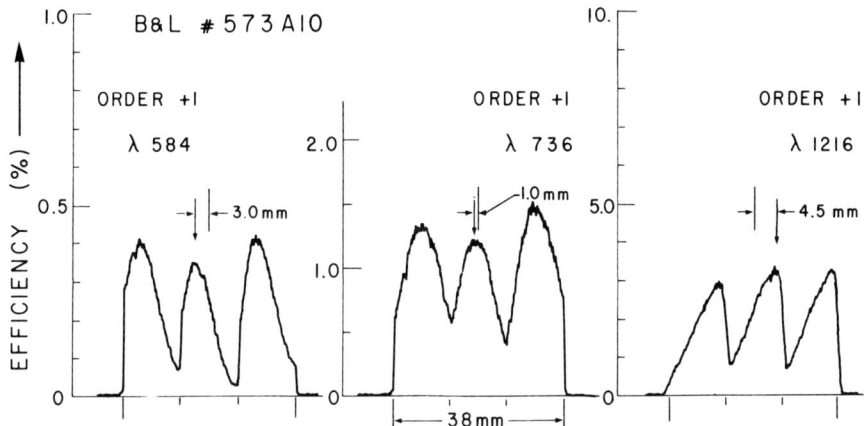

FIG. 10. Efficiency maps of a grating measured at near-normal incidence with a 40 cm radius of curvature, 600 grooves/mm, an oxidized aluminum coating, and blazed at approximately 736 Å. Maps were recorded at 584, 736, and 1216 Å, and show the tripartite nature of the ruling and how the blaze shifts as the wavelength departs from the blaze wavelength (*11*).

change in blaze becomes quite large, so that multipartite rulings are required. Because proper phasing between the sections during ruling is fortuitous, the resolving power of a multipartite grating is usually proportional to the total number of grooves divided by the number of sections.

Other factors may also contribute to the nonuniformity of efficiency across a grating surface. Figure 11 shows two efficiency maps, at 1216 Å, of a 1 m grating with 1200 grooves/mm. This grating is tripartite with a blaze angle of 2°45′, corresponding to a blaze wavelength of about 800 Å at normal incidence. The coating is platinum. The first-order efficiency, shown to the left, varies from about 8 to about 33%, with an average value of 22%. Only the center section shows the type of change in efficiency expected from the change in blaze. The left section consists of three distinct regions wherein something has occurred to change the slope of the groove facet. The right section also has three regions, the leftmost region showing approximately the expected behavior. The center region shows a minimum in efficiency, and the rightmost region a comparatively slow change in blaze. Measurements of a large number of gratings have shown that these anomalous effects are more the rule than the exception. They have never been satisfactorily explained.

Figure 11 also shows the zero order of the grating and illustrates quite clearly the complementarity of the two orders. Such complementarity is not always so obvious, especially if the diffracted energy is distributed more evenly throughout the other orders. In principle, the sum of the

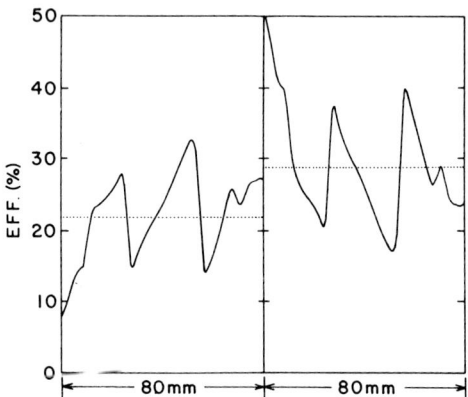

FIG. 11. Efficiency maps of zero and first orders at 1216 Å of a 1 m radius of curvature grating with 1200 grooves/mm, a blaze angle of 2°45′, and a platinum coating. The ruling is tripartite. The map shows unexpected changes in efficiency across the grating surface.

zero-order and diffracted intensities should equal the reflectance of the coating. In practice this is seldom the result because of scattering from the surface.

One possible reason for abrupt, or unexpected, changes in efficiency is the so-called *target pattern*, as shown in Fig. 12. This is a photograph in white light, of a sister grating of the one whose efficiency map is shown in Fig. 11. The target pattern appears as dark semicircular bands. No direct evidence exists to demonstrate that target patterns affect the VUV efficiency, probably because the height of the light patch that is used to measure the efficiency is too large to permit resolving the circular pattern.

Another reason for abrupt changes in efficiency is the existence of grating anomalies. If the angle of diffraction is 90°, the diffracted order proceeds parallel to the grating surface, which may cause a redistribution of radiation in the diffracted spectrum. Another type of anomaly is caused by a resonance set up in the groove itself, which is closely connected with surface plasmons. The effect of both of these anomalies is an abrupt change in efficiency and in polarization characteristics. Palmer *et al.* (*14*) have given a brief description of anomalies to which the reader is referred for both information and further references, and Hutley and Maystre (*15*)

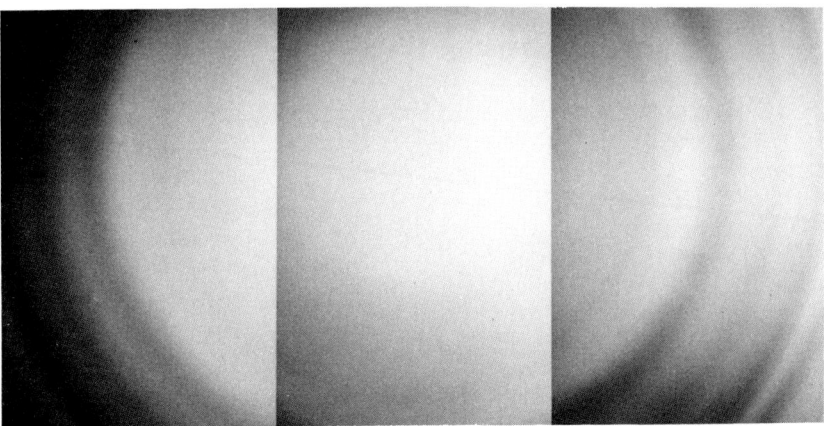

FIG. 12. Photograph of the surface of the grating of Fig. 11 in white light showing target patterns (dark circles) that may cause unexpected changes in efficiency across the grating surface. The photograph also illustrates the change in blaze in that the brightness of each panel is reduced from right to left.

have reported conditions under which diffraction gratings can be total absorbers.

Holographic gratings are recorded in a photoresist layer that has been "spun" on the appropriate substrate. The recording is done using a laser beam that is split; both resulting beams are expanded and finally recombined. In the volume where recombination takes place, interference fringes are formed. If the coated substrate is placed in this volume, the fringes can be recorded in the photoresist. The spatial distribution of energy in the fringe pattern follows a $\cos^2$ law. Therefore, when the photoresist is etched, the resulting surface consists of a number of hills and valleys that are approximately sinusoidal in profile and that are the grating grooves. The uniformity of such a grating depends on the uniformity of illumination over the area of the grating blank. If both recombining beams are uniform, the grating will have a uniform efficiency over its surface because all groove depths are approximately the same. However, if the illumination is not uniform, the groove depths will vary, being smaller where the illumination is weaker, resulting in a nonuniform efficiency.

Figure 13 (2) shows efficiency maps of the zero and two first orders of a holographic grating with an approximately sinusoidal groove profile. The wavelength is 1440 Å, the angle of incidence is 15°, the groove density is 2400 grooves/mm, and the overcoating is Al + MgF$_2$. This grating is exceptional in that the width of the recorded area is 26.8 cm and its radius of curvature is 85 cm, approximately f-3. The length of the rulings is about 8 cm. If this grating were to be ruled conventionally, a number of sections would be required to rule the entire surface. Here the two first orders are

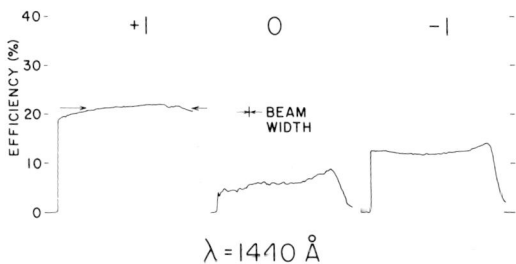

FIG. 13. Efficiency maps of the zero and two first orders of a concave holographic grating at 1440 Å. The radius of curvature is 85 cm, there are 2400 grooves/mm, an Al + MgF$_2$ coating, and the angle of incidence was 15°. The recorded area of the grating is 26.8 cm wide and the grooves are 8 cm long (2).

remarkably uniform across the recorded width, and the zero order changes by less than a factor of 2.

Groove profiles of holographic gratings need not be sinusoidal, but unless special recording techniques are used, they are generally symmetrical. As a result of this symmetry, the diffracted energy, at normal incidence, tends to be distributed approximately equally in the positive and negative orders. Figure 14 (16) shows the orders of a holographic grating at three wavelengths. The grating radius is 1 m, the coating is gold, and the groove density is 1200 grooves/mm. In contrast, blazed gratings have one or two strong orders and a number of very weak orders.

Blazed concave holographic gratings have been reported (17) that are quite efficient in the VUV. Figure 15 shows efficiency maps in the zero and two first orders of such a grating. This grating is coated with gold, has a 1 m radius of curvature, a groove density of 1200 grooves/mm, and was illuminated at an angle of incidence of 10°. The width of the recorded area is approximatly 5 cm. The negative first order is quite efficient compared to the positive first and zero orders, having an average efficiency more

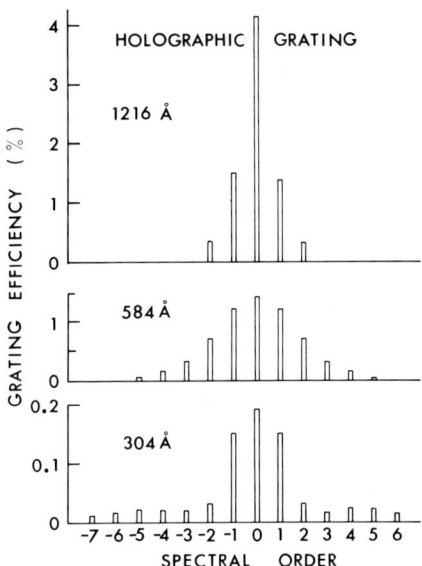

FIG. 14. Distribution of the diffracted energy from a holographic grating in different orders. The grating has sinusoidal groove profiles. The radius of curvature is 1 m, with 1200 grooves/mm, a gold coating, and the measurements were made at near-normal incidence. Measurements were made at the center of the grating only (16).

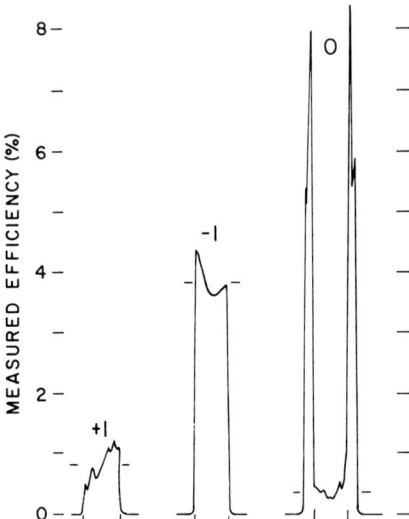

Fig. 15. Efficiency maps of a blazed, concave, holographic grating at 412 Å. Radius of curvature is 1 m, 1200 grooves/mm, the coating is gold, and the angle of incidence is 10° (*17*).

than four times that of the positive first order and about eight times that of the zero order. The two large efficiency maxima on either side of the zero order are specular reflections from parts of the grating surface where no rulings were recorded. The actual zero order lies between the two small vertical lines at the zero level of efficiency. The strong first order is not as uniform across the 5 cm width as was the grating of Fig. 13 across its 26.8 cm width. This nonuniformity may be caused by the special technique used to blaze the grating, which requires one of the beams to go through the grating blank.

At grazing incidence, gratings have the same nonuniformities as at normal incidence. Figure 16 (*18*) shows efficiency maps of a ruled and a holographic grating. The grating radii are 1 m, the coatings are gold, the groove densities are 1200 grooves/mm, the angle of incidence in each case was 80°, and the wavelength for the measurement was 192 Å. The rule grating is blazed for about 160 Å at this angle of incidence, and the first order shows, by its asymmetry, that the measurement was not made at the blaze wavelength. The second order, which has about 0.1 of the average efficiency of the first order, is quite nonuniform. At the zero order, the average efficiency is slightly larger than that of the first order.

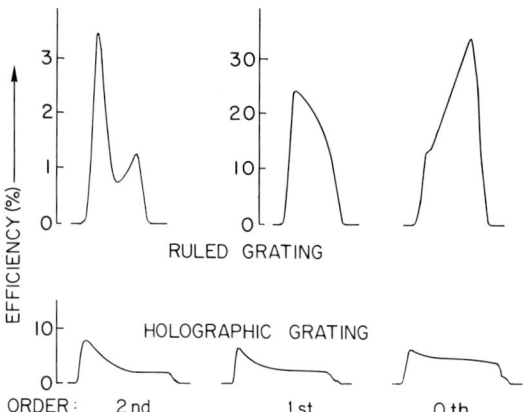

FIG. 16. Comparison of efficiency maps of a conventional (ruled) grating and a holographic grating with sinusoidal groove profiles at an angle of incidence of 80° and at 192 Å. Both gratings had gold coating, 1 m radii of curvature, and the conventional grating was blazed for about 200 Å at 80° angle of incidence (*18*).

This high efficiency in the zero order appears to be a characteristic of blazed gratings at grazing incidence, even at the blaze wavelength.

The holographic grating shows much less change in efficiency across its surface than the ruled grating—slightly more than a factor of 2 in the first order, and about a factor of 3 in the second order. This holographic grating had a sinusoidal groove profile, and measurements of the groove depth using a Talysurf machine, showed that it changed by a factor of about two across the grating; from about 500 Å, where the efficiency is highest, to 250 Å.

2. Efficiency vs. Wavelength

In order to determine grating efficiency as a function of wavelength, efficiency maps are obtained for the desired wavelengths and the average values are plotted as a function of wavelength. The wavelength dependence of these average values is the main interest of spectroscopists, who generally use the entire ruled, or recorded, area to obtain their spectra. Averaging can be done with a planimeter, or by any convenient method, and the procedure need not concern us here.

Figure 17 (*17*) shows the wavelength dependence of the zero and two first orders of a ruled tripartite grating that has a 1 m radius, 1200

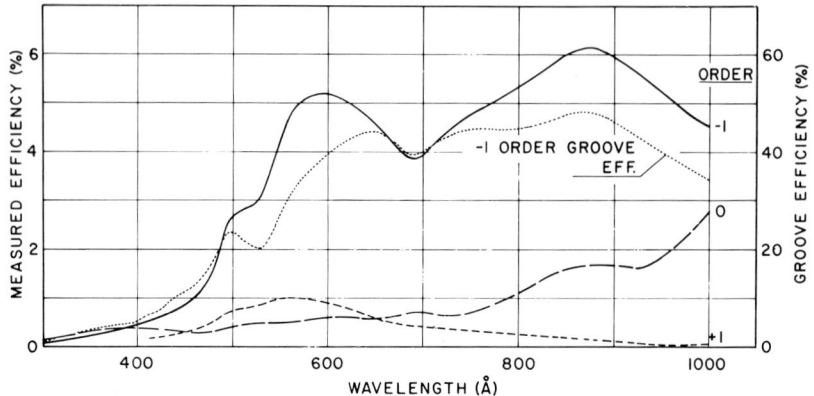

Fig. 17. Wavelength dependence of the efficiency of a 1 m, concave, gold-coated conventional grating from 200 to 1000 Å measured at 10° angle of incidence (*17*).

grooves/mm, a gold coating, and was measured at an angle of incidence of 10°. The negative first order is quite strong, compared to the other two orders, and reaches its maximum value of 6.2% just short of 900 Å. The zero order is small at the short wavelengths and rises gradually to longer wavelengths, reaching a value of about 2.7% at 1000 Å. The positive first order has a maximum value of 1% at about 560 Å and drops to very small values toward the longer wavelengths. Both first orders have maxima at 560 Å which is a characteristic of the gold coating rather than the groove shape. If the groove efficiency is plotted, as was done for the negative first order, the maximum at 560 Å disappears. Groove efficiency is defined as the measured efficiency divided by the specular reflectance of the coating. For the grating in Fig. 16, the maximum groove efficiency should occur at the blaze wavelength, about 800 Å. It is located at 870 Å and has a value of approximately 62%.

For an ideal grating there is only one maximum in the groove efficiency curve, and it is located at the blaze wavelength. However, because of the change in blaze across a concave grating, target pattern, etc., there will be fluctuations in the groove efficiency curve and the location of the blaze wavelength will not always be clear.

Figure 18 (*17*) shows the wavelength dependence of efficiency in the zero and two first orders of a concave blazed holographic grating that has the same groove density, radius, and coating as the grating of Fig. 17. A large maximum in the negative first order occurs at about 560 Å because of the reflectance spectrum of gold. The groove efficiency for this order

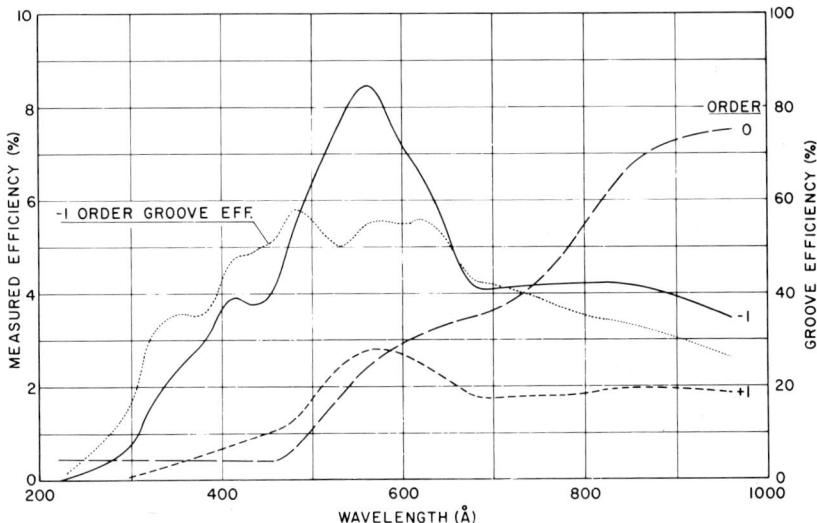

Fig. 18. Wavelength dependence of efficiency of a blazed, 1 m concave, gold-coated holographic grating from 200 to 1000 Å measured at 10° angle of incidence (*17*).

also has fluctuations but has a somewhat better defined, although broad, maximum than the grating of Fig. 17, which extends from 450 to 650 Å. Because of the broadness of this maximum, the reflectance maximum of the gold at 560 Å is somewhat more emphasized than that shown in Fig. 17. The holographic grating also has a higher maximum efficiency of about 8.5% at 560 Å. Its zero order, however, is not as small as that of the ruled grating and rises to a value of about 7.5% at close to 900 Å. Also, its positive first order is larger than that of the ruled grating. These differences in efficiencies may be due to a slightly different groove profile of the holographic grating. Although no verification exists, all indications are that the grooves do not have sharp edges as do those of the conventional grating, and this difference in groove profile may be responsible for the increase in zero and positive first orders.

Figure 19 (*18*) shows the wavelength dependence of the efficiencies in the zero, first, second, and third orders of the concave holographic grating of Fig. 16. Efficiency measurements were made at the same angle of incidence (80°) on both sides of the grating normal to verify whether or not the grooves are symmetrical. If they are, the measured efficiencies should be the same. Cross hatching helps to identify the two measurements for a given order. The differences in efficiencies are within the limit of error of

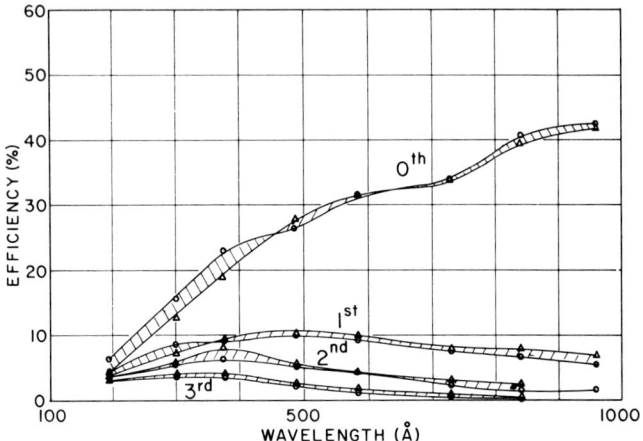

FIG. 19. Wavelength dependence of the efficiency of a 1 m concave holographic grating with sinusoidal groove profiles at 80° angle of incidence. The coating is gold and there are 1200 grooves/mm (*18*).

the measurement, and so we conclude that the groove profiles are indeed symmetrical. The first, second, and third orders show good uniformity as a function of wavelength, but their values are within an order of magnitude of each other, which means that overlapping orders might be a problem in interpreting spectra. In addition, the efficiencies are not very large.

Figure 20 (*18*) shows the zero-, first-, and second-order efficiencies of the conventional (ruled) grating of Fig. 16 (solid lines) and compares these results with the zero- and first-order efficiencies of the holographic grating of Fig. 19 (dashed lines). Here the average values for the holographic grating efficiencies are shown. Note that the conventional grating has a well-defined blaze at about 200 Å, and that the second-order value, where measured, is 10% or less than that of the first order.

The zero orders of both gratings are large, especially toward the longer wavelengths. Such behavior is characteristic of grazing incidence mountings in the VUV.

3. Efficiency in Conical Diffraction

Recently the use of gratings in conical diffraction mountings has become of interest because of the increase in efficiency over that of a conventional mounting for wavelengths less than 500 Å. The conical

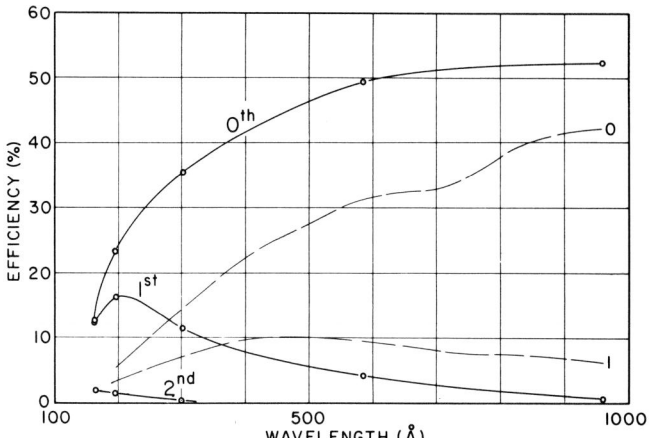

FIG. 20. Wavelength dependence of the efficiency of a 1 m concave, conventional grating measured at 80° angle of incidence (solid lines). The grating has 1200 grooves/mm and a gold coating. Shown for comparison are the zero and first orders of the holographic grating of Fig. 19 (dashed lines) (18).

mounting is a mounting wherein the incident wave vector is not perpendicular to the grating grooves as it is in the conventional mounting, but almost parallel to them. The geometry and properties of such mountings have been discussed in detail by Neviere et al. (19) and are not given here. Figure 21 (19) compares the efficiency of a plane grating used in both the conical and conventional mounting. The abscissa shows both the wavelength scale and the angle of incidence θ''. Curve 1 shows the measured efficiency in the first order from about 250 to 1050 Å using the conical mounting. Curve 4 represents the measured first-order efficiency in the conventional mounting but at the blaze condition, i.e., it shows the maximum efficiency of the grating, in the conventional type of mounting, as a function of wavelength.

At 1000 Å the conical mounting has little advantage over the conventional mounting as far as efficiency is concerned, but at 500 Å it shows an efficiency about $3\frac{1}{2}$ times greater than that measured for the conventional mounting at the blaze condition, useful in the VUV at wavelengths less than about 500 Å. These results indicate that conical diffraction will be quite useful with synchrotron, or other highly collimated, radiation in the VUV. Curves 2 and 3 show efficiency measurements at 30 and 60° from the conical position, respectively, and show that maximum efficiency is obtained with conical diffraction.

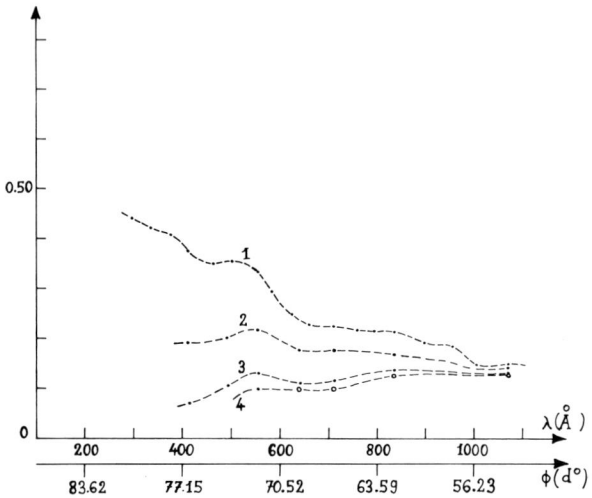

FIG. 21. Wavelength dependence of the efficiency of a gold-coated plane grating with 1800 grooves/mm, and a blaze angle of 9°19′ in the conical position (curve 1), in the conventional position at the blaze condition (curve 4), and at 30 and 60° from the conical position (curves 2 and 3, respectively) (*19*).

Only plane gratings can be used in this configuration because of the large aberrations arising from the use of spherical surfaces at large angles of incidence.

4. Measurement of Stray Light from Gratings

Stray light from gratings arises from a number of sources. Ghost spectra are spurious lines at different distances from the parent line and are caused by periodic errors in spacing of the grating grooves. Diffuse scatter, sometimes called general, or unfocused, stray light originates from small, randomly spaced, roughness on the grating surface and, as the name implies, is light scattered more or less uniformly into a 2π solid angle. Diffuse scatter and ghosts are seldom a problem in the VUV with good conventional or holographic gratings. More serious problems are caused by small random errors in groove spacing, or by roughness of the groove surface or both. Such errors give rise to a continuous background spectrum that lies in the plane of dispersion and is often referred to as focused stray light (FSL), or sometimes as "grass." An example is shown in Fig. 22 (*2*), where the spectra from two conventional gratings, identical

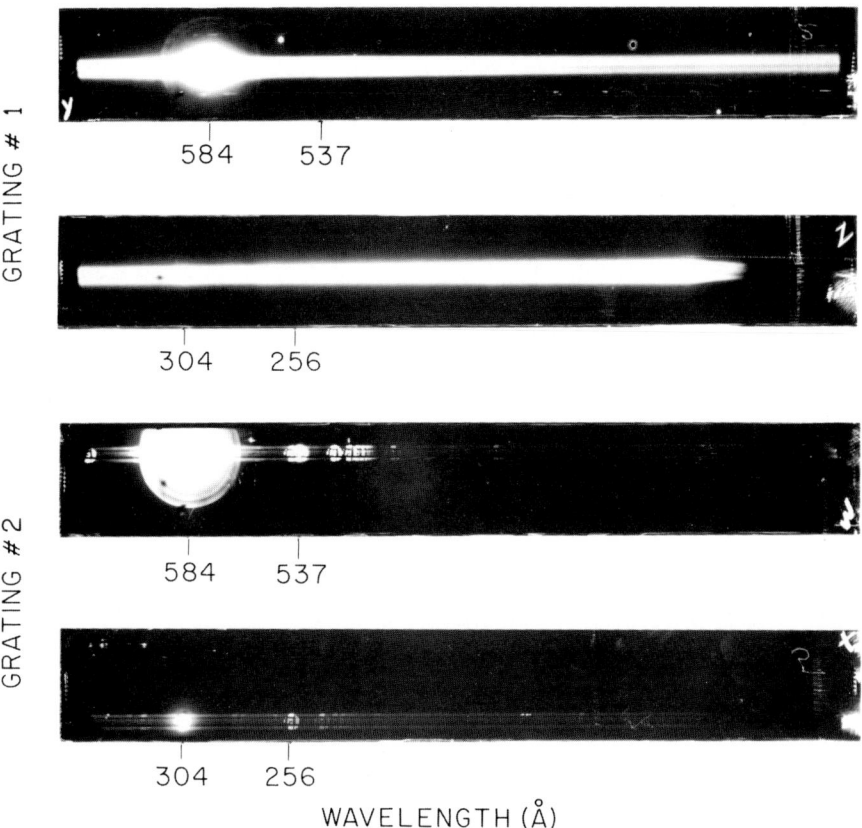

FIG. 22. Comparison of focused stray light (FSL) from two gratings (2).

other than for their FSL characteristics, are shown. These gratings were for use in Wadsworth mountings intended to record the XUV spectrum of the sun. The sun was simulated by a resolution mask, which is imaged on the film in the various lines produced by a dc glow discharge in helium. Each grating had 3600 grooves/mm, a gold coating, and a 4 m radius of curvature. A thin aluminum film filter (20) was placed just in front of the photographic film to eliminate the radiation for all wavelengths longer than 800 Å. Therefore, the pseudocontinuum obtained is actually focused stray VUV radiation. Grating 1 had an unacceptably high FSL level. In fact, it is almost impossible to distinguish the first two members of the series beginning at 304 Å. Grating 2 also has an appreciable FSL but it is far smaller than that of 1.

Because all of the grooves of holographic gratings are recorded simultaneously, and their spacing is governed by the interference field of two recombining coherent light beams, the probability of random, or periodic, errors in groove spacing is negligible. Hence these gratings are essentially free of ghosts or FSL. Conventional gratings, however, are ruled one groove at a time, thus providing ample opportunity for both random and periodic errors in spacing. By equipping ruling engines with interferometers that control the motion and position of the carriage carrying the grating during ruling, such errors are minimized but not eliminated. Consequently, it sometimes becomes desirable to measure the intensity and spectral distribution of the FSL of gratings, either for studies of the gratings themselves or to choose the best grating for a particular application.

FSL measurements in the VUV can be difficult, especially for modern conventional gratings that have a low FSL level. Figure 23 (*11*) shows results obtained from such a measurement in which the grating, light source, and detector were approximately on the Rowland circle. The center of the grating only was illuminated, and the grating was rotated about the vertical axis through its apex while the detector was held motionless.

To interpret such a measurement quantitatively, the fraction of incident radiation per angstrom of dispersion must be known. If such a measure-

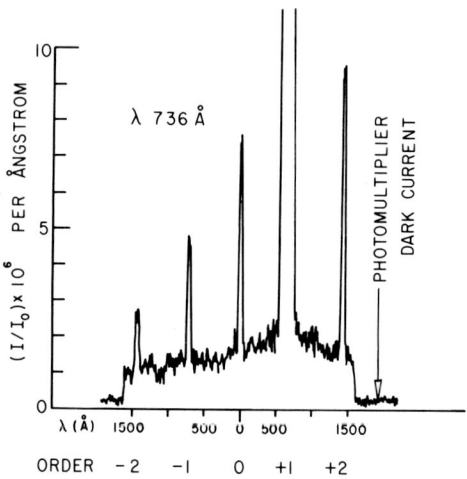

FIG. 23. Measurement of FSL between orders at 736 Å of the grating of Fig. 10 (*11*).

ment is attempted with no focused condition, one can no longer associate a definite spectral width with the width of the slit in front of the detector, and the magnitude of $I/I_0/\text{Å}$ cannot be determined. Relative measurements can be made, however, and are useful if the object of the measurements is to choose a grating with the minimum stray light or to compare gratings.

Another type of relative FSL measurement, useful if the light source is a continuum or a quasi-continuum of very closely spaced emission lines, has been reported by Johnson and Tousey (21). A grating is mounted in a spectrograph and illuminated by light from a carbon arc crater. Blackening of photographic film below the air cutoff at approximately 1850 Å is compared with blackening just above the cutoff. The greater the ratio of density at wavelengths greater than 1850 Å to the density at wavelengths less than 1850 Å, the weaker the FSL. This technique can also be used at shorter wavelengths. For example, film blackening at 800 Å, when using a glow discharge in hydrogen, will give a rough idea of the FSL level throughout the hydrogen VUV spectrum since there is no emission from hydrogen at 800 Å. If, in addition, filters are used, such as LiF or MgF_2, the film blackening below each filter cutoff can be used to obtain an approximate wavelength distribution of the FSL.

Hunter (2) has discussed some of these techniques and has shown how FSL can affect reflectance measurements in the VUV.

References

1. B. W. Bach, Hyperfine Inc., Fairport, New York (private communication).
2. W. R. Hunter, *Proc. Int. Conf. Vac. Ultraviolet Radiat. Phys., 4th, 1974* p. 633 (1974).
3. M. C. Hutley, National Physical Laboratory, Teddington, Middlesex, England (private communication).
4. W. J. Choyke, W. D. Partlow, E. P. Supertzi, F. J. Venskytis, and G. E. Brandt, *Appl. Opt.* **16**, 2013 (1977).
5. G. Hass and W. R. Hunter, *in* "Physics of Thin Films" (G. Hass and M. H. Francombe, eds.), Vol. 10, p. 72. Academic Press, New York, 1978.
6. G. Hass, W. R. Hunter, and R. Tousey, *J. Opt. Soc. Am.* **46**, 1009 (1956).
7. W. R. Hunter, *Opt. Acta* **9**, 255 (1962).
8. E. G. Loewen and M. Neviere, *Appl. Opt.* **17**, 1087 (1978).
9. M. Neviere and W. R. Hunter, *Appl. Opt.* **19**, 2059 (1980).
10. W. R. Hunter, T. L. Mikes, and G. Hass, *Appl. Opt.* **11**, 1594 (1972).
11. D. J. Michels, T. L. Mikes, and W. R. Hunter, *Appl. Opt.* **13**, 1223 (1974).
12. W. R. Hunter and D. K. Prinz, *Appl. Opt.* **16**, 3171 (1977).
13. D. J. Michels, *J. Opt. Soc. Am.* **64**, 662 (1974).
14. E. W. Palmer, M. C. Hutley, A. Franks, J. F. Verrill, and B. Gale, *Rep. Prog. Phys.* **38**, 975 (1975).

15. M. C. Hutley and D. Maystre, *Opt. Commun.* **19,** 431 (1976).
16. T. Namioka and W. R. Hunter, *Opt. Commun.* **8,** 229 (1973).
17. W. R. Hunter, M. C. Hutley, P. R. Stuart, D. Rudolph, and G. Schmahl, *J. Opt. Soc. Am.* **66,** 1136 (abstr.) (1976).
18. W. R. Hunter, A. J. Caruso, and J. G. Timothy, Presented at the "Fifth International Conference on Vacuum Ultraviolet Radiation Physics," Sept. 5–9, 1977, Montpellier, France; also T. Namioka and W. R. Hunter, *J. Phys. (Paris)* **39,** C4–169 (1978).
19. M. Neviere, D. Maystre, and W. R. Hunter, *J. Opt. Soc. Am.* **68,** 1106 (1978).
20. W. R. Hunter, *in* "Physics of Thin Films" (G. Hass, M. H. Francombe, and R. W. Hoffman, eds.), Vol. 7, p. 43. Academic Press, New York, 1973.
21. F. S. Johnson and R. Tousey, *J. Opt. Soc. Am.* **38,** 1103 (abstr.) (1948).

Progress, Problems, and Applications of Molecular-Beam Epitaxy

COLIN E. C. WOOD

*Department of Electrical Engineering and National Research
and Resource Facility for Submicron Structures
Cornell University
Ithaca, New York*

I. Introduction	36
II. Some Considerations in System Design	36
III. *In Situ* Assessment Techniques	40
1. Surface Analytical Facilities	40
2. Real-Time Analysis	41
IV. Initial Stages of Growth	48
1. Substrate Preparation	48
2. Buffer Layers	49
V. Doping	52
1. Unintentional Impurities and Deep Levels	52
2. N-Type Dopants and Their Problems	57
3. P-Type Dopants and Their Problems, Including Ion Implantation Doping	63
VI. III–V Binaries Other Than GaAs	66
VII. Ternary III–V Alloys	69
1. III, III'–V Alloys	70
2. III–V, V' Alloys	72
3. Quaternary Alloys	74
VIII. Device Applications	75
1. Majority Carrier Devices	75
2. Future Majority Carrier Device Applications	77
3. Minority Carrier Devices	78
4. Multilayer Structures	82
5. Heterojunction Structures	84
6. Application of MBE Periodic Structures	86
IX. Technological Developments	86
1. Metal Semiconductor Systems	86
2. Insulator–Semiconductor Systems	89
3. Selected-Area Epitaxy	89
X. Projections	92
XI. Conclusions	95
References	96

I. Introduction

Many successful scientific and practical applications of molecular-beam epitaxy (MBE) have appeared in the literature in recent years and have now established it as a viable technique despite the initial (in many cases unnecessarily high) capital investments. A testimony to its viability is the number of establishments currently utilizing this technique, at the time of writing estimated to be 65 worldwide (24 of which are in the U.S. and ~20 in Japan).

This chapter is an attempt to give an overview of some of the more interesting aspects and recent developments, and to present a somewhat subjective but critical appraisal of the fundamental limitations and as yet unsolved practical problems that are still to be overcome if the full potential of this versatile tool is to be fully exploited. Because of the author's subjectivity, certain facets have received far less than their due attention and others far more.

The mechanics of MBE are not treated in full here. To obtain background information on this and early work in the field of MBE, the reader is referred to excellent reviews and articles that have already appeared in the literature.

For a comprehensive coverage on work prior to 1976 reviews by Cho and Arthur (*1*), Chang *et al.* (*2*), Chang and Ludeke (*3*), Joyce and Foxon (*4*), Chang (*5*), Farrow (*6*), and Cho (*7*) should be read.

For information on design criteria for MBE systems, papers by Luscher and Collins (*8*), Massies *et al.* (*9*), and Williams (*10*) are recommended.

The subject matter covers only compounds of the elements of the third and fifth column of the periodic table, and while it is recognized that excellent work on both silicon and compounds of the type IIB–VI, IIA, IIA′–VI and IVB, IVB′–VI has been reported, the reader is referred to relevant articles for this material. Specifically, articles by Smith and Pickhardt (*11, 12*) and others (*13–16*) on II–VI compounds, Walpole *et al.* (*17*) and others (*2, 18–25*) on IV–VI compounds and those by Ota (*26, 27*), Shiraki (*28*), and others (*29, 30*) on Si MBE are recommended.

II. Some Considerations in System Design

While the specific design of MBE systems is variable (*8, 10*), depending upon the materials to be grown and the *in situ* analysis to be carried out, there are certain features that most systems should have in common.

Some important points are generalized below. A basic good compromise design is also diagrammatically represented in Fig. 1.

Stainless-steel vacuum systems offer the best versatility and ruggedness, and for practical reasons cylindrical vacuum systems are used. Oil diffusion, ion, and turbo molecular pumps have all been employed for MBE systems. If phosphorus is to be used as a source material, ion

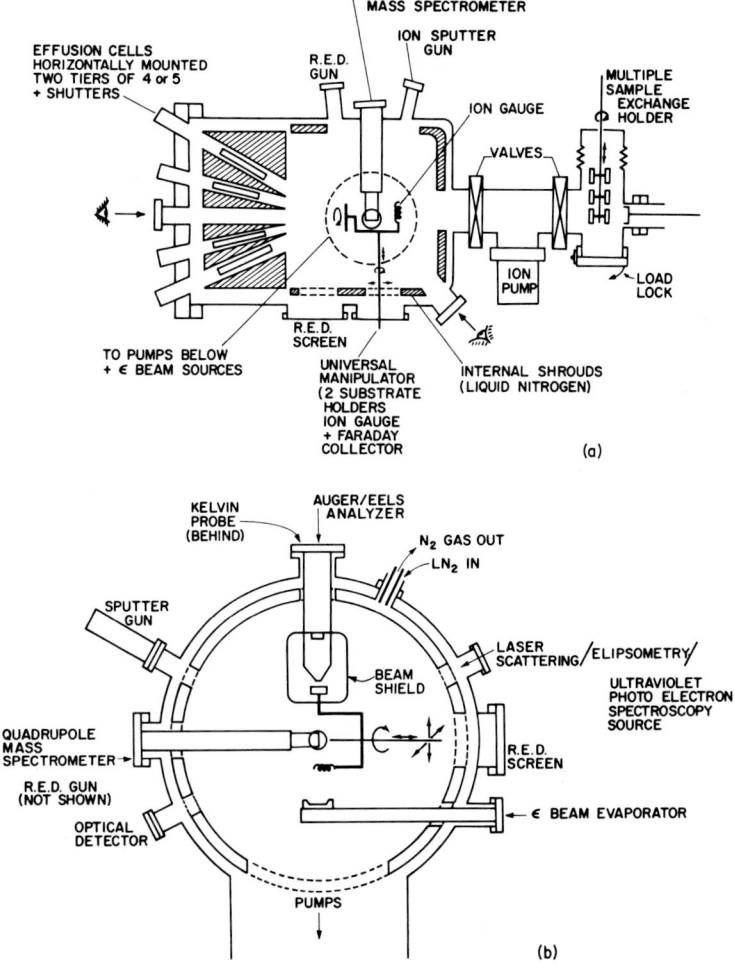

FIG. 1. Diagrammatic representation of a typical modern MBE system. (a) Plan cross section; (b) vertical cross section.

pumps offer the safest, oil- and maintenance-free pumping capability. In all cases titanium sublimation pumping with large-area liquid-nitrogen-cooled surfaces (>20,000 liter sec^{-1}) in the growth chamber is extremely desirable for efficient, rapid reduction of residual CO and H$_2$O species. The efficacy of auxiliary closed-cycle He pumping is still being tested and will be reported later. However, significant reduction in volatile hydrocarbon species has been observed by the author.

Source assemblies should be housed inside liquid-nitrogen-cooled shrouds, with their respective orifices as close as practically possible, and preferably isolated to prevent intercontamination and thermal cross talk. They should also be mounted off the vertical axis to prevent contamination from materials accumulated at the top of the chamber (*31*) over many growths.

In addition to the source shroud and titanium sublimation pump shield, as large an area of liquid-nitrogen-cooled surface as possible should be included in the growth chamber, preferably behind the substrate, to reduce background gas contamination.

Source assemblies should be made with materials that have very low vapor pressure, such as tantalum, graphite, alumina, and boron nitride, and made in such a way that no volatile interreaction products are produced at the operating temperatures, e.g., Al$_2$O$_3$ and C, Al$_2$O$_3$ and Ta (*32*). Contrary to popular opinion, the use of BN cells offers no advantage over graphite except when Al is to be evaporated. In fact, considerable difficulty is found in measuring BN effusion cell temperatures accurately.

The substrate holder should be made with considerations similar to cell assembly fabrication. However, it is necessary to use Mo blocks or plates as good thermally conducting surfaces that do not alloy readily with the In or In/Ga mixtures used to mount substrates. A very efficient large-area nonradiation type heater design that has been used for several hundred growth experiments is shown in Fig. 2. The choice of source to substrate distance depends upon the compromise of layer uniformity and growth rate that is required.

The uniformity in layer thickness or doping level (see Fig. 3), assuming axial cells with point source effusion, is given by a cosine distribution:

$$J_{\text{EDGE}} = J_{\text{CENTER}} \cos^4 \theta$$

where J_{CENTER} can to a first approximation be calculated from the Knudsen equation

$$J = 1.11 \times 10^{22}[aP/r^2(MT)^{1/2}] \text{ molecules/cm}^2/\text{sec}$$

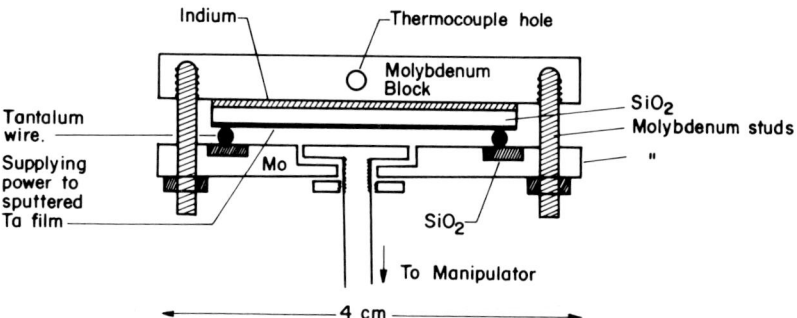

FIG. 2. Efficient conduction-type large-area substrate holder–heater.

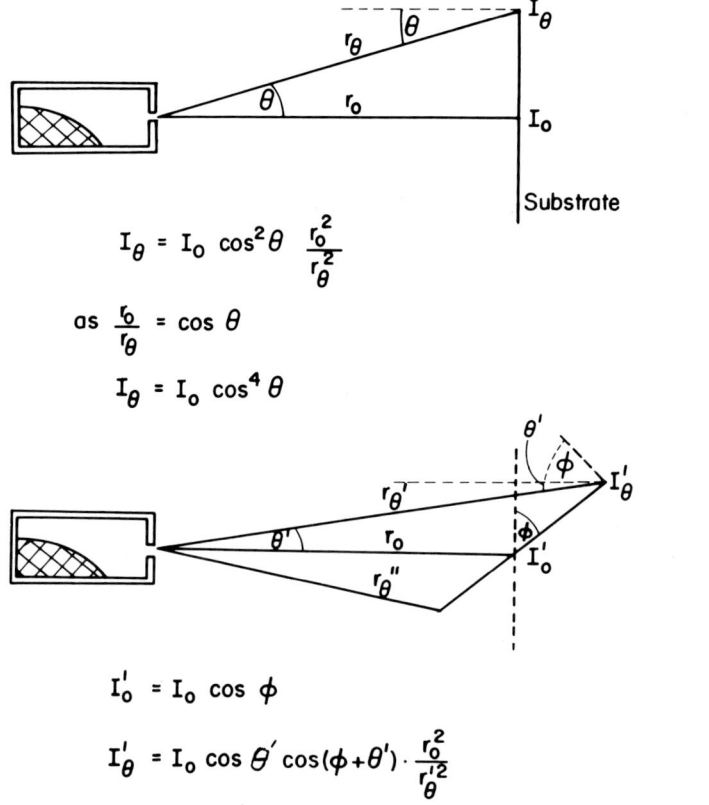

$$I_\theta = I_0 \cos^2\theta \; \frac{r_0^2}{r_\theta^2}$$

as $\frac{r_0}{r_\theta} = \cos\theta$

$$I_\theta = I_0 \cos^4\theta$$

$$I_0' = I_0 \cos\phi$$

$$I_\theta' = I_0 \cos\theta' \cos(\phi+\theta') \cdot \frac{r_0^2}{r_\theta'^2}$$

FIG. 3. Point source flux distribution across a substrate mounted axially (upper) and nonaxially (lower).

where a is the orifice area, P the vapor pressure (mm), r the source–substrate distance (cm), M the molecular wt, T the temperature (Kelvin). For nonaxial cells there is a further nontrivial correction.

In practice, however, significant deviations from cosine distributions occur, as practical orifices do not approximate point sources.

Thermocouples should be of tungsten/rhenium alloys to avoid contamination from volatile species observed in other type of thermocouple (e.g., chromel–alumel) (*33*).

Sample insertion should, whenever possible, be through a vacuum interlock to avoid disturbing the vacuum integrity.

III. *In Situ* Assessment Techniques

A basic advantage of MBE over most other forms of epitaxy is the ability to include facilities for monitoring, control, and surface analysis in the growth chamber, or in chambers separated by an isolation valve.

1. Surface Analytical Facilities

There is no rigid geometry for arranging surface analytical facilities but care must be taken that (a) there is no associated contamination from them, and (b) they should not be adversely affected by the beams.

An ion gauge, movable into the substrate position, is invaluable for neutral-beam monitoring and should be included either on the substrate heater carousel or on a separate motion drive.

There is no ideal position for a mass spectrometer analyzer, but it should be in the main chamber and preferably on the central axis of the source–shroud assembly, immediately behind the substrate. This allows both background gases and beam species (by raising the substrate holder) to be monitored directly without undue recourse to geometrical corrections.

These and other facilities enable either real-time or immediate postgrowth assessment of structural, chemical, and electrical properties related to growth parameters. Information thus gained can then be rapidly fed back for growth parameter control.

The scope of this chapter does not allow a detailed account of the various assessment techniques. However, in the next section mention is made of applications, progress, and some of the problems that have been elucidated by them.

2. Real-Time Analysis

While quadrupole mass spectrometry (QMS) is not strictly a surface technique, it is the most important single analytical facility in MBE systems. Background analysis enables leak detection, readings of residual vacuum gases (both inherent to the system and associated with heating effusion cells and substrate heaters) and the effect of ancilary liquid nitrogen, helium cryo-, and titanium sublimation pumping to be evaluated.

During growth, the nature of species effusing from cells or leaving substrate surfaces can also be monitored. In addition, relative effusion–flux ratios can be estimated and in certain cases used in computerized control of MBE growth (34, 35) (see Fig. 4). In quantitative applications, however, attention must be paid to a particular instrument's mass discrimination and variations of electron multiplier characteristics. A rapid, simple, and reliable method uses an ion guage mounted on the substrate–holder carousel (36) in conjunction with a mass spectrometer.

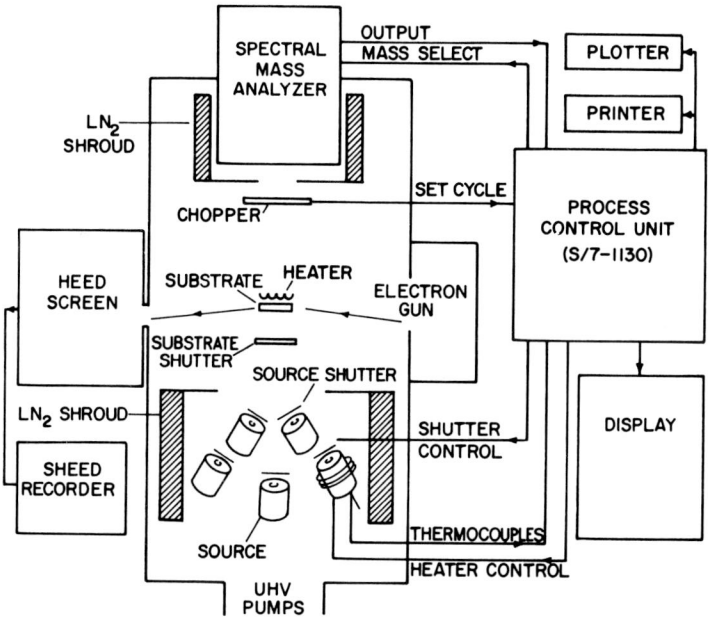

FIG. 4. Schematic diagram of a computer-controlled MBE system. After Chang et al. (3).

This combination allows direct measurement of beam fluxes and has been used to determine the spatial distribution of flux intensity over substrate areas.

The second most widely applied technique is that of glancing incidence ($\sim 2°$) reflection electron diffraction (RED) first reported in the context of MBE by Cho (37). Energies are usually in the range of 2.5 (MEED) to 40 KeV (HEED) (2, 21). Real-time display of diffraction patterns before, during, and after growth gives invaluable information about crystal surfaces.

From the sharpness of patterns, information is gained on the disorder, surface smoothness, crystallinity, and orientation of specific surfaces, and in certain cases mechanisms of growth (37). Surface reconstruction patterns can be also related to stoichiometry (38, 39) and to some extent atomic structure of surfaces (1).

It should be mentioned here that streaked patterns, which are commonly interpreted as arising from very flat and ordered surfaces, can in fact reflect combinations of nonmonochromatic electron energies, low beam coherence lengths, and thermal as well as atomic disorder. Dove (40) and Ludeke (41) showed that an ideally ordered coincidence lattice gives patterns comprised of a series of spots lying on a circle that passes through the undiffracted incident beam image.

Many reconstruction patterns have been reported for GaAs under various conditions of substrate temperatures, beam fluxes, and surface indices (42–46). However, for most purposes of practical growth the (100) plane is chosen for the technologically important reason that it has orthogonal (110) type cleavage planes. Growth temperatures for this compound are usually in the range 520–620°C and under these conditions three distinct surface structures $C(2 \times 8)$, (1×1), and $C(8 \times 2)$ exist (see Fig. 5), corresponding to arsenic excess, arsenic-stabilized, and gallium-stabilized surfaces respectively. Most other III–V compounds display equivalent structures, but GaSb shows only a $C(2 \times 6)$ structure under Sb-rich and (2×3) under Ga-stabilized conditions (see Fig. 6) as do its alloys with GaAs for Sb/As > 4 (41), and presumably InSb and AlSb, although diffraction patterns of these materials have not as yet been reported.

Other routine applications of RED are in determining the removal of oxide prior to growth by thermal treatment, (555 $\pm$ 5°C for GaAs) (36), the recovery of crystallinity during annealing of sputter-cleaned surfaces (6, 46), and in determining the nucleation properties and interfacial crystal quality of heteroepitaxial layers and single/poly growth (47).

It must be stressed here that although almost essential, RED produces only spatially integrated statistical information, so that absolute interpre-

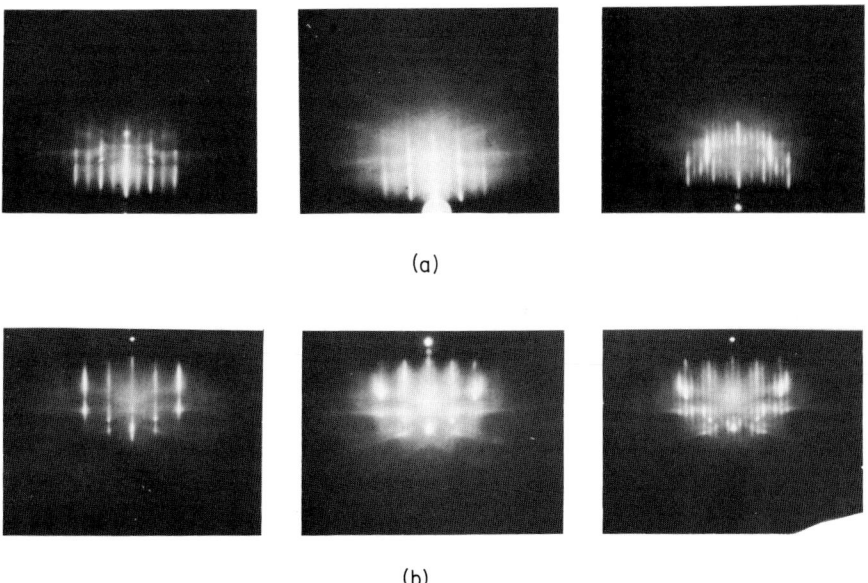

FIG. 5. 10 keV RED patterns of (100) GaAs. (a) [110] Azimuth As rich, As stable, Ga rich; (b) [1$\bar{1}$0] azimuth As rich, As stable, Ga rich.

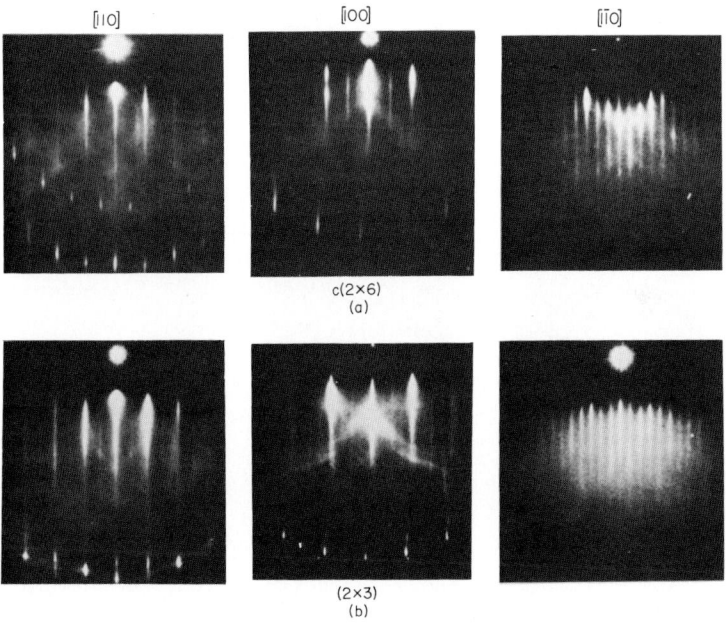

FIG. 6. 20 keV RED patterns of (a) $c(2 \times 6)$ and (b) (2×3) surface reconstruction for [110], [100], and [1$\bar{1}$0] azimuths. After Ludeke (41).

tations of crystal quality, surface smoothness, and details of atomic rearrangements must be to an extent subjective unless reinforced by TEM, SEM replication (2), or other techniques (48).

Certain steps have been made to increase the reliability of interpretations of RED patterns in which the variation of intensity of streaks is determined in a fashion (SHEED) analogous to emission spectroscopic densitometry (40) (see Fig. 7).

Auger electron spectroscopy (AES) has found extensive use as one of the few relatively nondestructive techniques for examination of surface chemical contamination (1), accumulation of dopants (38, 39, 49) (Fig. 8), semiquantitative estimation of alloy composition (50–53), interface sharpness (50, 51, 53), and interdiffusion (54). The usefulness of AES as a technique stems from the fact that most Auger electrons have escape depths on the order of one to three atomic layers. The absolute sensitivity of this technique is unfortunately more than 0.1% of a monolayer and quantitative Auger is exceedingly laborious. However, as a tool for estimation of relative changes in composition, etc., it has been used to

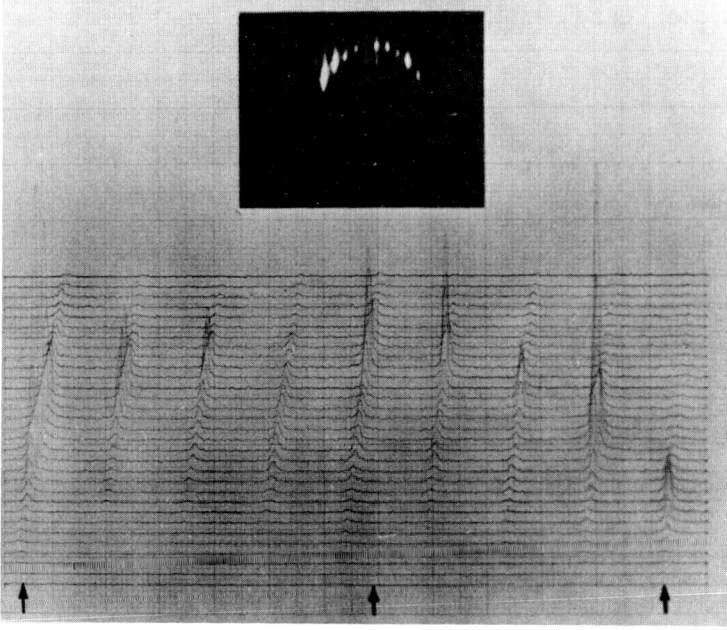

FIG. 7. Scanning high-energy reflection electron diffraction (SHEED) trace with conventional 1/4 order RED pattern inset (20 keV). Arrows mark position of bulk spacing. From Dove et al. (40).

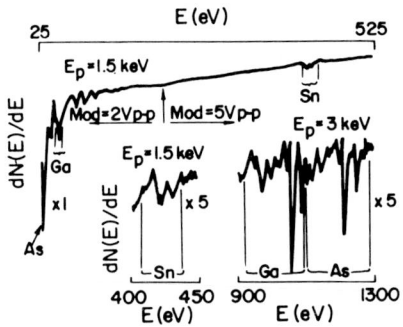

FIG. 8. Auger electron spectrum of 0.5 μ thick tin-doped MBE GaAs film ($N_D - N_A = 2 \times 10^{18}$ cm^{-3}) grown at 820 K. After Wood and Joyce (36).

significant advantage to advance many of the understandings of MBE-related studies either by itself or with sputter ion etching (49, 53) or secondary ion mass spectrometry (SIMS) (49).

The geometrical restrictions of low-energy electron diffraction (LEED) preclude its use as a real-time analytical technique. However, more detailed information can be obtained than from REED, as a reciprocal lattice projection is obtained directly on hemispherical phosphor-coated optics. For the reason that interpretation is more difficult and usually requires sophisticated computer analysis, LEED has been used only for simple lattice symmetry and dimension studies in III–V MBE (55) (see Fig. 9).

Many other *in situ* analytical facilities have been less extensively employed in connection with MBE; the following are some of the more interesting ones. Raman spectroscopy is more sensitive than AES for the determination of residual surface carbon on substrates prepared for epitaxy (56) and depth profiles of alloy compositions (57). *In situ* secondary ion mass spectrometry (SIMS) is a useful but destructive method for studies of dopant accumulation (49), alloy composition (45, 58) and impurity incorporation (59). Laser scattering, another non-destructive technique, has been employed real-time for monitoring of dynamic surface topography associated with nucleation and growth of epitaxial InP (60) and GaInAs alloys (48) grown on different orientations of InP substrates (see Fig. 10). Electron energy loss spectroscopy, with low beam-energy Auger-type analyzers, has been extensively used by Ludeke in the study of intrinsic surface state energies of GaAs (61), GaAlAs (62), GaSb and InAs (41), and those related to oxygen

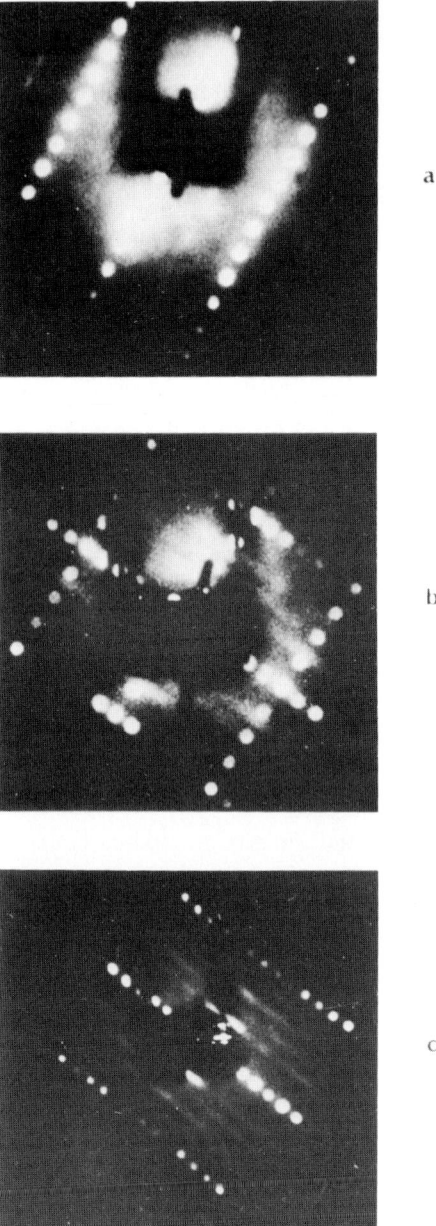

FIG. 9. LEED patterns of (100) GaAs with (a) $c(8 \times 2)$, (b) (4×6), and (c) (1×6) reconstructed surfaces.

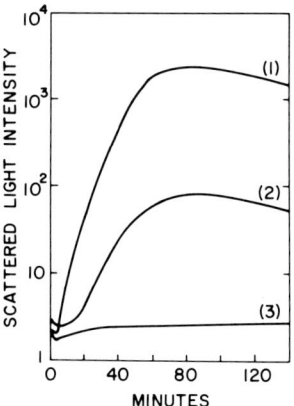

FIG. 10. Nonspecular scattered laser light intensity for growth of (1) $Ga_{0.59}In_{0.41}As$, (2) $Ga_{0.39}In_{0.61}As$, and (3) $Ga_{0.46}In_{0.54}As$ layers on InP substrates. From Miller et al. (60).

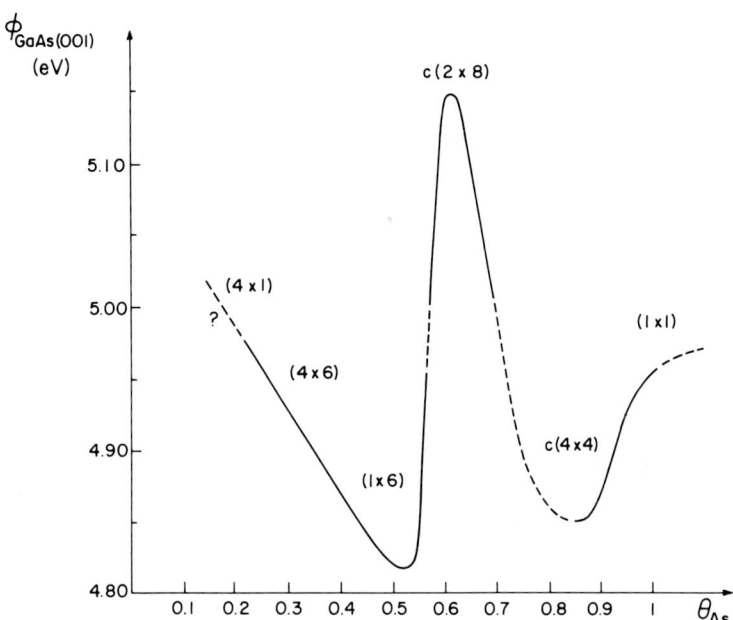

FIG. 11. Work function (ϕ) measurements of (100) GaAs with different proportional arsenic coverages. From Massies et al. (65).

exposure (63) and induced by antimony (64). Results of similar studies were also reported by Massies et al. (65).

The latter author (66) has shown by *in situ* Kelvin probe experiments that the work function of room temperature (100) GaAs surfaces is relatively independent of carrier type (Fermi level position) but dependent on the surface stoichiometry (Fig. 11) (65).

Very few of the known postgrowth electrical, optical, and structural assessment techniques have not at some time been applied to layers grown by MBE. A systematic summary of this work is outside the scope of this chapter, but some specific results are included in the text.

IV. Initial Stages of Growth

1. Substrate Preparation

This is one of the most intractable problems yet found in the whole MBE process. The main problem is to achieve a damage-free, atomically clean, stoichiometric surface.

Following chemical treatments such as Br_2/methanol solutions, water treatments, and various sulfuric acid–hydrogen peroxide–water mixtures (typically 7:1:1), substrates are normally mounted on molybdenum heater blocks with In or In/Ga alloys and loaded into the growth system. Auger analysis typically reveals that these surfaces are rich in oxygen with traces of carbon. Heating (in the case of GaAs to $555 \pm 5°C$) (36), under a flux of the relevant group V element, decomposes the oxide and in most cases oxidatively removes carbon, within the detection limits of AES (1) resulting in a fairly strong reconstructed surface as seen by RED (Fig. 12) (36). For substrates less stable than GaAs [e.g., InP (32)], thermal treatment leads to severe stoichiometric deviations unless unacceptably high group V fluxes are used to stabilize the surface. Calawa *et al.*

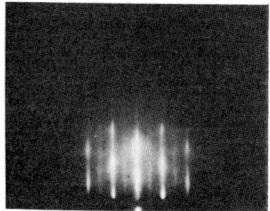 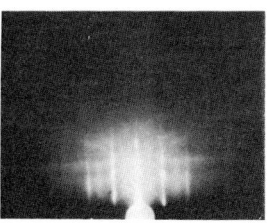 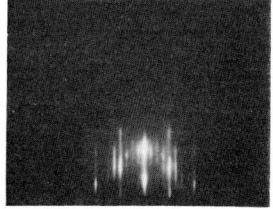

FIG. 12. 10 KeV RED patterns of GaAs surfaces heat-cleaned with an arsenic flux $\sim 10^{16}$ $cm^{-3}\ sec^{-1}$; [110] azimuth, (a) As rich, (b) As stabilized, (c) Ga rich.

have found it possible to stabilize InP surfaces by incident As fluxes during heat cleaning at ~500°C prior to growth of GaInAs alloys (67).

There are often residual traces of carbon even after such treatments, and alternative methods such as sputter-ion cleaning at elevated temperatures under group V fluxes have been used for its removal. However, stoichiometric problems normally arise by preferential loss of the group V element on annealing. Large numbers of In metal droplets form on sputtered and annealed InP, which interfere with subsequent epitaxy or metallizations (68).

A third method reported successful by the author (69) used water vapor ($\sim 10^{-6}$ Torr) pressures as a transport agent for carbon at elevated substrate temperatures (~580°C).

Thermal treatments are in most cases successful for GaAs but significant oxygen concentrations in excess of 10^{11} cm^{-2} have been found at substrate–layer interfaces by secondary ion mass spectrometry (SIMS) analysis (70). SIMS has also identified interfacial pile-up of chromium from chromium-doped semi-insulating (SI) substrates (71). The surface accumulation is presumably associated with relaxation of the high Cr concentrations in excess of the thermodynamic solubility during heat treatment (Fig. 13).

2. Buffer layers

For certain applications, notably FETs, the substrate is required to have a very high resistance ($>10^7$ Ω cm) in order to reduce shunt

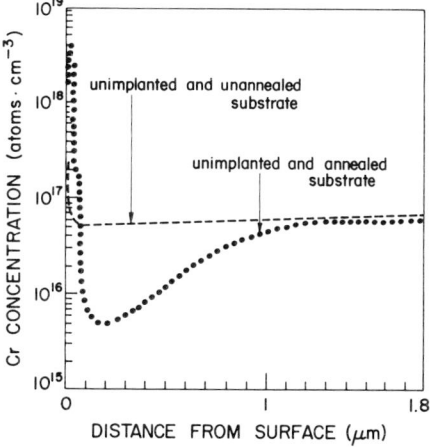

Fig. 13. SIMS profile of chromium in untreated and heat-treated (900°C for 20 min) semi-insulating GaAs substrates. From Huber et al. (71).

conduction, etc. However, direct growth of active layers on bulk SI substrates, in general, produces inferior interface properties (*69, 72*). Stoichiometric deviations and impurity accumulation at the substrate surface during heat treatment (*70*) or sputtering and impurity complex formation with As vacancies, etc. (*73*), prior to growth of the epilayer can all be responsible for nonideal interfaces.

Rapid diffusion of accumulated impurities (Cr, O?, etc.) during MBE (see previous section) can compensate the first-grown part of the film. Such increases in deep acceptor densities reduce carrier mobilities and can severely alter the expected free-carrier profile. Concentrations of impurities remaining at layer–substrate interfaces can also cause capacitance and I–V hysteresis loops in FET devices (Fig. 14) (*46*).

The usual practice to overcome substrate–layer interfacial problems is to grow high-resistivity buffer layers (*46*).

Nominally undoped MBE GaAs epilayers are normally p type in the region 10^{15}–10^{13} cm^{-3}. p–n junctions associated with growth of 10^{17} cm^{-3} n-type FET active layers on such layers cause depletion in the buffer (typically ~3 μm at 10^{14} cm^{-3}).

An active layer grown on a semi-insulating substrate may also have an associated depletion width ~0.1 μm (Fig. 15) associated with excess traps pinning the level near midgap, which must be considered when a specific

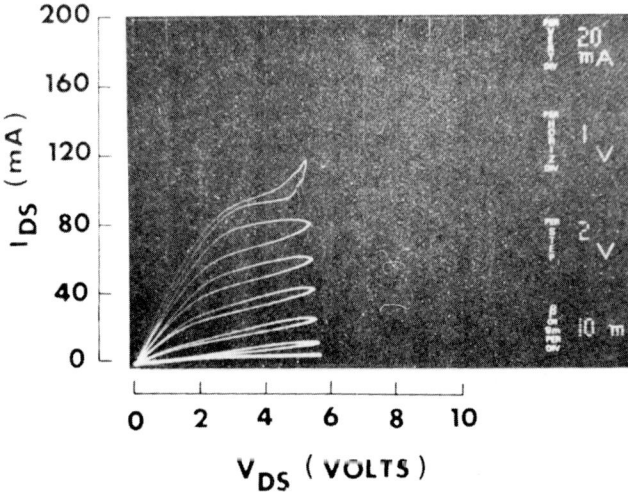

FIG. 14. Transfer characteristics of FETs produced on MBE GaAs layers without a buffer layer, showing looping. After Cho *et al.* (*46*).

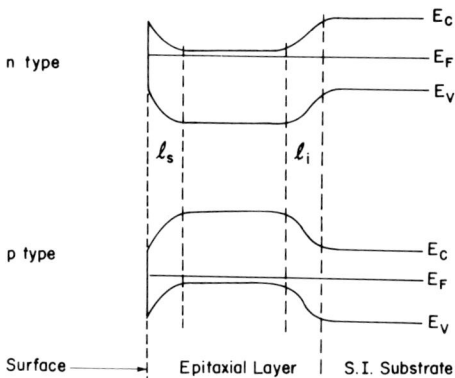

FIG. 15. Simplified band diagrams showing surface and interfacial band bending regions associated with pinning of the Fermi level by surface states and excess substrate traps, respectively, upper n type, lower p type.

thickness doping ($n \times l$) product is required (74) and especially when doping levels are to be estimated from Hall measurement for $N_D - N_A < 10^{17}$ when $l \lesssim 1$ μm (or 10 μm for 10^{15}).

There are disadvantages with p buffer layers in that the p–n junction depletion may not be stable to the voltage transients in FET device operation.

Alternative approaches for good-quality, high-resistance MBE buffer layers have been attempted. The first of these involves incorporation of chromium, a deep acceptor (0.75 eV below E_c) that overcompensates residual impurities. Morkoc has recently prepared semi-insulating MBE GaAs by use of Cr doping (75).

In the second method, low substrate temperatures were used during growth of buffer layers (76). At $T_s < 460°$C (76, 77) (Fig. 16) (78) GaAs resistivities approach 10^6 Ω cm. Electrical deep centers produced by this technique were found to be stable to the higher temperatures employed in subsequent active-layer growth and are currently being evaluated as buffer layers for FET devices (76, 79). Oxygen-doped GaAlAs alloys have recently proven interesting as potential insulators in MIS structures (80, 81). This type of material could conceivably be used for buffer layers (80). Benefit would arise for both the deep-trap nature of oxygen in $Ga_{1-x}Al_xAs$ alloys, and also the discontinuity in the conduction band edge, which confines carriers to conduct through the interfacial GaAs channel region only.

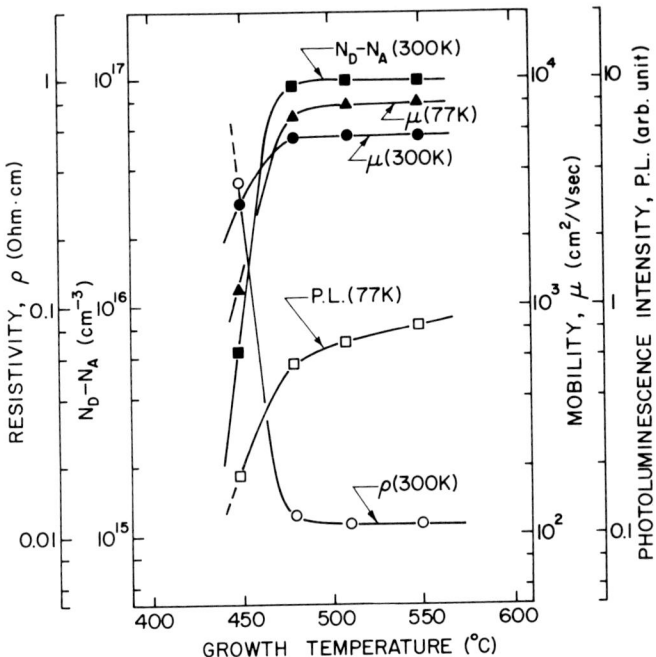

FIG. 16. Substrate temperature dependence of mobility, photoluminescence intensity (P.L.), resistivity (ρ), and free-donor density ($N_D - N_A$) of MBE GaAs layers. From Murotani et al. (78).

V. Doping

1. Unintentional Impurities and Deep Levels

The electrically active centers in the forbidden gap of semiconductors that are separated by more than a nominal 100 meV from either band edge are generally termed "deep levels." This broad classification includes several types of centers, among which are simple substitutional impurities such as certain transition metal atoms (82), stoichiometric defects such as gallium vacancies (83), and complexes of the two types.

Although certain levels have been identified (83–87) by transient capacitance, transient current, and other forms of spectroscopy, most have eluded systematic study.

Every form of epitaxy introduces traps that are related to the method (84, 85). However, different workers (87) report different levels for nominally the same process, leading to the conclusion that they are in the main dominated by, or complexed with, substitutional impurity atoms.

The interest in deep levels arises from their electron- and hole-trapping ability whereby they behave as compensating deep donors and acceptors, respectively. The presence of high concentrations of such centers causes several problems:

(1) Control of intentional doping levels becomes difficult when trap densities approach the required carrier levels.

(2) Deep centers in the forbidden band increase leakage currents across Schottky barriers and p–n junctions by reducing the effective barrier height (see Fig. 17).

(3) The presence of electron trap levels at the interface of microwave FET devices can cause looping and noise performance to be increased. If not stable with time or temperature, they will also cause drift in pinch-off voltages (*88*).

(4) Finally, minority traps act as nonradiative recombination centers, reducing lifetimes in optical devices such as LEDs and double heterojunction lasers (DHLs), and reducing the diffusion length in photocathode devices.

For these reasons alone it is important that the identification and elimination of carrier trapping centers is not ignored.

The first study of electron traps in MBE GaAs (*89*) showed the presence of at least nine such centers, two of which (M_2 and M_8) dominated under gallium-stabilized conditions and three of which dominated under arsenic-rich growth (M_1, M_3, and M_4) (see Fig. 18).

The obvious conclusion from this result is that stoichiometry of the surface plays a major role in the incorporation of impurities or in the formation of complexes with impurities that do become incorporated (e.g., $X–V_{As}$ or $X–V_{Ga}$).

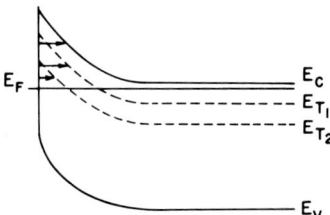

FIG. 17. Simplified band diagram showing "trap-tunneling" mechanism of reverse-biased Schottky barrier leakage.

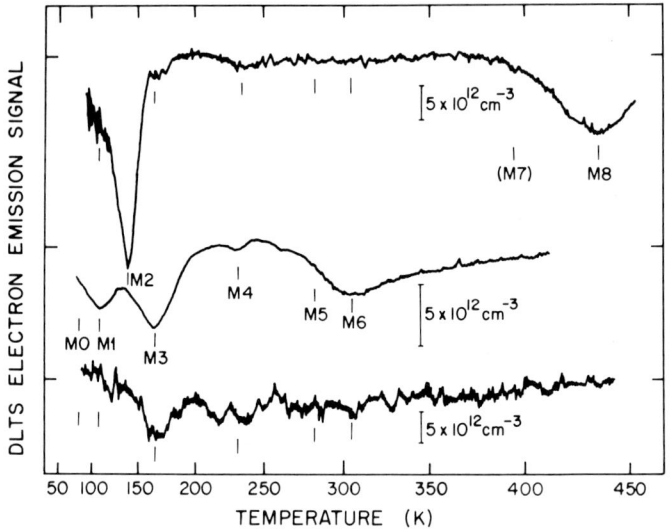

FIG. 18. DLTS spectra of electron traps in MBE GaAs showing two As-rich samples with lowest trap concentration (lower two traces) and a Ga-rich sample (upper trace). The rate window is 51 sec^{-1}. From Lang *et al.* (*89*).

The present low concentrations ($\lesssim 1 \times 10^{12}$ cm^{-3}) (*79*) are below those that give problems in majority carrier devices but it is not certain that the same is true for minority carrier devices. There is direct evidence, however, that the luminescence efficiencies of GaAs samples grown under gallium-stabilized conditions are very much higher than with arsenic excesses (*80*).

Lowering the substrate temperature significantly increases trap densities N_T (*90*). Below 480°C, levels not previously seen in MBE GaAs appear (Fig. 19), some of which have been seen in 1 MeV electron-beam-irradiated GaAs (*83*) (Fig. 20) and identified as a simple gallium vacancy (V_{ga}) ($E_A - 0.43$). We have seen two major electron traps at 300°C growth temperature, which we tentatively ascribe to the antisite arsenic species (As$_{Ga}$). The fact N_T increases with decreasing T_s during growth is the biggest single factor that limits the growth of electronic-quality GaAs to above 480°C. Similar effects are expected at differing values of T_s for other III–V compounds and alloys, although at $T_s < 360°$, InP becomes more conducting by increased defect density conduction centers.

Hole traps in MBE GaAs are very much less understood than electron traps for the simple reason that it is difficult to form a good Schottky barrier on p-type GaAs. The only other ways are to use p–n junctions

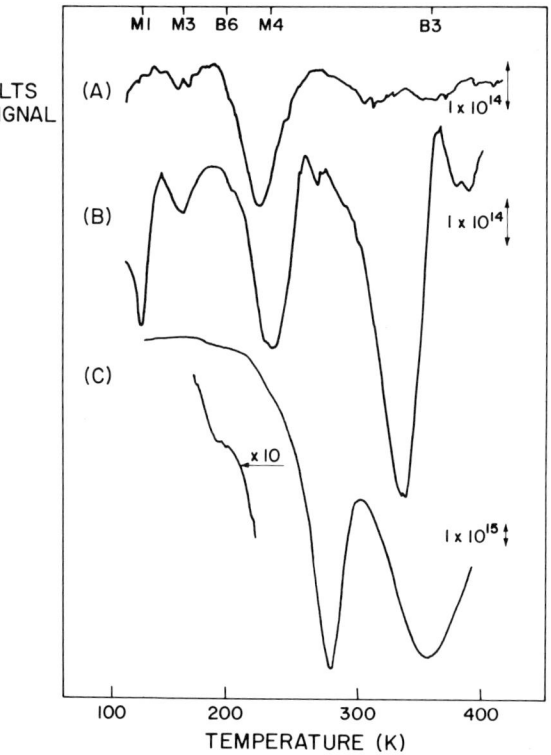

FIG. 19. Typical DLTS spectra of electron traps in MBE GaAs grown at (a) 500°C, (b) 400°C, and (c) 300°C, showing the increase in concentration and change of "fingerprint" traps. Note the existence of traps B_3 and B_6, which correspond to V_{Ga} in (c). The rate window is ~25 msec.

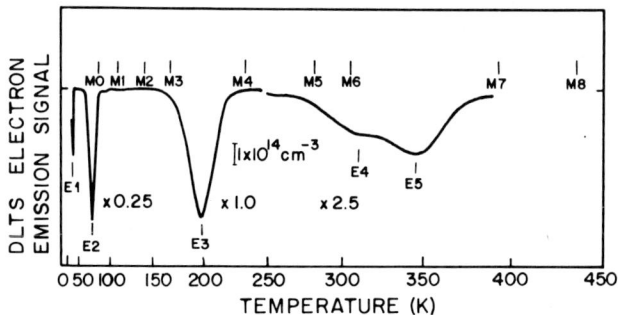

FIG. 20. DLTS spectrum of 1 MeV electron-irradiated GaAs. The "normal" MBE trap positions are shown above. From Lang et al. (89).

specially grown for the purpose or optical excitation, which produces spectra that are less easily understood. Martin *et al.* (*86*) reported the existence of two levels by the latter method in GaAs grown by the author, and recently other traps have been found in layers prepared in our laboratory at Cornell (*90*) (Fig. 21).

The increase of the Cu- and Cr-related traps with lower T_s again suggests that these impurities are complexed with gallium or arsenic vacancies and most importantly do diffuse rapidly at MBE growth temperatures.

Apart from deep-level centers, several investigations (*1, 91–93*) of nominally undoped MBE GaAs have been carried out. It is now accepted that nominally undoped GaAs grown in the range 480°C < T_s < 630°C

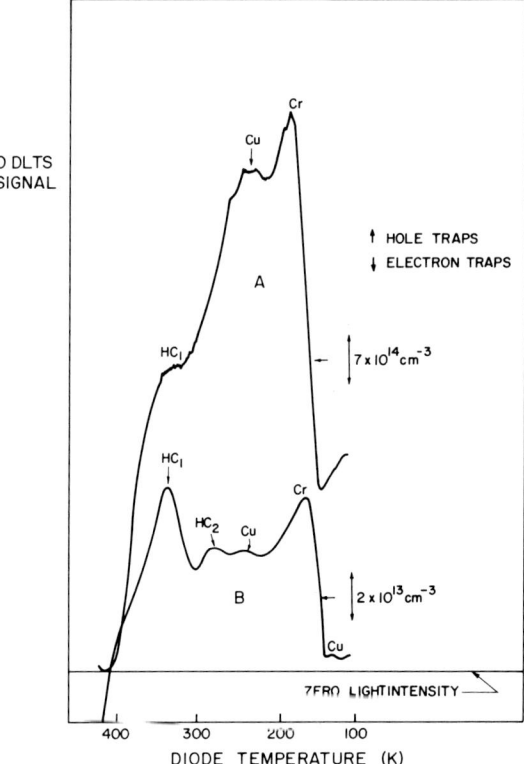

FIG. 21. Typical optical DLTS spectra of MBE GaAs grown at (a) 550°C; and (b) 430°C. Time constant, 20 msec; reverse bias, (a) 1.5 V, (b) 3.0 V; neodymium YAG laser.

usually has residual p-type character in the range (5×10^{13}–10^{15} cm^{-3}) when an elemental arsenic beam source is used. When high-purity (nominally undoped) bulk GaAs is used as the As beam source (As$_2$ + a little Ga), layers are n type with $N_D - N_A \sim 10^{15}$–10^{16} cm^{-3} as a result of silicon from the starting GaAs (*1*).

The most common residual acceptor impurities have been identified by photoluminescence (*77, 93–95*) as carbon with $E_a \sim 27$ meV binding energy, manganese (*94, 95*) with $E_a \sim 113$ meV, and germanium with $E_a \sim 40$ meV (*77, 94–96*), although there is still controversy about the identity of "residual manganese" photoluminescence peaks, which could also be ascribed to a C_{As}–vacancy complex (*97*). Other impurities such as Al, Cr, Ni, W, and Mo have been identified (*96*) by mass spectrometry and are associated with inadequate shielding of furnace components, chromel–alumel thermocouples, and W and Mo furnace windings.

Adequate care in reducing source contamination, improved pumping of residual CO and H$_2$O species, and the use of sample exchange vacuum interlocks has resulted in very much lower residual acceptor densities. Evidence for this conclusion comes from recent values of liquid-nitrogen temperature Hall mobilities in excess of 100,000 cm^2 V^{-1} sec^{-1} (*75*) for lightly tin-doped samples and above 60,000 in the author's laboratory (*98*).

When a vacuum interlock is not used, Calawa (*95*) has shown that background hydrogen pressures $\sim 10^{-6}$ Torr reduce incorporation of carbon and oxygen (Fig. 22).

Once parameters important in the preparation of layers of a particular compound or alloy have been defined, and these parameters achieved in practice, electrical properties must then be controlled. To this end, the incorporation of dopant atoms must be understood. Most work on the incorporation of dopants from neutral-atom beams has been empirical and mainly in studies of GaAs. However, recent years have seen doping studies of a more fundamental nature and on III–V compounds and alloys other than GaAs.

2. N-Type Dopants and Their Problems

Of the elements that are potentially useful n-type dopants in GaAs, the group IV elements have been used almost exclusively until recently.

a. Group IV Elements

Carbon. As mentioned above, carbon is the most common background impurity in MBE, where it behaves as an acceptor. The incorporation

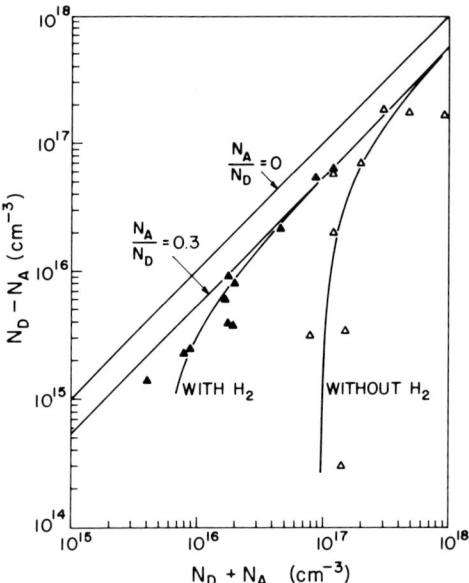

FIG. 22. Effect of background pressures of hydrogen during growth on the compensation ratio properties of MBE GaAs. The net donor concentration $N_D - N_A$ is plotted as a function of total ionized impurity concentration $N_D + N_A$ at 77 k. From Calawa (95).

mechanisms are still not clear but it is thought to originate from residual background CO molecules.

Silicon. Silicon is a well-behaved donor impurity in GaAs although it is slightly amphoteric as evidenced by its 10 K photoluminescence peak at 1.492 eV (99). It has an effective unity incorporation coefficient but is very nonvolatile. Effusion cell temperatures used for Si doping are higher than for Ge or Sn (being typically 1100°C for 10^{17} cm^{-3}), which can produce higher background gas loads and hence rather more heavily compensated layers than those doped with Sn. This problem may preclude its use for Al-containing III–V compounds and alloys. Good GaAs free-donor mobility results for Si doping have been reported, however (*1, 3, 100, 101*) (Fig. 23)

Germanium. Like silicon, germanium has unity incorporation coefficient (no surface accumulation); however, it does exhibit marked amphoteric character (*1, 77, 102, 103*) in GaAs, although most entirely donor like in InP (*103*).

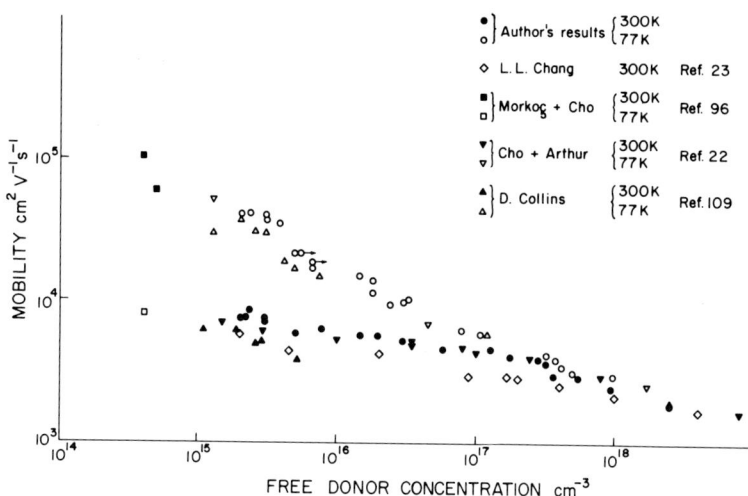

FIG. 23. Free-electron mobility for MBE GaAs as a function of donor concentration.

Cho demonstrated under arsenic rich conditions (104) growing from Ga and As$_2$ beams that Ge was incorporated predominately as a donor and under gallium-stabilized conditions as an acceptor. Wood confirmed this result using Ga and As$_4$ molecules (77) and showed that the donor/acceptor ratio was not only continuously dependent upon the As$_4$/Ga flux ratio but also on the substrate temperature (Fig. 24). This latter result has been demonstrated also by Smith (12) and used by Dohler (105) for preparation of some novel n.i.p.i. structures (see later). Above ~2 × 10^{19} atoms Ge cm^{-3}, MBE GaAs becomes heavily p-type under all conditions (105a).

Tin. Tin is the most commonly used n-type dopant in MBE growth of GaAs and other III–V binaries and alloys.

With the exception of GaAs$_{1-x}$Sb$_x$ for $x > 0.8$, tin behaves predominantly as a donor. In GaAs it has been used to prepare the highest n-type liquid-nitrogen mobilities (75, 96).

Cho, however, demonstrated quite early (102) that tin accumulated at the surface during growth, a process confirmed by Wood (69) and by Ploog (49) using SIMS and Auger electron spectroscopy. In a study of doping transients as a function of step changes in As$_4$/Ga flux ratio and substrate temperature, the author (36) showed the rate of incorporation to be controlled by a surface reaction limited by the concentration of surface

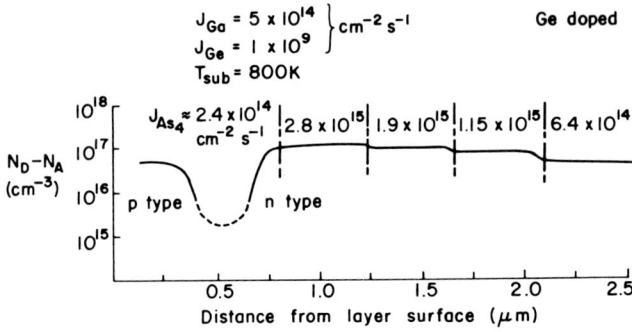
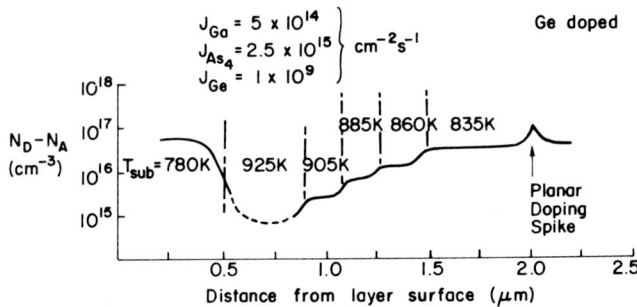

FIG. 24. Effect of arsenic to gallium flux ratio (upper trace) and substrate temperature (lower trace) on the site occupancy of Ge in MBE GaAs. From Wood et al. (77).

gallium vacancies (see Fig. 25), a result recently confirmed by Murotani et al. (76). The nonunity incorporation coefficient means that only at low substrate temperatures (below 520°C) and with high As_4/Ga flux ratios (typically >10/1) can sharp drops in the doping profile be achieved.

The surface accumulation means that predeposition of tin must be used (36) in growth of layers on semi-insulating or n^+ substrates if interface carrier dips are to be avoided (Fig. 26). The use of the exponential decrease in carrier concentration obtained solely from predeposited tin concentrations has recently been used to advantage in preparing layers for power FETs with increased linearity of transconductance (g_m) (98) (Fig. 27).

Lead. Lead is a very large atom and is not detectably incorporated in GaAs under normal conditions of MBE growth even when the lead flux exceeds that of the gallium. In fact, gallium arsenide can be grown through several monolayers of lead in a pseudo-liquid-phase epitaxy

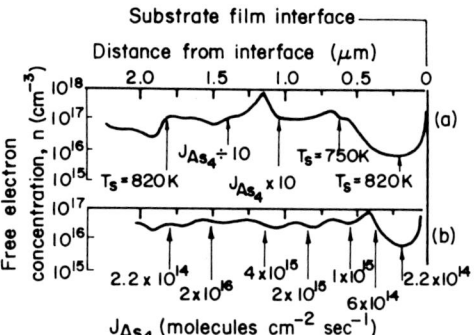

FIG. 25. Effect of changes in As$_4$/Ga flux ratio and substrate temperature on the incorporation of tin (from a constant flux) during MBE growth of GaAs. From Wood and Joyce (*36*).

system (*106*). (Lead reevaporates with a surface lifetime of a few seconds at ~500°C.)

b. Group VI Elements. With the exception of oxygen, all the group VI elements are well-behaved donors in GaAs grown by liquid or vapor phase epitaxy (*107*). However, the elements have vapor pressures that are too high to allow their use as controlled dopant sources in MBE (*108*).

This problem is aggravated by the practice of baking vacuum systems (~250°C) to obtain lower pressures prior to growth. Another problem associated with the use of elemental group VI sources is the complexity of species that evaporate from the solids (e.g., S_2, S_4, S_8). Arthur (*1*), using elemental tellurium as a dopant source, found a surface reaction with

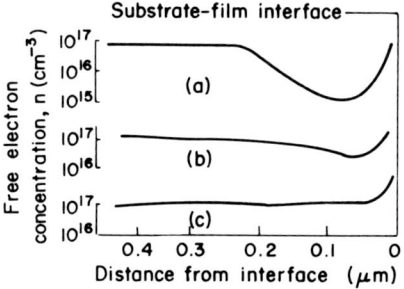

FIG. 26. Interfacial free-carrier concentration profiles for tin-doped MBE GaAs layers grown at constant rate. (a) $J_{As_4} = 2.2 \times 10^{14}$ cm^{-2} sec^{-1}; (b) $J_{As_4} = 2 \times 10^{16}$ cm^{-2} sec^{-1}; (c) as (b) with 3×10^{12} tin atoms deposited prior to growth. From Wood and Joyce (*36*).

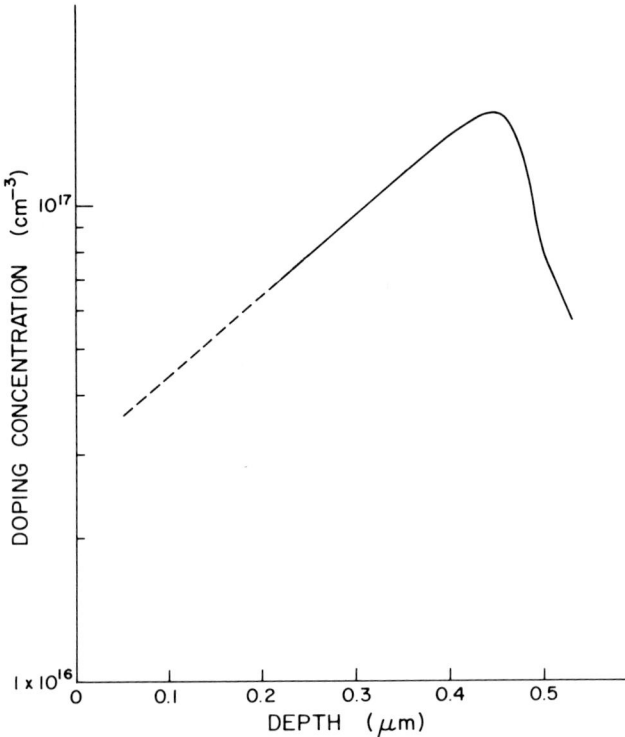

FIG. 27. Exponential free-electron profile of a GaAs power FET active layer produced by predeposition of tin dopant. From Wood et al. (98).

gallium at higher fluxes to form GaTe, which remained on the surface and doped further GaAs as uncontrollably high levels.

Making use of the virtual insolubility of lead in GaAs under normal MBE conditions, it has been shown possible to use a surface exchange reaction with lead sulfide and lead selenide as captive monatomic dopant sources (106). These lead chalcogenides have vapor pressures that are low enough, even at bakeout temperatures, to allow their use. Excess lead is necessary in these sources, however, to reduce dissociational evaporation.

The incorporation of S and Se by this process is extremely efficient ($\geq 50\%$) and little dependent upon the As_4/Ga flux ratio or growth rate. No surface accumulation was observed and very sharp changes in doping level were obtained (Fig. 28). However, above 550°C reevaporation becomes increasingly important.

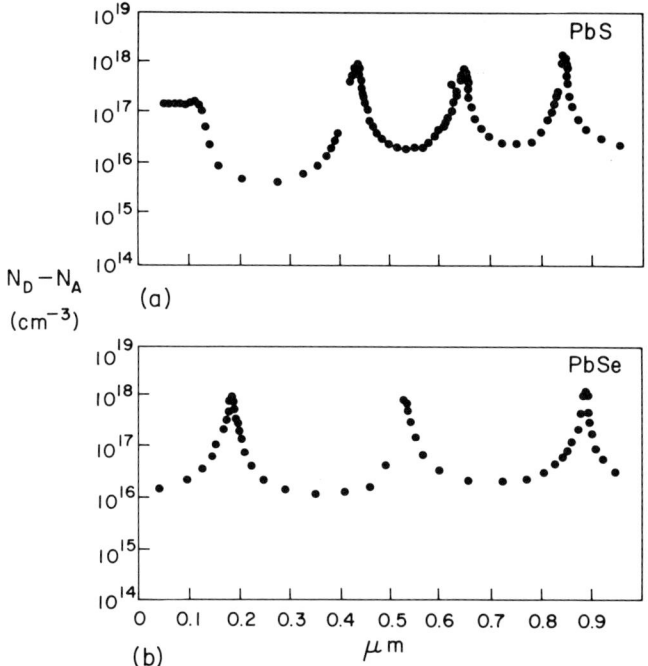

FIG. 28. Capacitance–voltage-generated free-donor profiles of 150 Å wide spikes in GaAs doped (a) with sulfur from lead sulfide and (b) with selenium from lead selenide. From Wood (*106*).

As these elements are not amphoteric, they should find use in III–V compounds and alloys in which group IV elements produce p-type layers (e.g., GaAs$_{1-x}$Sb$_x$ for $x > 0.2$).

Smith (*109*) has extended the idea of "captive sources" using SnS molecular or double doping. Collins (*110*) similarly has used SnTe with encouraging results for MBE GaAs FET layers that did not require the predeposition of tin to achieve abrupt free-donor profiles (Fig. 29).

3. P-Type Dopants and Their Problems, Including Ion Implantation Doping

These still present significant problems. The conventional acceptor impurities zinc and cadmium have very low incorporation coefficients (K_i) in MBE GaAs and Ga$_{1-x}$Al$_x$As (*111*). However, low-energy simultaneous

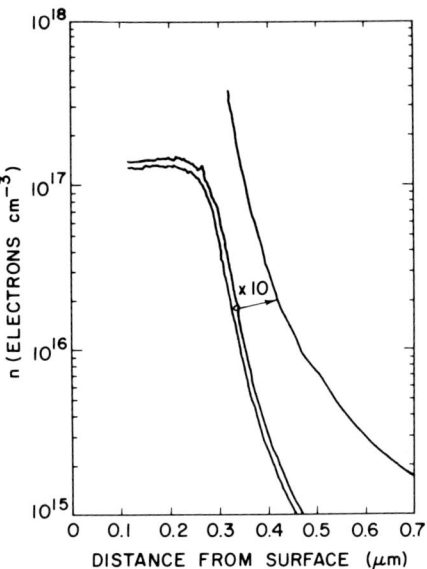

FIG. 29. Capacitance-voltage generated free donor profiles of SnTe-doped MBE GaAs FET layers. Note the abrupt doping level rise at the interface. From Collins (*110*).

Zn ion implantation (Fig. 30) during epitaxy has improved the effective electrical incorporation coefficient of this element up to ~0.01% (*112–114*) of the zinc effused into the system (~3% incorporation × ~0.3% ionization efficiency).

Magnesium. K_i for magnesium is reputedly high (*115*), but its effective electrical incorporation coefficient K_e has only been reported up to ~2 ×

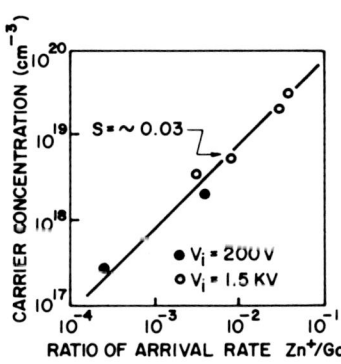

FIG. 30. Dependence of carrier concentration on the ratio of Zn^+ to Ga arrival rates. From Naganuma (*113*).

10^{-3} (*113*) in GaAs. However, K_e is increased with Al content in $Ga_{1-x}Al_xAs$ alloys (*116*). Use of high-Mg fluxes to obtain high doping levels can cause degradation of the surface smoothness (*1*).

Manganese. Manganese also suffers from adverse surface degradation problems for fluxes above that required to increase $N_A - N_D$ above $\sim 10^{17}$ cm^{-3} and has a higher binding energy (90–100 meV) (*95*) and also a high dependence of electrical activity on surface stoichiometry during growth.

Germanium. Germanium, as mentioned above, is an amphoteric dopant requiring low As_4/Ga flux ratios and high values of T_s to become predominantly acceptorlike ($E_a \sim 40$ meV). However, films grown under these conditions do have good photoluminescent efficiencies despite their inevitable compensation (*77*).

Beryllium. Despite its extreme toxicity hazard, beryllium has to date shown the most promising p-type doping properties (*117, 118*) with $K_e \approx 1$. Acceptor levels up to 6×10^{19} cm^{-3} with higher hole mobilities than other acceptors above 10^{18} cm^{-3} (*119*) (Fig. 31) have been proven possible. However, Be is extremely sensitive to CO and H_2O background gases, which result in neutral Be–O type centers (*118*). This last problem currently puts an effective lower limit on the controllable acceptor density around 10^{16} cm^{-3}. Beryllium has a very high resistance to diffusion in

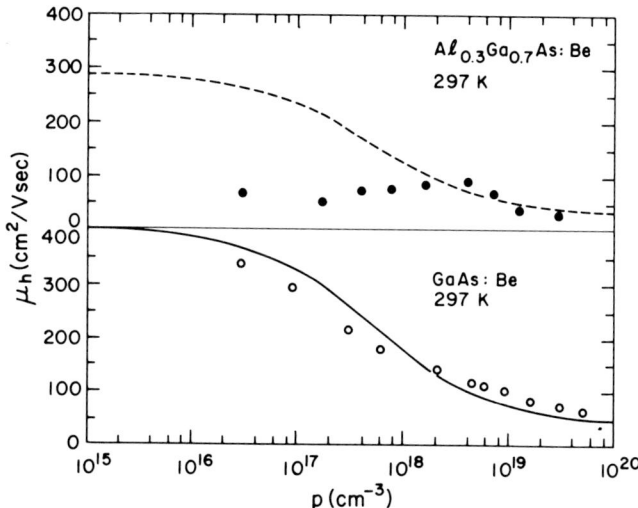

FIG. 31. Hole mobility for Be-doped MBE GaAs and $Ga_{0.7}Al_{0.3}As$ layers. The solid line represents the highest mobilities measured in GaAs grown by other techniques. The dashed line represents a corresponding estimate for $Ga_{0.7}Al_{0.3}As$. From Ilegems (*117*).

MBE GaAs (*117, 118*) and AlGaAs (*120*) below 900°C and has been successfully used in hybrid LPE–MBE double heterojunction GaAs/GaAlAs lasers (*121*).

Isoelectronic doping. Nitrogen behaves as efficient isoelectronic radiation centers in many indirect-gap III–V compounds and alloys (e.g., *122*), and as such the use of ionized nitrogen implantation simultaneously with MBE growth to dope $GaAs_{1-x}P_x$ should be mentioned in this section (*123*). Isoelectronic trap densities greater than 10^{18} cm^{-3} were achieved with $K_I \approx 0.06$.

VI. III–V Binaries Other Than GaAs

Early work on binaries other than GaAs has been reviewed by Cho (*1*) and Chang (*5*). This section discusses only recent developments and individual results of specific interest.

MBE growth of all III–V binaries follows the same basic trends reported in Cho (*1*) and Chang (*5*). Two-dimensional growth rates are dependent upon the arrival rate of the group III elements providing the flux of the group V element is in excess ($J_{V_4} > 0.5xJ_{III}$) and that desorption of the group III element is not significant. One notable exception to this general trend is that of InSb (see below). There are significant differences in the ranges of T_s that allow two-dimensional growth. In general, above the congruent sublimation temperature (T_{cs}), surface topographies degrade by competitive decomposition and growth. Above T_{cs} significantly increased values of J_V/J_{III} are required to retain stoichiometry.

Reported values of T_{cs} for (100)-oriented surfaces are GaAs ~640°C (*38, 124*), InP ~360°C (*125*), GaP ~670°C (*126*), and InAs ~370°C (*127*). For orientations other than (100), information is limited to (111) InP ~415°C (*60*) and (111) and (110) GaAs ~520°C (*87*).

Aluminum binaries have indirect gaps with low carrier mobilities that limit their application in the main to optical devices, although as buffer layers (*128*) and in multilayer structures (see below) some technological advantages are found in the use of AlAs. They are also exceedingly difficult to grow pure because of the high reactivity with oxygen-containing background gases. This gettering process severely degrades electrical quality of both Al-containing binaries and alloys.

AlAs is the only binary III–V compound of aluminum that has been grown by MBE and was subsequently used as a uniformly thin coating with potential MOS applications (*129*). Chang (*130*) showed by X-ray secondary interference scattering studies that it was possible to grow coherent multilayer AlAs/GaAs structures. This and analogous GaAlAs structures are covered in the section on multilayer growth. Recently, reactive MBE has been used to prepare GaN on sapphire substrates from beams of NH_3 and Ga (*130a*).

Little has been published on MBE GaP since reviews by Chang (*5*) and Cho (*1*). Epitaxy of GaP is, not surprisingly, very similar to that of GaAs in most respects: for example, $T_{cs} \simeq 670°C$, P_4/Ga flux ratios $> 1/2$, similar RED patterns, and surface treatments for epitaxy.

GaSb MBE growth has only recently received attention. Its use in a closely lattice-matched multilayer structures with InAs, with interesting interfacial band structures, is considered later. Layers of GaSb can be grown at temperatures similar to GaAs (460–600°C). However, n-type doping of MBE GaSb is not possible with the group IV elements, which all behave as acceptors (*131*). Another anomaly of this compound (and suspected of all Sb-rich III–V binaries and alloys) is the third- and sixth-order surface reconstructions (*41*) in the [110] direction of (100)-oriented gallium- and antimony-stabilized surfaces, respectively, above 350°C (see Fig. 6). Current work on n-type GaSb utilizes elemental Te doping, but poor control is obtained (*132*). This problem may be overcome by use of compound dopants such as PbS, PbSe (*105*), or SnTe (*110*).

InP growth has been reported by several authors (*60, 125, 133–133b*). The thermal instability of this compound necessitates sputter-ion cleaning and annealing under P_4 fluxes or thermal removal of oxides under As overpressures (*67*). RED patterns are analogous to GaAs for both (100) and (III) (*60, 125*) surfaces. McFee *et al.* (*60*) in an elegant experiment, grew on curved surfaces demonstrating the orientation dependence of surface quality and concluded the best growth temperature was T_s ~450°C, on (III) surfaces misoriented between 2 and 4° toward the [211] direction. Matsushima *et al.* (*133*) concluded that tin doping improved the morphology of InP grown at 240°C on (100) GaAs.

Doping with Sn and Mg has been found relatively successful for this compound (Be less so) to levels in excess of 10^{19} cm^{-3}. However, measured room temperature electron mobility figures $\mu_n \sim 3000$ cm^2 V^{-1} sec^{-1} at $\sim 2 \times 10^{16}$ cm^{-3} have been disappointing. Contrary to the case of GaAs, however, lowering of substrate temperatures <350°C for InP (vs. ~460°C for GaAs) increased film conductivity, which was associated with the introduction of stacking faults and a photoluminescence peak at ~60 meV above the valence band edge.

InAs has applications as high cutoff diodes, galvanomagnetic sensors, and IR detectors. It was first grown by MBE on GaAs (despite the 7.4% lattice mismatch) by Yano *et al.* (*134*) and Meggit *et al.* (*135*), and on InAs by Ludeke (*41*) and Chang *et al.* (*131*).

Showing "GaAs RED patterns" (*41, 134*) InAs can be grown epitaxially between 225 and 530°C (*135*). However, the choice of substrate temperature to a large extent decides the free donor concentration (5 × 10^{16} and 2 × 10^{18} cm^{-3} at T_s-530 and 330°C, respectively) (*135*), which is not related to impurities but to excess As incorporation (*135*). Both high growth rate and As/In flux ratios increased donor levels (*136*). However, in the best layers grown, residual carrier densities could be explained by donors concentrated at the interface (*134, 135*) (Fig. 32). Free-electron mobilities (μ_{n300}) up to 22,000 cm^2 V^{-1} sec^{-1} for $N_D - N_A \sim 3 \times 10^{16}$ cm^{-3} (*134*) and hole mobilities μ_{p300} up to 670 cm^2 V^{-1} sec^{-1} for nominally undoped p-type films have been reported (*137*). Yano found Mg was a useful p-type dopant up to $\sim 6 \times 10^{18}$ cm^{-3} with $\mu_{p300} \sim 150$ cm^2 V^{-1} sec^{-1} and $\mu_{p77} \sim 220$ cm^2 V^{-1} sec^{-1}. Grange recently found that H_2 has little effect but O_2 degrades the electrical properties of MBE InAs (*136*).

In the only article to date on MBE InSb, Baba *et al.* (*138*) treated the effects of T_s and J_{Sb}/J_{In} on the resulting film stoichiometry. At low T_s

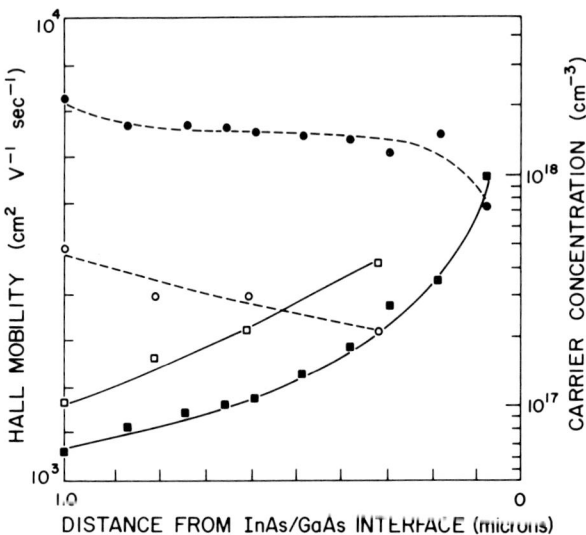

FIG. 32. Variation of carrier concentration and mobility with depth for two InAs epilayers grown on GaAs. ●, Mobility of buffered layer; ■, carrier concentration of buffered layer; ○, mobility of unbuffered layer; □, carrier concentration of unbuffered layer. From Grange *et al.* (*136*).

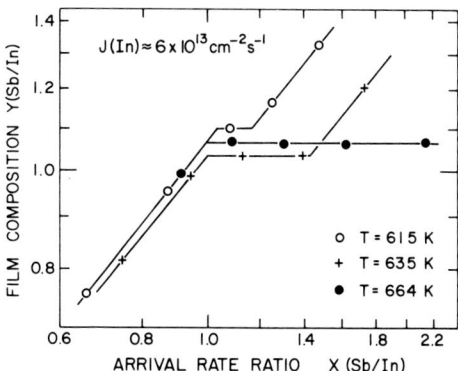

FIG. 33. Auger electron spectroscopic determination of deposited compositions of InSb films vs. flux ratio at different substrate temperatures. From Baba et al. (138).

values and high J_{Sb}/J_{In} ratios, the excess antimony desorption rate can be slower than the arrival rate, resulting in two-phase growth (Fig. 33). It was thus found necessary to choose T_s below the InSb melting point (525°C) and above ~390°C to obtain stoichiometric, single-crystal epitaxy. The flux ratio conditions for single-phase growth in terms of T_s-dependent maximum Sb flux (n_{T_s}) was described by

$$n_{T_s} = 1.6 \times 10^{29} \exp\left(\frac{-2.3 \times 10^4}{T_s}\right) \text{ atoms cm}^{-2} \text{ sec}^{-1}$$

No electrical data were reported by Baba. However, μ_{77} values as high as 40,000 cm² V⁻¹ sec⁻¹ (only ~1/2 bulk values) have been obtained in InSb grown by the three-temperature method (139).

VII. Ternary III–V Alloys

Ternary III–V compound alloys can be conveniently divided into two basic types: (1) III_{1-x}–III'–V and (2) III_{1-x}–V–V'. The former can normally be grown with attention only to the relative group III beam flux ratios providing sufficient excess group V element is used, there is no miscibility gap or lattice symmetry change, and the substrate temperature is not so high that desorption of one of the group III elements becomes significant. In general, the trends of T_{cs} and lattice parameter tend to follow Vegard's law. In the second type, competition for the group V site is a complicated problem depending not only on relative flux ratios, but also on T_s and presumably to some extent the absolute growth rate (J_{III}).

1. III, III'–V ALLOYS

$Ga_{1-x}Al_xAs$ alloys have received much more attention than any other III–V ternary alloy and are discussed further under device applications. However, it must be stressed that, in common with other Al-containing compounds, they suffer from severe background contamination. Yet because of their unique matching properties to the GaAs lattice over the whole composition range and their large band gap they are still probably the most useful type 1 ternary compound for optical-device applications.

In the growth of lattice-matched GaInAs on (100) InP substrates, previously reported by Tateishi et al. (140), the alloy compositional variation across the substrate was overcome by Miller and McFee (48) by a single cell containing mixed indium in gallium solutions. Recently, Ohno et al. (140a) have used a concentric double cell containing elemental charges of Ga and In with relevant orifice areas for growth of spatially uniform matched InGaAs/InP films. This advance can be employed in the growth of other $Ga_{1-x}In_x$–V alloys and probably for GaAl–V alloys as well. Sakaki et al. (137) found the expected monotonic dependence of mobility and bandgap with composition (Fig. 34), which makes this material interesting for FET active layers where $x \sim 0.53$ (lattice match to InP substrates) and for lasers where $x \sim 0.53$ and $x \simeq 0.18$.

Hiyamizu et al. (141) showed the effect of composition in this alloy on the electrical and optical inactive (dead) layer thickness* (Fig. 35) when grown on (100) GaAs (72) and used step grading of buffer layers to reduce the effect. More recently this author (141) used "saw-tooth" grading (see Fig. 36) to produce a buffer layer for 77 K $Ga_{0.84}In_{0.16}As$ homojunction lasers at 0.960 μm. The junctions in this case were prepared by zinc diffusion from $ZnAs_2$, producing threshold currents as low as 3 kA/cm². No doubt much more work will be carried out on this alloy system in the near future.

Recent studies by Scott et al. (142) of $Ga_{0.49}In_{0.51}P$ lattice-matched to GaAs have drawn attention to the decompositional problem of alloy compounds containing both In and P, which reduces the value of T_s at which this material can be usefully prepared. Foxon and Joyce (52) has treated the case of GaInAs and GaInP and found little surface accumulation of In in either compound.

---

* This thickness is defined as the distance from the GaAs substrate/GaInAs film interface to the inflection points in the mobility or photoluminescence plate.

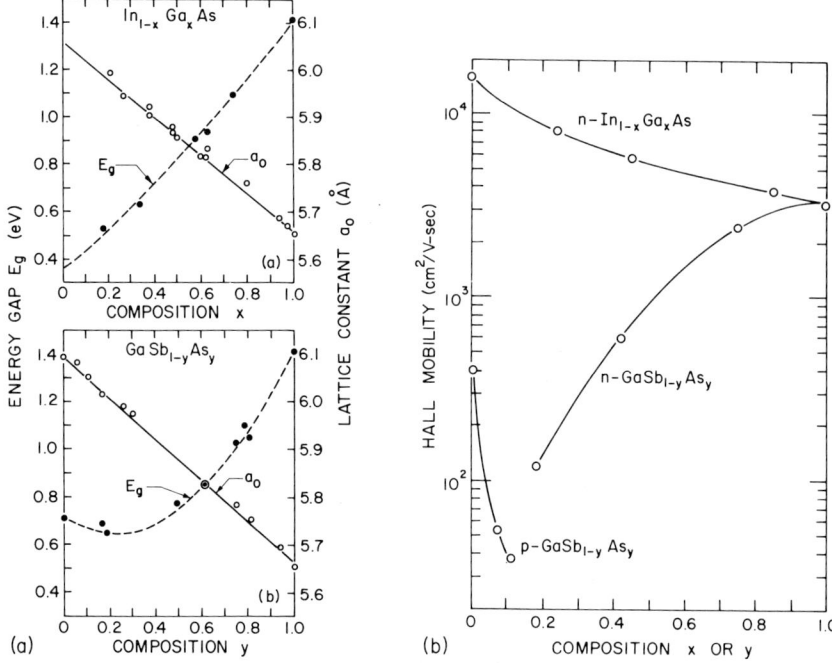

FIG. 34. (a) Experimental band gap and lattice parameter determinations for MBE alloys of the systems $Ga_xIn_{1-x}As$ (top) and $GaAs_ySb_{1-y}$ (bottom). From Sakaki et al. (137). (b) Corresponding mobilities for the tin-doped (3×10^{17}) alloys. From Chang et al. (131).

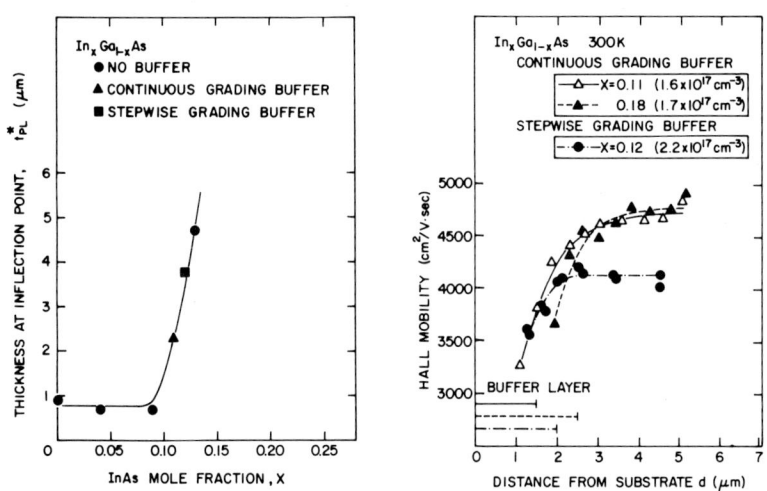

FIG. 35. Thickness of layer at the inflection point of photoluminescent intensity vs. mole fraction of InAs in $Ga_{1-x}In_xAs$ alloys (left). (Note the rapid increase in t above $x = 0.1$.) Depth profiles of electron mobility in $Ga_{1-x}In_xAs$ films for different buffer layers (right). From Hiyamizu et al. (141).

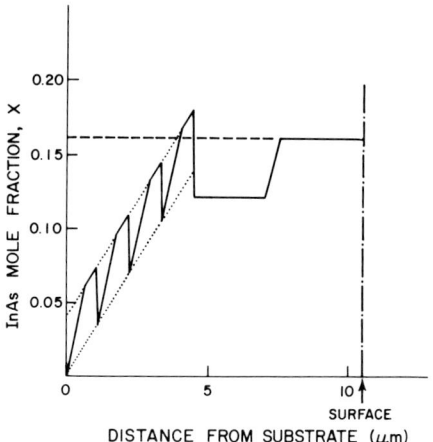

FIG. 36. Sawtooth compositional graded buffer layer used for improved $In_{0.16}Ga_{0.84}As$ laser efficiency. From Hiyamizu et al. (141).

Nothing at the time of writing has been published on MBE growth of InGaSb or Al ternaries other than GaAlAs, but it is expected that the general trends discussed above will be followed.

One major problem that must be mentioned in any discussion of ternary or quaternary compounds is the choice of substrate materials. The problem, not unique to MBE, limits the determination of many electrical properties to layers that can be grown on GaP, GaAs, or InP, which are the only III–V binaries that can yet be prepared semi-insulating in bulk form. Growth of non-lattice-matched layers suffer from electrically and optically interfacial dead layers, and even when matched at the growth temperature, differences in thermal expansion coefficients between layer and substrate can give rise to similar effects. This last problem is discussed for the specific examples of GaInAs and GaAsSb by Chang and Segmuller (143).

2. III–V, V' ALLOYS

Of the possible alloys in the second type of ternary systems (i.e., III–V, V') GaAsP and GaAsSb have received the most study, although recently requirements for long-wavelength photo-detectors have focused attention on alloys of the type InAsSb and InSb(Bi).

The most notable problem in these systems concerns the control of composition. Unlike the type 1 ternaries, there are flux ratio and substrate-temperature-dependent competitions for the group V sites during growth. For both the $GaAs_{1-x}P_x$ system (*1, 58, 140, 144*) (Fig. 37) and the $GaAs_{1-x}Sb_x$ system (*145*) (Fig. 38) (and presumably in the analogous aluminum and indium alloys as well as the III–$P_{1-x}Sb_x$ alloys) the higher Z element has the greater relative incorporation coefficient (K_i). The relative values K_{i1}/K_{i2} for the two alloys studied (*144, 145*) decreased in magnitude with increasing T_s, but reasonable control only appears to be obtained if the ratio of higher Z group V element flux to the group III element flux is kept below that required for stoichiometric growth at that particular T_s.

Little has been recently published in the GaAsP alloys system save that of nitrogen ion doping by Matsushima *et al.* (*123*) mentioned earlier.

The lattice constant of $GaAs_{1-x}Sb_x$ alloys grown on GaAs substrates follows Vegard's law across the whole composition range (*137*), showing no miscibility gap (*146*). However, in common with $Ga_{1-x}In_xP$ alloys (*130*) the differential thermal mismatch causes matching composition to be different at different growth temperatures (~550°C) (*146a*). The surface structure of this alloy system for high values of $x/1 - x$ has been found to

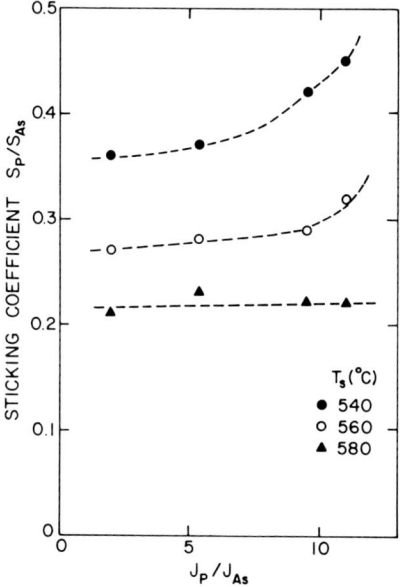

FIG. 37. Sticking coefficient ratio S_P/S_{As} as a function of incident fluxes J_P/J_{As} at various substrate temperatures for GaAsP growth by MBE. From Matsushima and Gonda (*58*).

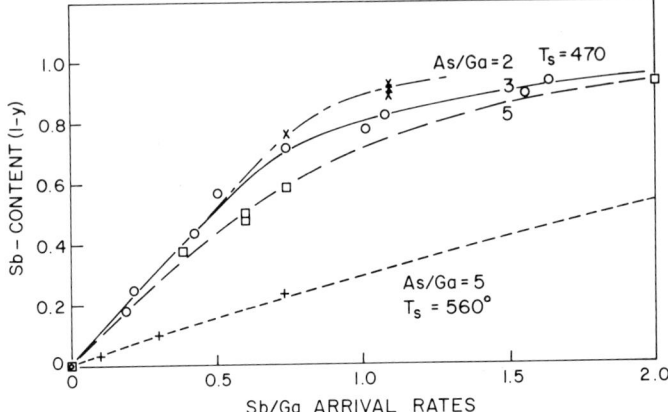

FIG. 38. Dependence of Sb content in $GaSb_{1-y}As_y$ films on the Sb flux for various As fluxes (normalized to the Ga flux) at 470 and 560°C. From Chang et al. (145).

be different from GaAs by Ludeke (41) and by Maruyama et al. (147). The former reports $(2 \times 3) \times (2 \times 6)$ patterns and the latter diffuse elongated streaks in the <114> direction, which were interpreted as indicating surface atom displacements.

Chang et al. (131) in studies of $GaAs_{1-x}Sb_x$ found that tin behaved amphoterically in this alloy (see Fig. 34). It is not therefore known if the electron mobility follows x monatomically. The predominant carrier types are donors for $x < 0.82$ and acceptors for $x > 0.86$. This composition range also corresponds to the changeover of the surface reconstruction from the GaAs-like arsenic-rich $C(2 \times 8)$ and Ga-stabilized $C(8 \times 2)$ patterns to the Sb-stabilized $C(2 \times 6)$ and the Ga-stabilized (2×3) patterns (41) (see Fig. 6).

Sakaki (137) studied the systems InGaAs and GaAsSb and showed monatonic downward bowing of the band gaps toward the center of the composition range for InGaAs and toward $x = 0.2$ in the system $GaAs_xSb_{1-x}$. Cho (148) graded $GaAs_xSb_{1-x}$ buffer layers from $x = 1$ to $x = 0.9$ during growth at 580°C for light-emitting diode layers (see Section VII).

3. QUATERNARY ALLOYS

The only reported MBE growth of a quaternary III–V layer was that by Cho of lattice-matched $Ga_{1-x}In_xAs_{1-y}P_y$ on InP. However, few details or results were recorded (*148*) save that initial photoluminescence results were poor compared to MBE $Ga_{0.9}As_{0.1}Sb$ LEDs. No doubt the near future will see more work on these potentially important alloys.

VIII. Device Applications

The many devices that depend upon two-dimensional semiconductor structures can be divided into two basic classes: majority and minority carrier devices. The first usually exploit the favorable electron mobilities of the direct gap III–V semiconductors for microwave and magnetoelectric applications, and the latter the ability to tailor the band gap and hence the wavelength for optical emission and detection applications. Devices that have been fabricated from MBE layer structures are described under these two basic classifications below.

1. Majority Carrier Devices

The rapid increase in volume of and in information in communications, radar, and computation over recent years has greatly increased the demand for microwave and millimeter wave devices. The requirement for higher frequency carrier wave generation, modulation, signal detection, and amplification as well as logic elements, has placed demands on the sophistication, and in many cases reduced the physical dimensions, of devices required to perform these and other related operations. The ability of MBE to control layer thickness and free-carrier profiles down to angstrom dimensions has led to its use for several important device applications.

GaAs field effect transistors are typically used as low-noise microwave signal detectors and microwave signal generators. Both low-noise and high-power FETs require n-type layers less than 1 μm thick grown on, and physically supported by, nonconducting substrates. The product of active layer thickness and doping level required is typically $2-4 \times 10^{12}$ cm^{-2}, which corresponds to $\sim \frac{1}{2}$ that depletable by the gate electrode before breakdown.

In low-noise FETs, the doping profile required is defined such that the transconductance g_m should be as high as possible, especially near the gate pinch-off voltage. Now, as the mobility of carriers is dependent upon the quality of the epitaxial region closest to the substrate interface, it has been found advisable in most cases to first grow buffer layers prior to active layers to avoid deleterious substrate-related interfacial effects (*46*).

Low-noise MBE FET structures were first reported with DC g_m values ~4 mmhos by Naganuma et al. (*149*). However, transfer characteristics showed loops above 2 mA drain current and no microwave results were reported. Subsequently, Cho and Chen (*150*) reported low-noise microwave signal gains ~10 dB with corresponding noise figures ~3 dB at 6 GHz with $g_m \sim 28$ mmhos. Values of 3.3 dB noise with 5.6 dB gain at 10 GHz and 35 mmhos transconductance were also obtained by the author (*69*). Later results of low-noise FETs with ~2 dB noise and 12 dB gain at 6 GHz were given by Cho et al. (*46*). Most recently, results of low-noise FETs have been reported (*151*) using very thin (~1000 Å) heavily doped (3.5×10^{17} cm^{-3}) MBE GaAs layers with 0.3 μm gate lengths. In these devices, a gain of 15 dB was obtained associated with noise figures as low as 1.5 dB at 8 GHz.

In reference (*46*), the first MBE power FET results were also given: 0.43 W/mm gate width at 1 dB compression and 4.4 GHz operating at 35% power-added efficiency. Recently, power FET results that compare favorably with vapor epitaxial layers were obtained (*152*) using Si doping to obtain sharp doping profiles at the buffer–active layer interface: 38% power added efficiency at 8 GHz with 1 dB compression output power of 1.4 W. Murotani et al. (*78*) reported low substrate temperature buffer growth for power FETs with figures of merit, ~0.7 W/mm gate width at 6 GHz, and a linear gain of 10.4 dB.

To date, all power FET layers have been grown with nominally flat doping profiles. This leads to nonconstant g_m and hence nonlinearity of I_{drain} with V_{gate}. The problem of linearity is one that can be improved successfully by tailoring the doping profile (*153*). The author has recently shown significant improvement in linearity by exponentially increasing free-donor densities in the active layer toward the buffer layer by predepositional doping (*154*). The use of low–high and hyperabrupt profiles has also been found to increase linearity (*96*).

Other problems such as high source-to-drain series resistance (R_s) in both low-noise and power devices have been significantly eased by use of (tunnel contact) surface layers doped above 10^{19} (*155*) and by use of thin (~100 Å) degenerately As-doped epitaxial Ge overlayers (*156, 156a*). The latter structure also has been found to be stable at higher temperatures.

Variable-capacitance devices, microwave varactors and mixer diodes find application in signal detection, frequency division, and mixing. They share two important requirements in that the capacitance should bear a predescribed relationship over a large range with applied bias and that the RF resistance should be small. The first of these is usually met if a power law relationship holds between doping and distance in the layer from the junction (Schottky or p–n), i.e., $n_l \propto l^m$.

Schottky barrier hyperabrupt varactors were first reported (*157*) on MBE layers with $m = 1.5$ giving the expected $V - \phi \propto c^{-1/2}$ characteristic (where ϕ is the Schottky barrier height) that showed a resonant frequency linearly related to the applied bias voltage. A similar relationship ($m = 1.2$) was obtained by Covington and Hicklin (*158*) for a p$^+$–n hyperabrupt varactor. Series resistances (R_s) offered by nondepleted regions, contacts, and n$^+$ substrates were ~1.3 Ω for a 30 μm diam diode in (*158*) and 1.8 Ω in (*157*). Cutoff frequencies in these devices were estimated to be ~100 and 40 GHz, respectively.

Cryogenically cooled mixer diodes fabricated from MBE GaAs (*159*) were found to have exponential I–V characteristics and shot-noise-limited temperatures of ~55 K. R_s values ~1.5 Ω for 15 μm diam mixer diodes with Schottky ideality factors of 1.06 and cutoff frequencies ~42 GHz were reported by Wood et al. (*77*). Meeks et al. (*160*) achieved R_s values of ~7 Ω with ideality factors ~1.1 for similar devices with 2 μm diam electroplated gold Schottky barriers on tin-doped MBE GaAs layers.

In IMPATT diode devices, a high doping spike is introduced inside a low-doped drift region. Avalanching occurs in the high-field region. The resulting "plasma domain" drifts across the low-doped region with a characteristic transit time. Recombination at the anode contact allows the avalanche field again to be built up. Although intrinsically noisy because of the avalanche process, very good DC–microwave conversion efficiencies can be achieved. Avalanche mode microwave power generation has been achieved in IMPATT devices made from MBE GaAs (*161*). At 11 GHz a 44 dB noise measure with 2.8 W power output at 18% efficiency was achieved.

2. Future Majority Carrier Device Applications

Millimeter wave Gunn devices have not received attention by MBE, but it is felt that there are promising applications in this area. Device efficiency can be optimized by controlling a 6–7% slope in the donor level (*162*) through 1–2 μm layers, which is relatively easily achieved by MBE.

Significant improvements of FET output impedance can be expected with high indirect bandgap (e.g., $Ga_{1-x}Al_xAs$) buffer layers (*163*). Enhanced carrier confinement and flatter I–V curves should result as carriers that do penetrate into this type of buffer layer will suffer from much reduced mobility.

Use of thin oxygen-doped $Ga_{1-x}Al_xAs$ surface films could also enable reduced gate leakage in forward-biased conditions in MISFETs (e.g., *164*).

Future use of "two-dimensional electron gas" phenomenon, associated with modulation-doped heteroepitaxial conduction band discontinuities (*165*), will no doubt be found in majority carrier devices. Finally, the use of "lamellar doping" (*166*) will also be investigated for certain higher harmonic generating FETs and enhanced mobility devices.

3. Minority Carrier Devices

Most applications of minority carrier devices are related to light emission, detection, or amplification. Two basic types of light emission are possible with direct-gap semiconductors. First, simple recombination of minority carriers injected across forward-biased p–n junctions produces photon emission with energies close to that of the band gap. Second, if such injected electrons and the resultant light are both confined by higher bandgap and higher refractive index layers, respectively, efficient stimulated emission is produced (lasing) at high enough injection currents. In both cases, by choosing the compound or alloy with the correct bandgap, it is possible to decide a priori what wavelength the emission will be.

High-intensity LEDs have numerous display applications and are potentially important for optical communications. Two MBE-grown III–V alloy heterojunctions structures have been successfully made into LED devices. n GaAs$_{0.9}$Sb$_{0.1}$ on (100) GaAs layers were grown using compositional grading to reduce lattice mismatch problems. The heterojunction structure was used to prepare a 1.0 μm wavelength diode by Zn diffusion through SiO$_2$ masks (*148*). The resulting Burrus (*167*) type diode (see Fig. 39) had an external quantum efficiency ~0.1% between 60 and 150 mA pulsed input current. Double heterojunction (electron confinement) Burrus-type LEDs have also been made from n-type Ga$_{0.7}$Al$_{0.3}$As structures (*168*). These junction devices about 0.05% efficient at $\lambda_{max} \sim 0.87$ μm, were improved by Zn diffusion.

Recently, GaAsP alloys that were grown by MBE and doped with ionized nitrogen have shown nitrogen-center-related photoluminescence at wavelengths ~6200 Å.

There are many other possible combinations of substrate and layer structures useful for various wavelength LEDs, which have not yet but will no doubt be fabricated in the near future.

An extensive amount of work has been carried out at Bell Laboratories on MBE-grown GaAlAs heterojunction systems for laser applications and at IBM (*2*) for transferred electron device effects. Notable achievements

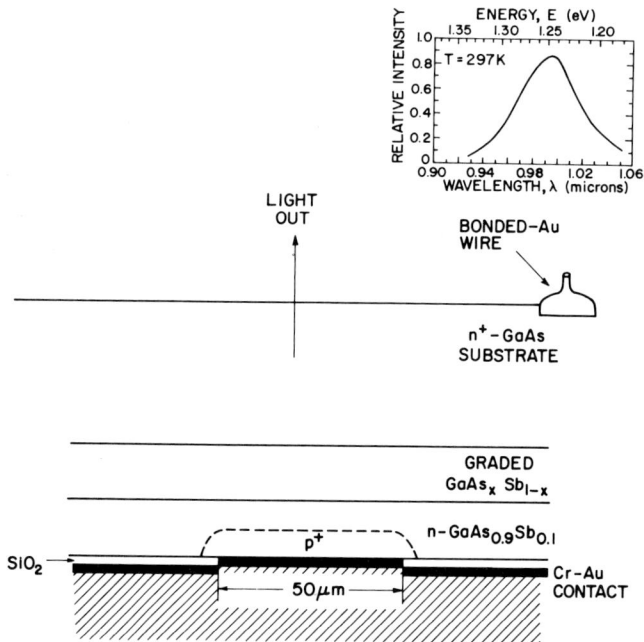

FIG. 39. Schematic of planar $GaAs_{0.9}Sb_{0.1}$ p-n junction LED with spontaneous emission shown inset. From Cho *et al.* (*148*).

in GaAlAs double heterojunction lasers (DHLs) have previously been reviewed (*1, 3, 6*). Little more will be added here except to draw attention to certain key articles. The continuous operation of 8820 Å $Ga_{0.65}Al_{0.35}As$/GaAs DHLs at room temperature (up to 100°C) as early as 1976 was reported by Cho *et al.* (*169, 170*) with threshold currents comparable with LPE diodes: 2.7×10^3 and 1.7×10^3 A/cm² for 380 and 100 μm cavity lengths and recently by Tsang with values as low as 800 A/cm² (*170a,b*). Distributed-feedback DHLs were grown over corrugated GaAlAs surfaces (produced by ion milling LPE GaAlAs) by MBE (*171*), with taper-coupled waveguides (*172–174*) (Fig. 40) and with embedded-stripe geometry (*175*) (Fig. 41). Multilayer structures have also shown interesting lasing properties (*176*) although the predicted lower thresholds have not materialized (see section on multilayer structures). Recently, 77 K lasing in zinc-diffused $Ga_{0.47}In_{0.53}As$/GaAs structures have been achieved (*137*) with a sawtooth graded buffer layer. Confinement regions in this structure were $Ga_{0.84}In_{0.16}As$ and $Ga_{0.82}In_{0.18}As$. Emission was at 0.96 μm with $I_{threshold}$ as low as 3 kA/cm².

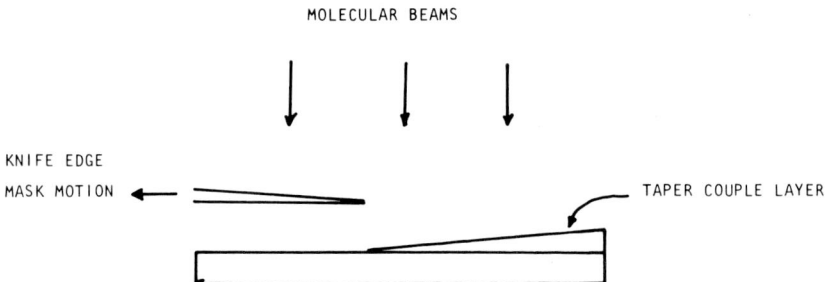

FIG. 40. Taper-coupled laser structure. From Reinhart et al. (173).

Pulsed 300 K lasing at 1.65 μm was also shown (103) using Cd-diffused MBE InP/Ga$_{0.47}$In$_{0.53}$As/InP structures. In this example, a mixed Ga/In effusion source was used to achieve lattice-matched compositions. The multiple-mode operation was found to be dependent upon device current, which had threshold values as low as 3.2 kA cm^{-2} with external quantum efficiencies ~20%. The latter composition has been obtained using computer-controlled effusion rates of Ga and In ($x \pm 2.5\%$) (177). No lasing action was reported in this case, although luminescent efficiencies were "high," centered at 1.6 μm.

Apart from the above electroluminescence devices, the ability to produce certain passive components, which may eventually allow complete integration of optical (IO) systems, has been demonstrated by MBE techniques. Among the more interesting are the preparation of taper

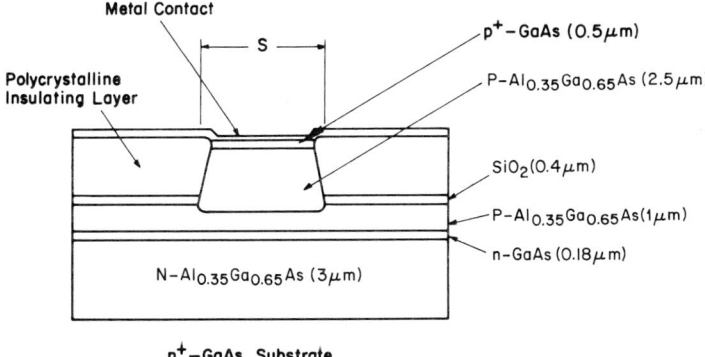

FIG. 41. Schematic of embedded-stripe GaAs/GaAlAs double heterojunction laser using selective epitaxy resist techniques. From Lee and Cho (175).

couplers by moving shutters during growth (*173*) or by combination of LPE and MBE (*178*). Single GaAs/AlAs (*172*) multilayers (*179, 180*) and three-dimensional waveguides (*181*) (Fig. 42) have also received attention directed toward low-loss integrated optical operation (*182*). Earlier work on GaP waveguides produced second-harmonic generation (*183*).

Other minority carrier applications of MBE that have been demonstrated plausible are those of photocollection (*184*) and amplification (*185*). In the former, MBE p GaAs/n $Ga_{0.67}Al_{0.33}As$ heterojunctions were shown to have high (0.4) quantum efficiencies for 1.75 eV photons. Shallow homoepitaxial p–n junctions have been used in high-efficiency ($\sim$16%) solar cell devices (*185a*), although the production costs would probably be prohibitive for the areas of devices that would be required in practical solar cell applications.

The extreme flatness of MBE GaAs has been used to grow unintentionally doped p-type nucleating layers on (111)B GaP at 520°C. The 2 μm thick films were used as nucleating surfaces for growth of $>10^{18}$ Zn-doped vapor pressure epitaxial GaAs. The resulting structures were C_S–O_2-activated for photocathodes. Although the efficiencies of such devices were very low because of the lattice misfit between GaP and GaAs (6.4%) the resulting resolution and image quality was good (see Fig. 43).

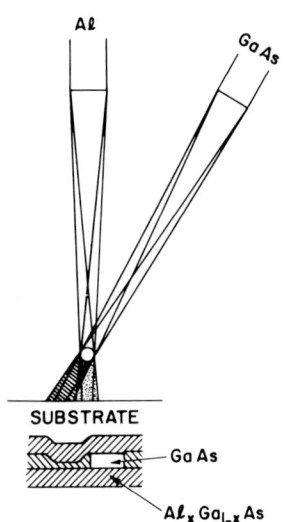

FIG. 42. Schematic of method for preparing three-dimensional waveguide structure. From Cho and Reinhart (*181*).

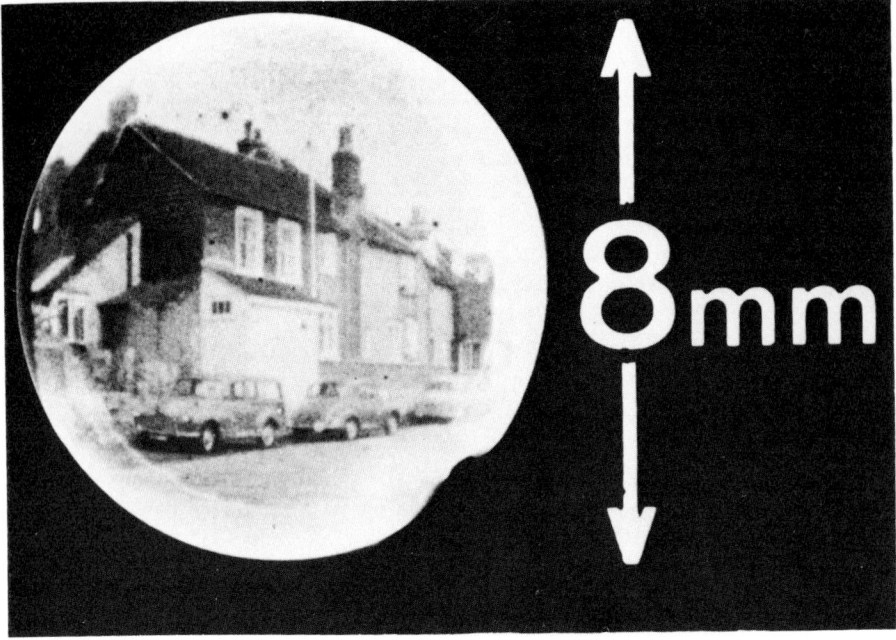

FIG. 43. Image from hybrid MBE/VPE GaAs/GaP photocathode structure. From Wood et al. (185).

4. MULTILAYER STRUCTURES

Periodic modulation of majority carrier type and bandgap both show interesting electrical (185b) and optical properties. It is in the preparation of this group of structures that MBE offers spectacular advantages.

In the former, the majority carrier type is alternated during growth of one compound or alloy, and in the latter the alloy composition or compound is alternated within limits allowed by lattice matching considerations. Cho first prepared n.i.p.i. structures with GaAs utilizing background donors for n-type layers and Mg for p-type (186). Dohler and Ploog (105) utilized the amphoteric nature of Ge incorporation in MBE GaAs to prepare n.i.p.i. structures in which the site occupancy of the germanium was periodically changed by variation of the substrate temperature during growth (187). Potentially interesting variations of effective bandgap and carrier mobility were also outlined for structures with doping superlattice periodicities about 1 order of magnitude greater than the heterojunction superlattices (typically 100–1000 Å) (*187, 188, 188a*).

Another potential application of the unique definition of doping structures is currently under investigation in the author's laboratory. By periodically terminating nominally undoped growth while dopant (in our case germanium) is deposited, dopant atoms can be confined to parallel atomic planes within the layer. Variations in the amount of dopant deposited and the separation of dopant lamellae show interesting anisotropic transport properties. Figure 44 shows C–V profiles of the first of such GaAs lamellar doped structures grown (*166*).

Surprisingly, the free electron mobilities in such lamellar doped films were as high as in uniformly doped films with equivalent average electron concentrations. This has allowed synthesis of complex free donor profiles

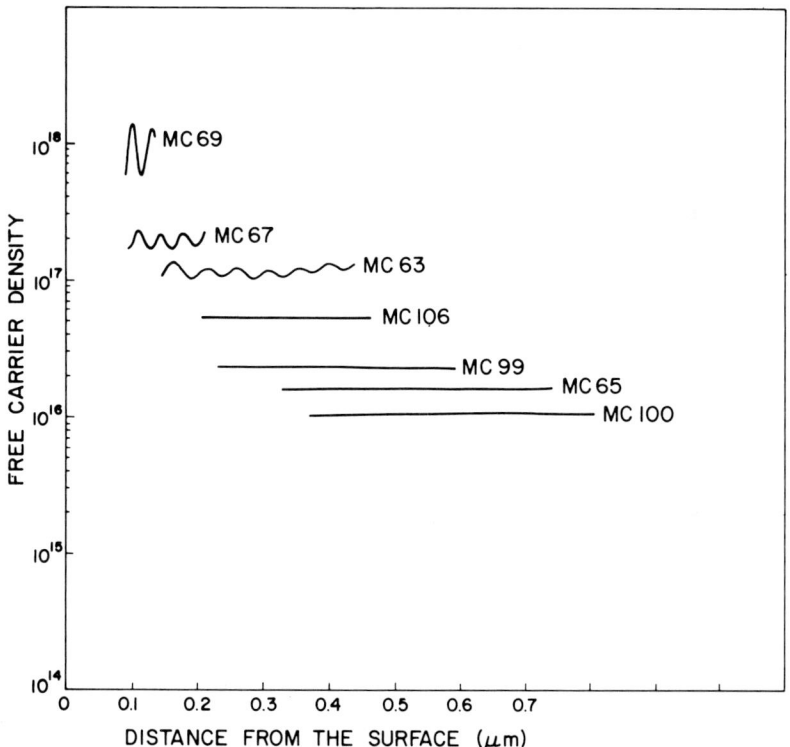

FIG. 44. Free-donor profiles (C–V) for lamellar-doped (Ge) GaAs MBE films. The Debye length effect is clearly seen in the layers with mean doping levels below 10^{17} cm^{-3}.

by either or both variation of the lamellar spacing or the individual planar dopant (Ge) concentrations. Attempts to prepare p-type profiles by lamellar Ge doping on Ga stabilized surfaces produced only heavily compensated n type films showing the nonunity incorporation coefficient of Ge on Ga stabilized surfaces.

Single lamellae with $>4 \times 10^{12}$ cm^{-2} free electron densities have been buried in buffer layers and used for constant capacitance enhancement and depletion mode FETs.

5. Heterojunction Structures

$Ga_{1-x}Al_xAs$/GaAs multilayer structures have been studied more extensively than any other since the first MBE demonstration (186). Discontinuities and new vibrational modes in phonon spectra, resulting from zone-folding effects, were observed in alternating GaAs/AlAs monolayer films by Raman spectroscopy (189, 190). Quantization of confined carrier motion in ~100 Å period GaAlAs/GaAs ultrathin structures gave rise to novel optical absorption phenomena (190–192) (see Fig. 45).

Stress associated with the interfacial mismatch can be less than in equivalent LPE structures as shown by exciton splitting and shifts in the 2 K photoluminescence spectra, and is independent of compositional grading (193, 193a). In this study annealing structures up to 850°C showed no significant interdiffusion. Chang (194) found that Mg center-associated

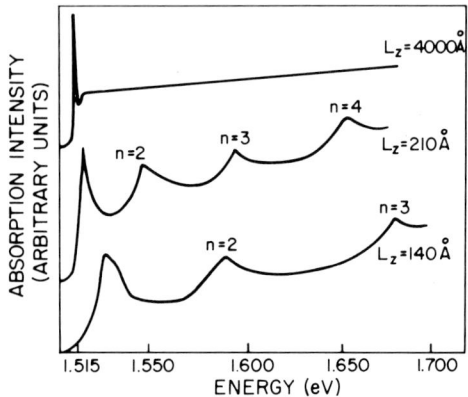

Fig. 45. Absorption spectra (2 K) of 400, 210, and 140 Å thick GaAs layers between $Al_{0.2}Ga_{0.8}As$ barriers. The quantum size effect manifests itself as an increase in the bulk exciton energy peaks in the thinner layers. From Dingle (191).

luminescence efficiencies rose by up to a factor 10 after 2 h bakes in arsenic atmospheres, and determined the activation energy of interdiffusion to be 4.3 and 3.6 eV, respectively, for GaAs and AlAs.

Imaging and diffraction mode transmission electron microscopy (*195*) demonstrated virtually perfect epitaxy with predicted periodicity and indicated a critical temperature (610°) above which surface roughening disrupted the ordered monolayer growth (*196*). Combination sputter etching AES (*197*), He ion backscattering, secondary electron microscopy (*198*), shallow-angle X-ray scattering studies (*130*), and TEM studies (*195*) have all been employed to characterize the interfacial properties of GaAs/GaAlAs multilayer structures.

A novel development of the work of Dingle, Stormer, and others (*162, 199–201*) arises from the conduction band discontinuity between modulation-doped GaAs and GaAlAs multilayer structures. Free electrons from donor atoms in thin doped GaAlAs layers populate the conduction band wells of undoped sandwiched GaAs layers (see Fig. 46). The dimensions of this effect are of the order of a few Debye lengths (L_D), but spectacularly high two-dimensional electron mobilities at 300 K are measured as the GaAs is relatively free from ionized impurity scattering centers (see Fig. 47).

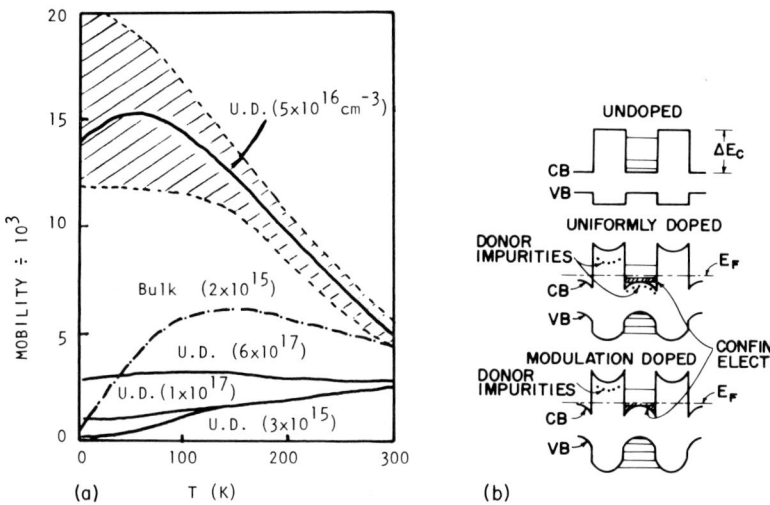

FIG. 46. (a) Mobilities vs. temperature for Si-doped GaAs/GaAlAs modulation-doped samples. Uniformly doped samples are shown for comparison (U.D.) from Dingle *et al.* (*165*). (b) Simplified band diagrams showing 2D electron gas regions.

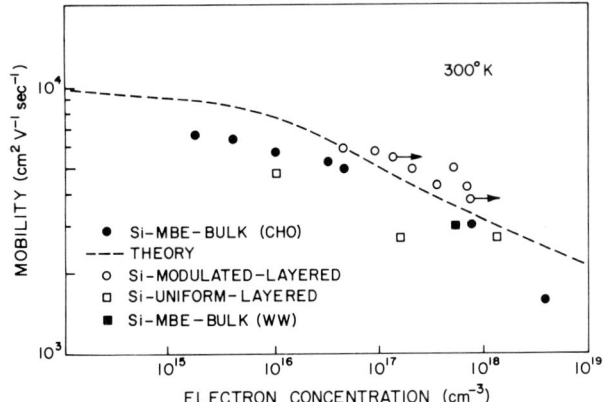

FIG. 47. 300 K mobilities of a range of Si-doped GaAs and Si-doped GaAs/Ga$_x$Al$_{1-x}$As superlattices. The solid circles and the dashed line (Brookes-Herring theory for $(N^+ + N^-/n) = 1$) are taken from Cho and Arthur (1).

Investigations of GaSbAs/GaInAs lattice-matched multilayer structures (202–204) show novel heterojunction properties in which the conduction band edge of the GaSbAs is very close in energy to the top of the GaInAs valence band. However, because of the difficulty in controlling the alloy composition for lattice matching, attention has recently switched to the simpler GaSb/InAs system alloys (205).

6. APPLICATION OF MBE PERIODIC STRUCTURES

Periodic GaAs/GaAlAs structures have shown useful wave-guiding properties (179, 180). Yeh et al. (206) studied optical wave propagation in periodic layered media. Laser oscillation (207) and optical second-harmonic generation (208–210a) have been observed in such periodic layers. Quarter-wave stacks (211) and interference filters (212) have also been prepared from GaAs/Ga$_{0.7}$Al$_{0.3}$As and GaAs/AlAs multilayers, respectively.

IX. Technological Developments

1. METAL SEMICONDUCTOR SYSTEMS

a. Schottky Barriers. Two technologically important extremes of metal semiconductor systems have been studied by MBE: rectifying or Schottky and nonrectifying or ohmic contacts. In the former, the Fermi level is pinned in the forbidden gap of the semiconductor by unfilled states in the metal–semiconductor interface (*213*).*

Whether the states are formed by surface steps, oxygen, or are metal induced differs with crystal orientation and the particular metal–semiconductor combination. As the barrier height of metal/(100) GaAs systems is almost independent of the electronegativity of the metal in most practical cases, heretofore it is felt that the pinning on (100) covalent III–V semiconductors is complete before metallization. MBE, however, offers the ability to deposit metal films on freshly prepared oxygen-free surfaces and hence to study virtually perfect Schottky barrier behavior.

Aluminum evaporation on arsenic-rich GaAs (100) surfaces was probably the first attempt to prepare epitaxial metal/semiconductor interfaces. Ludeke *et al.* (*214*) found initial growth of Al on GaAs at 400°C to be of mixed type with (110) Al and (100) Al on (110) GaAs. The latter dominated after ~25 Å with Al (100) parallel to (110) GaAs, which indicated the reconstruction of the underlying GaAs plays a deciding role in the epitaxial relationship. The same relationship was found when GaAs growth was subsequently resumed on the Al layer. Other GaAs orientations also gave epitaxial relationships, but the (100) surfaces produced the best topography.

Al layers sandwiched between GaAs epilayers exhibited back-to-back diode characteristics with breakdown voltages consistent with barrier heights ϕ_b ~0.6 eV. No further work has been reported on growth of semiconductors on epitaxial metals. However, it is the author's belief that this type of structure will find device applications in the future for ultra-high-speed electronic devices.

At room temperature, epitaxial Al deposition was single crystal with (100) Al on (100) GaAs (*215*). The resulting lattice mismatch was reduced by 45° rotation about the (100) axis with respect to the GaAs surface. Associated barrier heights of 0.66 and 0.72 eV were measured for arsenic-rich and gallium-stabilized surfaces, respectively, with ideality factors ~1.04.

Massies *et al.* (*216–218*) prepared Al, Au, and Ag contacts on various ion-sputtered and annealed surfaces of GaAs and InP. Epitaxial relationships for Al similar to those reported by Cho (*215*) were found for $T_s <$ 150°C and by Ludeke (*214*) at elevated temperatures ($T_s >$ 300°C). Silver contacts had reversed (cf. Al) epitaxial relationships with respect to

---

* The origin of these states is still debatable and the reader is referred to the proceedings of the PCSI Conferences held at Asilomar 1979 for further discussions of this controversial topic.

substrate temperature, i.e., (110) Ag‖(100) GaAs at ~150°C and (100) Ag‖(100) GaAs above 300°C. Silver Schottky barrier heights were ~0.6 and 0.67 eV on arsenic-rich and gallium-stabilized surfaces, respectively.

Epitaxial Ag/InP barrier heights were 0.43, 0.38, and ~0 for the series In-, As-, and P-stabilized surfaces. The essentially ohmic nature of the contact found for Ag/InP (P) was in agreement with that first reported by Farrow (6).

In a series of selective evaporations, Covington and Meeks (219) reported barrier heights of 0.81 for Ag and Au and 0.55 eV for Al epitaxial Schottky diodes formed on arsenic-rich MBE-grown GaAs surfaces at ~150°C.

b. Ohmic Contacts. Conventionally, ohmic contacts have been prepared by high-temperature alloying of evaporated metal combination overlays. The metal combination normally includes an element such as germanium (for n type) and zinc (for p type), which diffuses into the epilayer surface. A small amount of irregular dissolution (pitting) of the semiconductor also occurs, giving nonuniform interfaces. The combined effect is to produce a region under the metal in which the doping level is sufficiently high ($>10^{19}$ cm^{-3}) that the depletion region (associated with the Fermi level pinning at the interface) is sufficiently thin that it presents little or no barrier to tunneling. Relatively poor control of the alloying process leads to nonuniform R_s values and can cause eventual device failure by resistive heating or current crowding in pitted regions.

In 1976, the author demonstrated the efficacy of tin predeposition at normal growth temperatures to overcome interfacial doping level dips in MBE GaAs (69), and extended the application to the formation of ohmic contacts at lower substrate temperatures (220).

DiLorenzo and Cho (221) demonstrated the use of such nonalloyed ohmic contacts by tin-doping surface GaAs regions to $n > 5 \times 10^{19}$ cm^{-3} and found specific resistances of *in situ* grown contacts to be $\sim 2 \times 10^{-6}$ Ω cm^{-2}, in good agreement with theory (222). FET devices prepared with such contacts were consistently more uniform than equivalent alloyed contact devices (223). Tsang (224) prepared similar ohmic contact layers for p-type films by degenerate doping with beryllium.

A further practical problem not completely overcome by these "superdoped" surface contact regions is the temperature stability of device contact regions. The author has shown (77) that epitaxial growth of thin (~300 Å) degenerately doped ($>5 \times 10^{19}$ cm^{-3}) germanium films directly onto germanium superdoped GaAs surface regions gives specific contact resistances $<7 \times 10^{-8}$ Ω cm^{-2} (156, 156a), which exhibit high-temperature stability (225).

2. Insulator–Semiconductor Systems

A second type of system that is finding more technological employment in practical III–V devices is that of insulator–semiconductor (mis) structures (226). Uses are found as antireflection coatings for solar cells, for passivation of high-field device surfaces, for implant annealing and dopant diffusion, for beam–lead isolation (e.g., in mixer diodes), and as high dielectric constant films for MOSFETs.

The many techniques for preparing native oxides include anodic, thermal, and plasma oxidation, and refractory oxides/nitrides deposited by sputtering and vapor deposition. All these techniques have drawbacks and as yet there is no universal method for all applications. However, MBE has shown promise in several novel ways of preparing such films. Casey and Cho (227) used the gettering action of Al in the MBE growth of $Ga_{0.5}Al_{0.5}As$ alloys to prepare oxygen-doped ($E_a \sim 0.64$ eV) semi-insulating ($\sim 10^9$ Ω cm) layers for use as "lattice-matched" MIS structures (228). No hysteresis, capacitance dispersion, or interfacial trap densities ($>10^{11}$ cm^{-2}) were found in these studies.

Tsang (229) oxidized MBE AlAs layers, to form very uniform insulating amorphous oxides. Dry oxidation produced structures showing inversion and interface trap densities as low as 4×10^{11} cm^{-2} and little C–V hysteresis (229a). Breakdown fields were above 2×10^6 V/cm compared with 10^6 V/cm reported (230) for thermally grown native oxides on MBE GaAs. Low-temperature amorphous Al oxides grown *in situ* (231) were insulating, with $\sim 3 \times 10^{11}$ cm^{-2} interface states when prepared with an intermediate GaAlAs layer.

Work in this area by MBE and associated techniques has only recently been investigated and will no doubt become more important as the more fundamental problems of III–V epitaxy are solved.

3. Selected-Area Epitaxy

The ability, a priori, to define regions of electrically active films within a matrix of semi-insulating material, or on a semi-insulating surface, has many interesting and technologically important advantages for both discrete and integrated circuit device applications.

Four variations on this basic theme have already been demonstrated in MBE investigations: (1) growth on lithographically defined epiresist regions, (2) self-masking, (3) static proximity masking, and (4) moveable proximity masking.

In the first modification (232), a thin, thermally stable, noncontaminat-

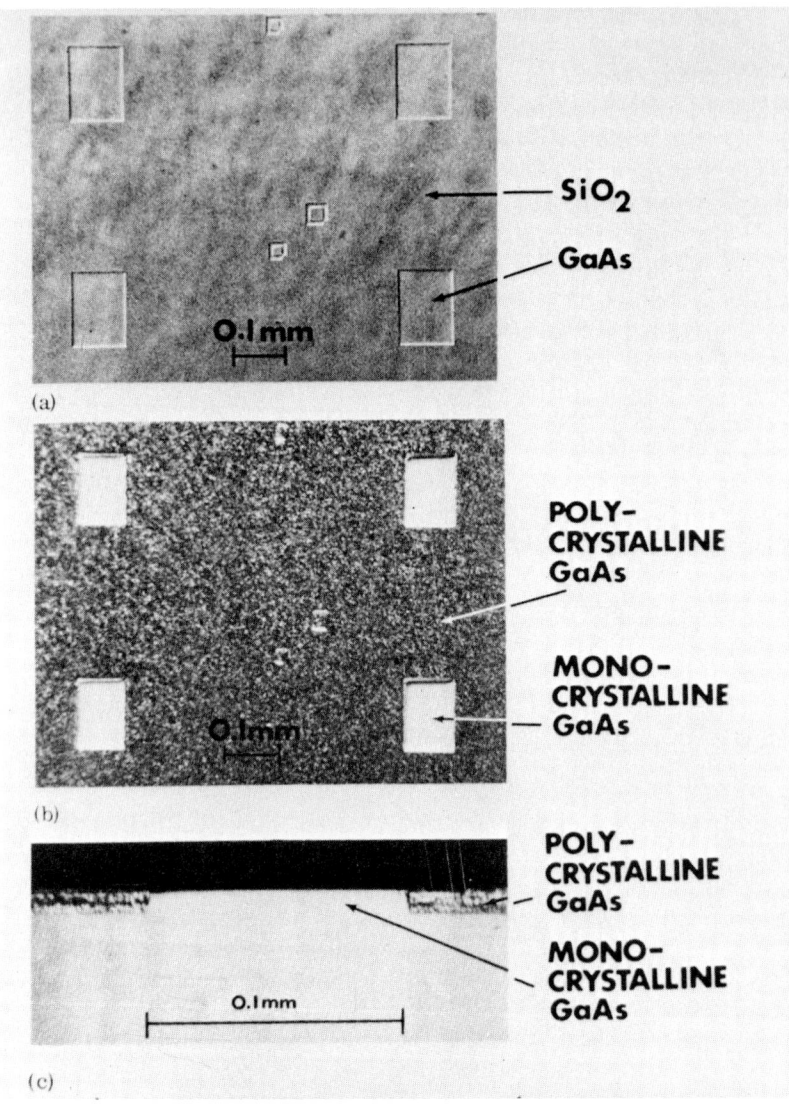

FIG. 48. Photomicrographs of (a) SiO$_2$ patterns on a GaAs substrate; (b) after 6 μm overgrowth of MBE GaAs, (c) enlarged (110) cleaved cross section of (b) showing featureless growth interface between epilayer and substrate, and (c) polycrystalline growth on the SiO$_2$ film.

ing material such as SiO_2 or Si_3N_4 or a native oxide was prepared on the substrate surface. Conventional lithography was used to "open up" free substrate surface regions. Growth over such structures gave good electronic-quality GaAs layers on the bare surface regions, in a matrix of polycrystalline material on the resist (see Fig. 48). The spatial resolution of this process is about 1 μm, but is very orientation dependent. The polycrystalline regions were highly resistive ($>10^6$ Ω cm).

If an epitaxy resist is used that can be selectively etched, the polycrystalline material can be removed by lift-off techniques. These have been used for producing inlaid n^+ contact regions and self-aligned metal contact regions (233).

The second modification requires conventional lithography on substrate or epilayer surfaces directly, followed by orientation-selective etching to produce undercut edges that act as masks for depositing structures in the etched regions (234) (see Fig. 49).

The third more conventional shadow technique of selective epitaxy was demonstrated by Tsang and Ilegems (235) by clipping tantalum or silicon masks in close proximity to the substrate to define areas of MBE GaAs growth (see Fig. 50). The problems associated with this technique relate to differential thermal expansion of the masks, which limits the spatial resolution to about 5 μm.

The fourth technique (a variation of the third) was to move the mask during growth. By varying hole shape and movement rate, it was shown possible to write three-dimensional patterns on the surface of a substrate with dimensions down to 10 μm (236) (see Fig. 51).

Preferentially etched structures for self-aligned masking have also been successfully used by Nagata et al. for GaAs (237) and $Ga_{1-x}Al_xAs$ (238). The lateral accuracy of epitaxial alignment was found to be ~0.1 μm and limited by preferential facet growth. At 550°C with $J_{Ga} \equiv 6000$ Å h^{-1} growth rate, the surface diffusion length of gallium atoms on GaAs was

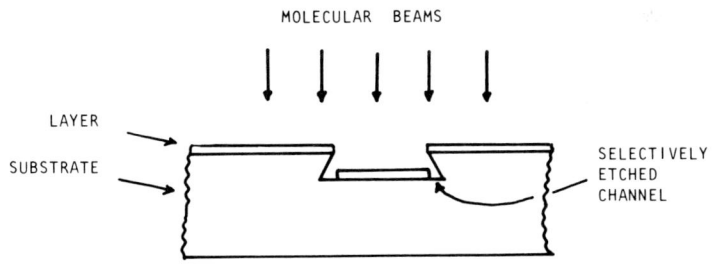

FIG. 49. Simplified schematic of selectively etched self-masking MBE growth.

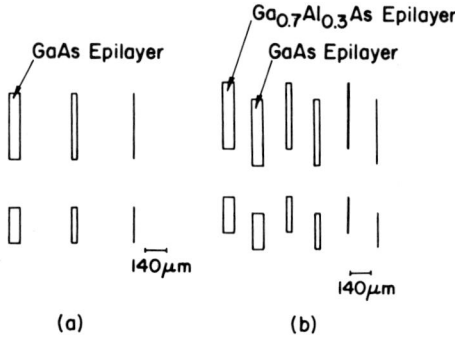

FIG. 50. Normalized optical micrograph of (a) single- and (b) two-level shifted GaAs ($Al_{0.3}Ga_{0.7}As$) selected-area epitaxial patterns.

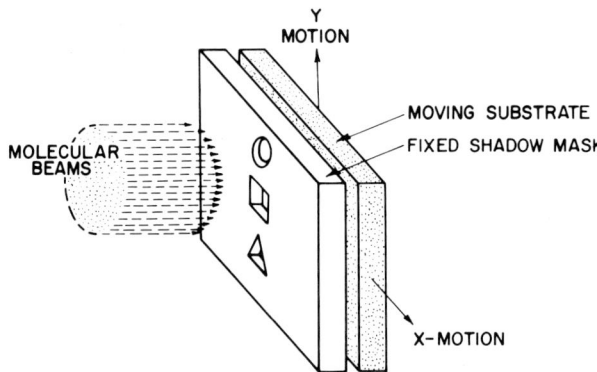

FIG. 51. Basic arrangement used for MBE writing of structures as narrow as 10 μm.

found to be ~200 Å under As-rich conditions and ~1900 Å under gallium-stabilized conditions (237).

Corrugated surface structures with 8 μm period were also used (238) to grow self-aligned GaAs/GaAlAs surface-oriented multistructures.

X. Projections

The recent explosive growth of interest in MBE is effectively illustrated by the fact that virtually all of the work covered by some 240 references cited here has been accomplished within the past five years. Many of the studies described are of a preliminary nature and indicate new directions that merit far more detailed investigation. In fact, this broad-brush appraisal shows that much consolidation of existing achievements needs to be carried out. In more technologically promising areas of growth this rather unrewarding task will no doubt be undertaken in laboratories where specific device applications are being considered for production. In this context there is now a need for a somewhat less sophisticated generation of machine design (and very definitely a need for a production machine) for use as fast turn-around prototypes for production evaluation.

Such equipment will dispense with the paraphernalia of research in the interests of reproducibility, ruggedness, and simplicity of operation, without compromising vacuum conditions or process control. Differentially pumped conveyor belt sample insert systems with tiers of evaporation cell banks for multiple growth could be envisaged. As the details of effusion cell design become better understood, it is conceivable that conveyor belt growth (moving substrates) will be achieved from linear arrays of, or racks of long, thin orifices (e.g., 20×0.1 cm) heated by clean, water-cooled external RF coils. Carrier profiles could be produced by allowing dopants to impinge only in specified regions of the sample train or by changing T_s during motion across the chamber. Cryopumping of such a continuous-growth system could be achieved by closed-cycle freon, liquid-nitrogen-type systems, or even water cooling. Expensive quadrupole mass spectrometers could be replaced by cheaper magnetic sector instruments with m/e resolution as low as 60 (using doubly charged cracking fragment peaks, e.g., As_1^{2+} at m/e-37.5, Ga^{2+} at m/e-34.5 and 35.5, or Sb_1^{2+}) for both major component and residual gas analysis. Lower energy RED systems could be incorporated (~ 2.5 keV) and movable ion gauges used to monitor flux rates, etc.

To the disdain of the purist, research in MBE techniques, as such, may be forced to resort to forms of reactive chemical substrate cleaning, e.g., with ozone, to remove the last traces of carbon that AES cannot detect. Methods must be developed for protecting surfaces of thermally unstable compounds for growth at higher temperatures by use of halogens or by hydrides (e.g., AsH_3). Indeed, we may find that vapor phase epitaxy by halide, hydride or even metal organic transport in UHV glass systems offers some advantages. Certainly the effect of hydrogen pressures has been found helpful under certain circumstances.

Use of metals more reactive to oxygen than Al that form volatile oxides (e.g., Cs) should be investigated for reducing the high deep-center concentrations of III-V alloys containing this metal. The use of lead (or similar) as a dynamic two-dimensional liquid phase may result in the ability to use the high LPE segregation coefficient of conventional p-type dopants such as Zn or Cd more efficiently and may even be found effective in reducing background-gas-associated carbon and oxygen contamination by screening out CO and H_2O background gases.

The use of multiple-cell compartments with different orifice sizes is currently under investigation in the author's laboratory for long-term stability of group III element flux ratios for growth of lattice-matched III_{1-x}, III'_x–V compounds and may be found useful for $III-V_{1-x}$, V'_x and quaternary compounds where control of the group V ratios is difficult. The role of background and intentional dopant incorporation process as a function of substrate orientation, and in more detail on compounds other than GaAs, needs investigation.

Regarding types of materials studied, more attention should be given to those with smaller bandgaps (to exploit the potential for long-wavelength detection and magnetic-field sensors), compounds and alloys such as InAsSb and InSb(Bi), InTlSb, or InTlSbBi as well as some of the more interesting indirect gap materials such as $Ga_{1-x}Al_xP$ and $GaP_{1-x}N_x$ alloys. The latter again directs attention toward the possible need for "mixed-discipline" growth using NH_3 or similar compounds. Some of the more interesting quaternary alloys (GaInAsP, for example) and alloys of the type III, III', III"–V and III–V, V', V" may be found feasible as growth parameter control becomes better understood.

New device applications are feasible with a knowledge of growth techniques for submicron structures. In particular the modulation-doped heterostructures in the GaAlAs/GaAs system appear to be promising for very fast two-dimensional electron gas devices (FETs) at lower operating temperatures. The use of multimonolayer heterojunction structures for integrated optical applications also looks exciting. Lamellar doping of compound semiconductor in the III–V family will be found to have intriguing and potentially important device properties. Focused dopant ion writing may become feasible for submicron integrated circuitry as well as local laser heating for definition of doped regions in low-T_s growth.

The ability to grow submicron $n^{2+}n^-n^{2+}$ structures will be interesting for ballistic devices in which the electron path length is greater than the mean free path between scattering centers even at room temperature. At liquid-nitrogen temperature where mean free paths can be >1 μm, GaAs devices operating below transfer voltage (~ 0.3 V) may pave the way to even faster computer logic systems.

Novel materials that will no doubt be grown are the alloys between GaAs and Ge, III–V, and II–VI, etc., and their properties evaluated. On a more mundane footing, the alloys of GaSb with GaAs and GaP will be more extensively studied for their interesting intervalley transfer properties and miscibility limits, respectively.

All the possibilities on this list may not find patrons or success, but the author is confident that many will find both and all will sooner or later be investigated.

XI. Conclusions

MBE does not offer to grow new materials that have not been grown before. It does, however, offer the potential to grow existing compounds with much more control of thickness (down to angstrom dimensions) and to be able to abruptly, reproducibly, and repeatedly change the doping type, tailor the doping profiles (both p and n), the composition of alloys, and to a certain extent the stoichiometry of the various growth processes.

The temperatures of epitaxy is lower than for conventional techniques, which is an advantage in the fight against impurity and alloy diffusion across interfaces and autodoping, although this last effect is to some extent offset by the optimum growth rates, which are significantly lower than with CVD or LPE.

The ability to characterize surfaces *in situ* and under certain circumstances *real time* is offering significant help toward understanding the various growth processes, although this aspect has to some extent taken second place to the more direct practical approach of MBE toward preparation of epitaxial structures suitable for devices.

The current state of the art is encouraging in that the densities of unintentional impurities (the most severely criticized aspect of MBE, apart from the initial capital investment) have now been reduced to levels below which they adversely affect electron devices.

There is still a long way to go in the preparation of minority carrier devices for use in electrooptical applications such as lasers and LEDs, which in many cases require the use of aluminum-containing alloys.

Other obvious advantages offered by MBE over those of CVD and LPE are of real technological importance and include the ability to prepare tunnel contact films, *in situ* ohmic and Schottky metallizations, selected-area growth, two-dimensional compounds, and it is expected in the future, better insulator–semiconductor structures.

To date, many of the devices prepared from MBE films show only slight improvements, and in certain cases slightly inferior properties, to those prepared by CVD or LPE. However, as the total effort × time product in

MBE is less than 10% of that invested in the other techniques, it is extremely encouraging that such device comparisons already can be made.

In conclusion it should be stated categorically that, although MBE does offer certain improvements over other epitaxial techniques where they overlap in capability, MBE should not be considered solely as a competitor but a complement in the ability to extend our understanding of epigrowth processes and surfaces, and to provide novel device features that are in many cases impossible by the alternative techniques.

Acknowledgments

The author is indebted to the many colleagues worldwide who willingly furnished and gave permission to reproduce material from preprints, reprints, and useful comments.

References

1. A. Y. Cho and J. R. Arthur, *Prog. Solid State Chem.* **10**, 157 (1975).
2. L. L. Chang and R. Ludeke, *in* "Epitaxial Growth" (J. W. Mathews, ed.), Part A, p. 37, Academic Press, New York, 1975.
3. L. L. Chang, L. Esaki, W. E. Howard, R. Ludeke, and G. Schul, *J. Vac. Sci. Technol.* **10**, 655 (1973).
4. B. A. Joyce and C. T. Foxon, *J. Cryst. Growth* **31**, 122 (1975).
5. L. L. Chang, to be published.
6. R. C. Farrow, *in* "Crystal Growth and Materials" (E. Kaldis and H. J. Scheel, eds.). North-Holland Publ., Amsterdam, 1976.
7. A. Y. Cho, *Jpn. J. Appl. Phys.* **16**, Suppl. 116-1, 435 (1977).
8. P. E. Luscher and D. M. Collins, *Prog. Cryst. Growth Charact.* (to be published).
9. J. Massies, P. Etienne, and N. T. Linh, *Rev. Tech. Thomson-CSF* **8**, No. 1 (1976).
10. D. Williams, *Int. Symp. Mol. Beam Epitaxy, 1st, Paris, 1978*.
11. D. L. Smith and V. Y. Pickhardt, *J. Appl. Phys.* **46**, 2366 (1975).
12. D. L. Smith and V. Y. Pickhardt, *Int. Symp. Mol. Beam Epitaxy, 1st, 1978*.
13. T. Yao, S. Amano, Y. Makita, and S. Maekawa, *Jpn. J. Appl. Phys.* **15**, 1001 (1976).
14. T. Yao, Y. Myoshi, Y. Makita, and S. Maekawa, *Jpn. J. Appl. Phys.* **16**, 369 (1977).
15. R. Ludeke, *Solid State Commun.* **24**, 725 (1977).
16. R. M. Ueda, *J. Cryst. Growth* **31**, 333 (1975).
17. J. N. Walpole, A. R. Calawa, T. C. Harman, and S. H. Groves, *Appl. Phys. Lett.* **28**, 552 (1976).
18. J. N. Walpole, A. R. Calawa, S. R. Chinn, S. H. Groves, and T. C. Harman, *Appl. Phys. Lett.* **29**, 307 (1976).
19. D. L. Smith and V. Y. Pickhardt, *J. Electron. Mater.* **5**, 247 (1976).
20. D. L. Smith and V. Y. Pickhardt, *J. Electrochem. Soc.* **125**, 2042 (1978).
21. D. K. Honke and M. D. Hurley, *J. Appl. Phys.* **49**, 4975 (1976).

22. D. K. Honek and S. W. Kaiser, *J. Appl. Phys.* **47**, 892 (1974).
23. J. N. Walpole, A. R. Calawa, T. C. Harman, and S. H. Groves, *Appl. Phys. Lett.* **28**, 552 (1976).
24. E. H. C. Parker and D. Williams, *Solid-State Electron.* **20**, 567 (1977).
25. D. Williams and E. H. C. Parker, *J. Phys. E* **10**, 1176 (1977).
26. C. A. Goodwin and Y. Ota, *IEEE Trans. Electron Devices* **ed-26**, 1796 (1979).
27. Y. Ota, *J. Electrochem. Soc.* **124**, 1795 (1977).
28. Y. Shiraki, Y. Katama, K. L. I. Kobayashi, and K. F. Komatsubara, *J. Cryst. Growth* **45**, 287 (1978).
29. J. C. Bean, *Appl. Phys. Lett.* **33**, 654 (1978).
30. U. Konig, H. Kibbel, and E. Kasper, *J. Vac. Sci. Technol.* **16**, 985 (1979).
31. N. T. Linh, private communication (1979).
32. R. C. Farrow, *Thin Solid Films* **55**, 303 (1979).
33. Chromel alumel thermocouples contain up to 2% Mn and as Cr, Mn and Nickel are deep acceptors (Mn not so deep) in III-V compounds it is not surprising that inferior results have been achieved with thermocouples made from these materials.
34. L. Esaki and L. L. Chang, *Thin Solid Films* **36**, 285 (1976).
35. P. Etienne, J. Massies, and N. T. Linh, *J. Phys. E* **10**, 1153 (1977).
36. C. E. C. Wood and B. A. Joyce, *J. Appl. Phys.* **49**, 4854 (1978).
37. A. Y. Cho, *J. Appl. Phys.* **41**, 2780 (1970).
38. J. R. Arthur, *Surf. Sci.* **43**, 449 (1974).
39. J. H. Neave and B. A. Joyce, *J. Cryst. Growth* **44**, 387 (1978).
40. D. B. Dove, R. Ludeke, and L. L. Chang, *J. Appl. Phys.* **44**, 1897 (1973).
41. R. Ludeke, *IBM J. Res. Dev.* **22**, 304 (1978).
42. B. A. Joyce and J. H. Neave, *J. Cryst. Growth* **43**, 204 (1978).
43. J. Massies, J. Chaplart, P. Devoldre, P. Etienne, and N. T. Linh, *Int. Symp. Mol. Beam Epitaxy, 1st, Paris, 1978.*
44. S. Gonda and Y. Matsushima, *J. Appl. Phys.* **47**, 4198 (1976).
45. K. Ploog and A. Fischer, *J. Appl. Phys.* **13**, 111 (1977).
46. A. Y. Cho, J. V. DiLorenzo, B. S. Hewitt, W. C. Niehaus, W. O. Schlosser, and C. Radice, *J. Appl. Phys.* **48**, 336 (1977).
47. A. Y. Cho and W. C. Bellamy, *J. Appl. Phys.* **46**, 783 (1975).
48. B. I. Miller and J. H. McFee, *J. Electrochem. Soc.* **125**, 1311 (1978).
49. K. Ploog and A. Fisher, *J. Vac. Sci. Technol.* **15**, 255 (1978).
50. K. Tateishi, M. Naganuma, and K. Takahashi, *Jpn. J. Appl. Phys.* **15**, 785 (1976).
51. T. Waho, S. Ogawa, and S. Maruyama, *Jpn. J. Appl. Phys.* **16**, 1875 (1979).
52. C. T. Foxon and B. A. Joyce, *J. Cryst. Growth* **14**, 75 (1978).
53. R. Ludeke, L. Esaki, and L. L. Chang, *Appl. Phys. Lett.* **24**, 417 (1974).
54. L. L. Chang and A. Koma, *Appl. Phys. Lett.* **29**, 138 (1976).
55. J. Massies, P. Etienne, and N. T. Linh, *Rev. Tech. Thompson*-CSF **8**, 5 (1976).
56. K. Ploog and A. Fisher, *Int. Symp. Mol. Beam Epitaxy, 1st, Paris, 1978.*
57. G. Abstreiter, E. Bauser, A. Fisher, and K. Ploog, *Appl. Phys.* (to be published).
58. Y. Matsushima and S. Gonda, *Jpn. J. Appl. Phys.* **15**, 2093 (1976).
59. K. Ploog, A. Fisher and F. Raisch, *Proc. Int. Vac. Congr., 7th, 1977* p. 1705 (1977).
60. J. H. McFee, B. I. Miller, and K. M. Bachmann, *J. Electrochem. Soc.* **124**, 259 (1977).
61. R. Ludeke and L. Esaki, *Surf. Sci.* **47**, 132 (1975).
62. R. Ludeke and L. Esaki, *Phys. Rev. Lett.* **33**, 653 (1976).
63. R. Ludeke, *Phys. Rev. Lett.* **39**, 1042 (1977).

64. R. Ludeke and A. Koma, *J. Vac. Sci. Technol.* **13**, 241 (1976).
65. J. Massies, P. Devoldre, P. Etienne, and N. T. Linh, *Proc. Int. Vac. Congr., 7th, 1977* p. 639 (1977).
66. J. Massies, P. Devoldre, and N. T. Linh, *Int. Conf. Phys. Semicond. Surf. & Interfaces, Asilomar, California* (1979).
67. A. R. Calawa, G. Davies, H. Ohno, and C. E. C. Wood, *Appl. Phys. Lett.* (to be published).
68. R. C. Farrow, *J. Cryst. Growth* **45**, 292 (1979).
69. C. E. C. Wood, *Appl. Phys. Lett.* **29**, 746 (1976).
70. A. M. Huber, private communication (1974); N. T. Linh, private communication (1979).
71. A. M. Huber, G. Morillot, N. T. Linh, P. N. Favenec, B. Deveaud, and B. Toulouse, *Appl. Phys. Lett.* (submitted for publication).
72. S. Hiyamizu, T. Fujii, K. Nanbu, and S. Maekawa, *Jpn. J. Appl. Phys.* **17**, Suppl. 17-1, 79 (1978).
73. W. Y. Lum, H. H. Wieder, W. H. Koschel, S. G. Bishop, and B. D. McCombe, *Appl. Phys. Lett.* **30**, 1 (1977).
74. A. Chandra, C. E. C. Wood, D. W. Woodard, and L. F. Eastman, *Solid-State Electron.* **22**, 645 (1979).
75. H. Morkoc and A. Y. Cho, *Pap., Workshop Compd. Semicond. Microwave Devices, Atlanta, Georgia* (1979).
76. T. Murotani, T. Shimano, and S. Mitsui, *J. Cryst. Growth* **45**, 302 (1978).
77. C. E. C. Wood, J. Woodcock, and J. J. Harris, *Proc. Int. Symp. Gallium-Arsenide Relat. Comp., 7th, 1978* I.O.P. Ser. 45 (1978).
78. T. Murotani, T. Shimano, and S. Mitsui, *J. Cryst. Growth* **45**, 302 (1978).
79. D. DiSimone, R. Stall, C. E. C. Wood, and L. F. Eastman, to be published.
80. H. C. Casey, Jr., A. Y. Cho, and E. H. Nicollian, *Appl. Phys. Lett.* **32**, 678 (1978).
81. H. C. Casey, Jr., A. Y. Cho, D. V. Lang, and E. H. Nicollian, *J. Vac. Sci. Technol.* **15**, 1408 (1978).
82. D. V. Lang and R. A. Logan, *J. Electron. Mater.* **4**, 1053 (1975).
83. D. V. Lang, R. A. Logan, and L. C. Kimmerling, *Phys. Rev. B* **15**, 4874 (1977).
84. A. Mircea and A. Mitonneau, *Appl. Phys.* **8**, 15 (1975).
85. C. E. C. Wood, *Int. Symp. Mol. Beam Epitaxy, Paris, 1st, 1978.*
86. G. M. Martin, A. Mittonneau, and A. Mircea, *Electron. Lett.* **13**, 191 (1977).
87. C. E. C. Wood, unpublished results.
88. J. Barrera, *Proc. Active Semicond. Dev. Microwave Int. Opt.*, (1975).
89. D. V. Lang, A. Y. Cho, A. C. Gossard, M. Ilegems, and W. Wiegmann, *J. Appl. Phys.* **47**, 2558 (1976).
90. R. Stall, C. E. C. Wood, P. D. Kirchner, and L. F. Eastman, *Electron. Lett.* **16**, 171 (1980).
91. D. W. Covington and E. L. Meeks, private communications in reference 10 of paper, *in* "Proceedings of Southeast Conference '78," p. 380. I.E.E.E. Region III, Atlanta, Georgia, 1978.
92. D. W. Covington and E. L. Meeks, *Int. Symp. Mol. Beam Epitaxy Paris, 1st, 1978.*
93. D. W. Covington, C. W. Litton, D. C. Reynolds, R. J. Almassy, and G. L. McCoy, *Proc. Int. Symp. Gallium-Ansenide & Relat. Compd, 7th, 1978*, I.O.P. Ser. 45, p. 171 (1978).
94. M. Ilegems, R. Dingle, and L. W. Rupp, *J. Appl. Phys.* **46**, 3059 (1975).

95. A. R. Calawa, *Appl. Phys. Lett.* **33**, 1022 (1978).
96. C. E. C. Wood and B. Clegg, unpublished results.
97. W. Y. Lum, A. R. Clawson, and H. H. Wieder, *J. Vac. Sci. Technol.* **14**, 1007 (1977).
98. C. E. C. Wood, D. DiSimone, and S. Judaprawira and L. F. Eastman, *J. Appl. Phys.* **51**, 2074 (1980).
99. A. Y. Cho and I. Hayashi, *Metall. Trans.* **2**, 777 (1971).
100. T. Shimanoe et al., *Int. Conf. Solid Films Surf.*, 4th, Tokyo, Japan (1978).
101. C. W. Litton, private communication (1978).
102. A. Y. Cho, *J. Appl. Phys.* **46**, 1733 (1975).
103. B. I. Miller, J. H. McFee, R. J. Martin, and D. K. Tien, *Appl. Phys. Lett.* **33**, 44 (1978).
104. A. Y. Cho and I. Hayashi, *J. Appl. Phys.* **42**, 4422 (1971).
105. G. H. Dohler and K. Ploog. *Prog. Cryst. Growth Charact.* (to be published).
105a. G. M. Metze, R. A. Stall, C. E. C. Wood, and L. F. Eastman, *Appl. Phys. Lett.* (to be published).
106. C. E. C. Wood, *Appl. Phys. Lett.* **33**, 770 (1978).
107. C. M. Wolfe and G. E. Stillman, in "Characterization of Epitaxial Semiconductor Films" (H. Kressel, ed.), p. 197. Elsevier, New York, 1976.
108. R. E. Hoenig and D. A. Kramer, *RCA Rev.* **30**, 285 (1969).
109. D. L. Smith, private communication (1978).
110. D. M.Collins, *Appl. Phys. Lett.* **35**, 67 (1979).
111. J. R. Arthur, *Surf. Sci.* **38**, 394 (1973).
112. M. Naganuma and K. Takahashi, *Appl. Phys. Lett.* **27**, 343 (1975).
113. M. Naganuma and K. Takahashi, *Electron. Eng. (Tokyo)* **94**, 15 (1974).
114. N. Matsunaga, T. Susuki, and K. Takahashi, *J. Appl. Phys.* **49**, 5711 (1978).
115. B. A. Joyce and C. T. Foxon, *J. Appl. Phys.* **16**, 17 (1977).
116. A. Y. Cho and M. B. Pannish, *J. Appl. Phys.* **43**, 5118 (1972).
117. M. Ilegems, *J. Appl. Phys.* **48**, 1279 (1977).
118. J. S. Roberts and C. E. C. Wood, unpublished results.
119. J. Vilms and J. P. Garrett, *Solid-State Electron.* **15**, 443 (1971).
120. W. V. McLevige, K. V. Vaidyanathan, B. Streetman, M. Ilegems, J. Comas, and L. Plew, *Appl. Phys. Lett.* **33**, 128 (1978).
121. M. Ilegems, H. C. Casey, Jr., S. Somekh, and M. B. Pannish, *J. Cryst. Growth* **31**, 158 (1975).
122. N. Holnyak, Jr., R. J. Nelson, J. J. Coleman, P. D. Wright, D. Finn, W. O. Groves, and D. L. Keune, *J. Appl. Phys.* **48**, 1963 (1977).
123. Y. Matsushima, S. Gonda, V. Makita, and S. Mukai, *J. Cryst. Growth* **43**, 281 (1978).
124. C. T. Foxon, J. A. Harvey and B. A. Joyce, *J. Phys. Chem. Solids* **34**, 1693 (1973).
125. R. C. Farrow, *J. Phys. D* **7**, 121 (1974).
126. R. C. Farrow, *Electrochem. Soc. Meet., 1975.* Extended Abstracts for Paper 214 (1975).
127. B. T. Meggit, E. H. C. Parker, and R. M. King, *Appl. Phys. Lett.* **33**, 528 (1978).
128. L. F. Eastman, private communication.
129. W. T. Tsang, *Appl. Phys. Lett.* **33**, 426 (1978).
130. L. Chang, A. Segmuller, and L. Esaki, *Appl. Phys. Lett.* **28**, 39 (1976).
130a. S. Yoshida, S. Misawa, Y. Fujii, S. Takada, H. Hiyakawa, S. Gonda, and A. Itoh, *J. Vac. Sci. Technol.* **16**, 990 (1979).
131. C. A. Chang, R. Ludeke, L. L. Chang, and L. Esaki, *Appl. Phys. Lett.* **31**, 759 (1977).
132. L. L. Chang, private communication (1979).

133. Y. Matsushima, Y. Hirofuji, S. Gonda, S. Mukai, and M. Kimata, *Jpn. J. Appl. Phys.* **15**, 2321 (1976).
133a. M. T. Norris and C. R. Stanley, *Appl. Phys. Lett.* **35**, 617 (1979).
133b. Y. Kawamura, M. Ikeda, H. Asahi, and H. Okamoto, *Appl. Phys. Lett.* **35**, 481 (1979).
134. M. Yano, M. Nogami, Y. Matsushima, and M. Kimata, *Jpn. J. Appl. Phys.* **16**, 2131 (1977).
135. B. T. Meggitt, E. H. C. Parker, and R. M. King, *Appl. Phys. Lett.* **33**, 528 (1978).
136. J. D. Grange, E. H. C. Parker, and R. M. King, *J. Phys. D* (submitted for publication).
137. H. Sakaki, L. L. Chang, R. Ludeke, C. A. Chang, G. A. Sai-Halasz, and L. Esaki, *Appl. Phys. Lett.* **31**, 211 (1977).
138. S. Baba, H. Horita, and A. Kinbara, *J. Appl. Phys.* **49**, 3632 (1978).
139. H. Freller, private communication.
140. K. Tateishi, M. Naganuma, and K. Takahashi, *Jpn. J. Appl. Phys.* **15**, 785 (1976).
140a. H. Ohno, C. E. C. Wood, L. Rathbun, and L. F. Eastman, *J. Appl. Phys.*, to be published.
141. S. Hiyamizu, T. Fujiki, K. Nanbu, S. Maekawa, and T. Hisatsugu, *Surf. Sci.* (to be published).
142. G. B. Scott and J. S. Roberts, *Proc. Int. Symp. Gallium-Arsenic & Relat. Comp., 7th, 1978* I.O.P. Ser. 45 (1978).
143. C. A. Chang and A. Segmuller, *J. Vac. Sci. Technol.* (to be published).
144. S. Gonda and Y. Matsushima, *J. Appl. Phys.* **47**, 4198 (1978).
145. C. A. Chang, R. Ludeke, L. L. Chang, and L. Esaki, *Appl. Phys. Lett.* **31**, 759 (1977).
146. T. Waho, S. Ogawa, and S. Muruyama, *Jpn. J. Appl. Phys.* **16**, 1875 (1977).
146a. L. M. Foster, *J. Electrochem. Soc.* **121**, 1662 (1974).
147. S. Maruyama, T. Waho, and S. Ogawa, *Jpn. J. Appl. Phys.* **12**, 1695 (1978).
148. A. Y. Cho, H. C. Casey, and P. W. Foy, *Appl. Phys. Lett.* **30**, 397 (1977).
149. M. Naganuma, K. Kamimura, K. Takahashi, and Y. Sakai, *Jpn. J. Appl. Phys.* **14**, 581 (1975).
150. A. Y. Cho and D. R. Chen, *Appl. Phys. Lett.* **28**, 30 (1976).
151. S. G. Bandy, D. M. Collins, and C. K. Nishimoto, *Electron. Lett.* (to be published).
152. M. Wataze, Y. Mitsui, T. Shimano, M. Nakatawi, and S. Mitsui, *Electron. Lett.* **14**, 759 (1978).
153. R. E. Williams and D. W. Shaw, *IEEE Trans. Electron Devices* ed. **25**, 600 (1978).
154. C. E. C. Wood, D. DeSimone, S. Judaprawira, and L. F. Eastman, *J. Appl. Phys.* **51**, 2074 (1980).
155. J. V. DiLorenzo, W. C. Niehaus, and A. Y. Cho, *J. Appl. Phys.* **50**, 951 (1979).
156. J. Devlin, R. Stall, C. E. C. Wood, and L. F. Eastman, *Electrochem. Soc. Meet. 115th Boston, Massachusetts* (1979).
156a. R. A. Stall, C. E. C. Wood, K. Board, and L. F. Eastman, *Electron. Lett.* **15**, 800 (1979).
157. A. Y. Cho and F. K. Reinhart, *J. Appl. Phys.* **45**, 1812 (1974).
158. D. W. Covington and W. H. Hicklin, *Electron. Lett.* **14**, 759 (1978).
159. M. V. Schneider, R. A. Linke, and A. Y. Cho, *Appl. Phys. Lett.* **31**, 219 (1977).
160. E. L. Meeks, G. N. Hill, C. W. Covington, and W. B. Day, *Int. Symp. Mol. Beam Epitaxy, 1st, Paris, 1978*.
161. A. Y. Cho, C. N. Dunn, R. L. Kuvas, and W. E. Schroeder, *Appl. Phys. Lett.* **25**, 224 (1974).

162. K. Kamei and L. F. Eastman, *IEEE Trans. Electron Devices* **ed-23**, 452 (1976).
163. A. Chandra and L. F. Eastman, *Annu. Workshop Compd. Semicond. Microwave Devices, Atlanta, Georgia, 9th, 1978*.
164. D. L. Lile, D. A. Collins, L. Messick, and A. R. Clawson, *Appl. Phys. Lett.* **32**, 247 (1978).
165. R. Dingle, H. L. Stormer, A. C. Gossard, and W. Wiegman, *Appl. Phys. Lett.* **33**, 665 (1978).
166. C. E. C. Wood, G. M. Metze, J. Berry, and L. F. Eastman, *J. Appl. Phys.* **51**, 383 (1980).
167. C. A. Burrus and B. I. Miller, *Opt. Commun.* **4**, 307 (1971).
168. T. P. Lee, W. S. Holden, and A. Y. Cho, *Appl. Phys. Lett.* **32**, 415 (1978).
169. A. Y. Cho, R. W. Dixon, H. C. Casey, Jr., and R. L. Hartman, *Appl. Phys. Lett.* **28**, 501 (1976).
170. A. Y. Cho, *Jpn. J. Appl. Phys.* **16**, Suppl. 16-1, 435 (1976).
170a. W. T. Tsang, MBE Workshop, Univ. Illinois, Urbana (1979).
170b. W. T. Tsang, Proc. IEDM, Washington, D.C. (1979).
171. H. C. Casey, Jr., S. Somekh, and M. Ilegems, *Appl. Phys. Lett.* **27**, 142 (1975).
172. J. L. Mertz and A. Y. Cho, *Appl. Phys. Lett.* **28**, 456 (1976).
173. F. K. Reinhart and A. Y. Cho, *Appl. Phys. Lett.* **31**, 457 (1977).
174. J. L. Mertz, R. A. Logan, W. Wiegmann, and A. C. Gossard, *Appl. Phys. Lett.* **26**, 337 (1975).
175. T. P. Lee and A. Y. Cho, *Appl. Phys. Lett.* **29**, 164 (1976).
176. J. P. Van Der Ziel, R. Dingle, R. C. Miller, W. Wiegmann, and W. A. Nordland, *J. Appl. Phys.* **26**, 463 (1975), and R. C. Miller, R. Dingle, A. C. Gossard, R. A. Logan, and W. A. Nordland, *J. Appl. Phys.* **47**, 4509 (1976).
177. H. Asahi and H. Okamoto, *Int. Conf. Solid Films Surf. Tokyo, Japan* (1978).
178. J. L. Mertz, R. A. Logan, W. Wiegmann, and A. C. Gossard, *Appl. Phys. Lett.* **26**, 337 (1975).
179. J. L. Mertz, A. C. Gossard, and W. Wiegmann, *Appl. Phys. Lett.* **30**, 629 (1977).
180. A. Y. Cho, A. Yariv, and P. Yeh, *Appl. Phys. Lett.* **30**, 471 (1977).
181. A. Y. Cho and F. K. Reinhart, *Appl. Phys. Lett.* **21**, 355 (1972).
182. J. L. Mertz, R. A. Logan, and A. M. Sergent, *J. Appl. Phys.* **47**, 1436 (1976).
183. J. P. Van Der Ziel, R. M. Mikulyak, and A. Y. Cho, *Appl. Phys. Lett.* **27**, 71 (1975).
184. H. Kroemer, W. Y. Chien, H. C. Casey, Jr., and A. Y. Cho, *Appl. Phys. Lett.* **33**, 749 (1978).
185. C. E. C. Wood, B. C. Easton, and D. Hood, unpublished.
185a. J. C. Fan, A. R. Calawa, R. L. Chapman, and G. W. Turner, *Appl. Phys. Lett.* **10**, 804 (1979).
185b. L. Esaki and R. Tsu, *IBM J. Res. Dev.* **14**, 61 (1970).
186. A. Y. Cho, *Appl. Phys. Lett.* **19**, 467 (1971).
187. G. H. Dohler and K. Ploog, *Int. Conf. Mol. Beam Epitaxy, Paris, 1st, 1978*.
188. K. Ploog, A. C. Fischer, and H. Kunzel, *Appl. Phys. Lett.* (to be published).
188a. C. E. C. Wood, S. Judaprawira, and L. F. Eastman, Proc. IEDM, Washington, D.C. (1979).
189. A. S. Barker, Jr., J. L. Mertz, and A. C. Gossard, *Phys. Rev. B* **17**, 3181 (1978).
190. R. Dingle, W. Wiegmann, and C. H. Henry, *Phys. Rev. Lett.* **33**, 827 (1974).
191. R. Dingle, *Festkoerperprobleme* **15**, 21 (1975).
192. R. Dingle, A. C. Gossard, and W. Wiegmann, *Phys. Rev. Lett.* **34**, 1327 (1975).

193. J. L. Mertz, A. S. Barker, Jr., and A. C. Gossard, *Appl. Phys. Lett.* **31**, 117 (1977).
193a. R. Dingle and W. Wiegmann, *J. Appl. Phys.* **46**, 4312 (1975).
194. L. L. Chang and A. Koma, *Appl. Phys. Lett.* **29**, 138 (1976).
195. A. C. Gossard, P. M. Petroff, W. Wiegmann, R. Dingle, and A. Savage, *Appl. Phys. Lett.* **29**, 323 (1976).
196. P. M. Petroff, A. C. Gossard, W. Wiegmann, and A. Savage, *J. Cryst. Growth* **44**, 5, (1978); see also P. M. Petroff, *J. Vac. Sci. Technol.* **14**, 973 (1977).
197. R. Ludeke, L. Esaki, and L. L. Chang, *Appl. Phys. Lett.* **24**, 417 (1974).
198. L. L. Chang, L. Esaki, W. E. Howard, and R. Ludeke, *J. Vac. Sci. Technol.* **10**, 11 (1973).
199. H. L. Stormer, R. Dingle, A. C. Gossard, W. Wiegmann, and R. A. Logan, *Proc. Int. Conf. Phys. Semicond., 14th,* (1978).
200. H. L. Stormer, R. Dingle, A. C. Gossard, W. Wiegmann and M. D. Sturge, *Solid State. Commun.* (to be published).
201. R. Dingle, H. L. Stormer, A. C. Gossard, and W. Wiegmann, *Proc. Int. Symp. Gallium-Arsenic & Relat. Compd., 7th, 1978* I.O.P. Ser. 45 (1978).
202. G. A. Sai-Halasz, R. Tsu, and L. Esaki, *Appl. Phys. Lett.* **30**, 651 (1977).
203. L. Esaki and L. L. Chang, *J. Vac. Sci. Technol.* **15**, 254 (abstr.) (1978).
204. L. L. Chang, *J. Vac. Sci. Technol.* **15**, 1478 (abstr.) (1978).
205. L. L. Chang, G. A. Sai-Halasz, N. J. Kawai, and L. Esaki, *Int. Conf. Phys. Compd. Semicond. Interfaces, Asilomar, California* (1979).
206. P. Yeh, A. Yariv, and A. Y. Cho, *Appl. Phys. Lett.* **32**, 104 (1978).
207. R. C. Miller, R. Dingle, A. C. Gossard, R. A. Logan, W. A. Nordland, Jr., and W. Wiegmann, *J. Appl. Phys.* **47**, 4509 (1978).
208. J. P. Van der Ziel and M. Ilegems, *Appl. Phys. Lett.* **28**, 437 (1976).
209. J. P. Van der Ziel and M. Ilegems, *Appl. Phys. Lett.* **29**, 200 (1976).
210. J. P. Van der Ziel, M. Ilegems, P. W. Foy, and R. M. Mikulyak, *Appl. Phys. Lett.* **29**, 775 (1976).
210a. J. P. Van der Ziel, M. Ilegems, and R. M. Mikulyak, *Appl. Phys. Lett.* **28**, 735 (1976).
211. J. P. Van der Ziel and M. Ilegems, *Appl. Opt.* **14**, 2627 (1975).
212. J. P. Van der Ziel and M. Ilegems, *Appl. Opt.* **15**, 1256 (1976).
213. V. L. Rideout, *Thin Solid Films* **48**, 261 (1978).
214. R. Ludeke, L. L. Chang, and L. Esaki, *Appl. Phys. Lett.* **23**, 201 (1973).
215. A. Y. Cho and P. D. Dernier, *J. Appl. Phys.* **49**, 3288 (1978).
216. J. Massies, P. Etienne, and N. T. Linh, *Surf. Sci.* **80**, 550 (1979).
217. J. Massies, P. Devoldere, and N. T. Linh, *J. Vac. Sci. Technol.* **15**, 1353 (1978).
218. J. Massies, P. Devoldere, P. Ettienne, and N. T. Linh, *Proc. 7th Int. Vacuum Congr. and 3rd Int. Conf. Solid Surfaces,* Vienna, p. 639 (1977).
219. D. W. Covington and E. L. Meeks, *Workshop Compd. Semicond. Devices, 1979* No proceedings (1979).
220. C. E. C. Wood, British Patent 15,291 (1977).
221. J. V. DiLorenzo and A. Y. Cho, *Int. Device Res. Conf.* (1978).
222. H. H. Berger, *J. Electrochem. Soc.* **117**, 507 (1972).
223. J. V. DiLorenzo, W. C. Hiehaus, and A. Y. Cho, *J. Appl. Phys.* **50**, 951 (1979).
224. W. T. Tsang, *Appl. Phys. Lett.* **33**, 1023 (1978).
225. J. Devlin, R. Stall, C. E. C. Wood, and L. F. Eastman, 7th Biennial Cornell Microwave Conference, Ithaca, New York (1979).
226. H. H. Wieder, *J. Vac. Sci. Technol.* **15**, 1498 (1978).

227. H. C. Casey, Jr., A. Y. Cho, D. V. Lang, and E. H. Nicollian, *J. Vac. Sci. Technol.* **15,** 1408 (1978).
228. H. C. Casey, Jr. and A. Y. Cho, *Appl. Phys. Lett.* **32,** 678 (1978).
229. W. T. Tsang, *Appl. Phys. Lett.* **33,** 426 (1978).
229a. W. T. Tsang, *Appl. Phys. Lett.* **34,** 408 (1979).
230. K. Ploog, A. Fischer, and R. Trommer, *J. Vac. Sci. Technol.* (to be published).
231. M. Hirose, A. Fischer, and K. Ploog, *Phys. Status Solidi A* **45,** K175 (1978).
232. W. C. Bellamy and A. Y. Cho, *J. Appl. Phys.* **46,** 783 (1975).
233. A. Y. Cho, J. V. DiLorenzo, and G. E. Mahoney, *IEEE Trans. Electron Devices* **ed-24,** 1186 (1977).
234. W. T. Tsang and A. Y. Cho, *Appl. Phys. Lett.* **30,** 293 (1977).
235. W. T. Tsang and M. Ilegems, *Appl. Phys. Lett.* **31,** 301 (1977).
236. W. T. Tsang and A. Y. Cho, *Appl. Phys. Lett.* **32,** 491 (1978).
237. S. Nagata and T. Tanaka, *J. Appl. Phys.* **48,** 940 (1977).
238. S. Nagata, T. Tanaka, and F. Fukai, *Appl. Phys. Lett.* **30,** 503 (1977).

Thin-Film IV–VI Semiconductor Photodiodes

H. HOLLOWAY

Research Staff
Ford Motor Company
Dearborn, Michigan

I. Introduction	106
II. Techniques for Vacuum Deposition of IV–VI Layers	108
1. Molecular Beam Epitaxy (MBE)	108
2. Hot-Wall Methods	111
3. Flash Evaporation and Sputtering	112
III. Properties of IV–VI Layers on Insulating Substrates	113
1. Layers on Alkali Halides	113
2. Layers on BaF_2 and SrF_2 Substrates	115
IV. Some Properties of IR Photodiodes	122
V. Thin-Film IV–VI Photodiodes	128
1. Development and General Characteristics	128
2. Quantum Efficiencies	132
3. Estimates of Resistance–Area Products, Lifetimes, and Detectivities	140
4. Detector Noise	149
5. Surface Effects and the Nature of the Metal-Barrier Devices	153
VI. Thin-Film Photodiodes for 3–5 μm Operation	157
VII. Thin-Film Photodiodes for 8–12 μm Operation	166
VIII. Unconventional Thin-Film Devices	168
1. The Junction Capacitance of Photodiodes	168
2. Pinched-off Photodiodes	172
3. Lateral-Collection Photodiodes	177
4. Self-Filtered Narrow-Band Photodiodes	186
5. Phototransistors	189
IX. Thin-Film Photodiode Arrays	191
X. Conclusions	198
References	199

I. Introduction

This chapter reviews recent work on the use of thin films of IV–VI compound semiconductors for the detection of infrared (IR) radiation. The field to be covered differs from that traditionally associated with IV–VI thin-film detectors because the emphasis is placed on single-crystal p–n junction devices instead of on polycrystalline photoconductors. This distinction may be clarified by consideration of the sequence of developments that has led to current IV–VI semiconductor photodiode technology.

The early work on IV–VI semiconductor IR detectors was with PbS, whose photoconductive applications were developed during the 1940s. Some use was made of vacuum-deposited layers (*1, 2*), but the major development was of polycrystalline layers that were precipitated from aqueous solutions (*3*). At a later stage similar techniques were applied to PbSe with consequent extension of the response to longer wavelengths (*4*). With both PbS and PbSe the attainment of useful photoconductors depends upon a poorly characterized partial oxidation that greatly improves the photoresponse over that of bulk crystals (*4*). The PbS and PbSe photoconductors remain valuable for applications where little or no cooling is used ($T \approx 200–300$ K), but they do not attain the high performance of p–n junction devices at lower operating temperatures. This performance limitation, together with a relatively slow response, sets the IV–VI photoconductors apart as an early branch from the evolution of high-performance IV–VI photodiodes.

More recent technological interest in the IV–VI semiconductors was stimulated by the discovery of band crossings in the pseudobinary alloys (Pb, Sn)Te and (Pb, Sn)Se (*5*). Thus, as the compositions of these alloys are changed by incorporation of SnTe or SnSe into PbTe or PbSe, the conduction and valence bands approach each other and eventually touch. This phenomenon made possible the attainment of arbitrarily small direct energy gaps by suitable choices of alloy composition and thereby permitted development of a new class of intrinsic detectors for the 8–12 μm atmospheric transmission band (*6*). The major emphasis of bulk-crystal IV–VI photodiode development has been on (Pb, Sn)Te for wavelengths near 10 μm, but similar crystal growth and diode fabrication techniques have been applied to other IV–VI semiconductors, for example, to PbTe for the 3–5 μm atmospheric transmission band (*7*).

The IV–VI semiconductors and their IR applications have been reviewed elsewhere (6, 8). The materials that are of most technological interest are the lead chalcogenides and their pseudobinary alloys with tin and other chalcogenides that permit adjustment of the energy gaps for specific applications. Some properties that are of particular interest are the near equality of the electron and hole effective masses, giving mobility ratios near unity, the relatively large dielectric constants [$\epsilon(\text{PbTe}) \approx 400\epsilon_0$ at 300 K], and the facility with which the compounds become nonstoichiometric. Chalcogen deficiency and metal deficiency give n- and p-type conductivity, respectively, and much of the device work has involved junction formation by stoichiometry change rather than by the incorporation of foreign dopants.

Epitaxial growth of IV–VI semiconductors on insulating substrates was demonstrated as early as 1948 (9) and was subsequently used as a preparative technique for studies of the optical and transport properties of these materials, particularly by Zemel and co-workers (e.g., 10). During this phase of thin-film studies growth was mostly on cleaved alkali halide substrates and the epitaxial layers seem to have been generally too imperfect for p–n junction fabrication. (Some success was achieved by Schoolar (11) with PbS layers on NaCl substrates. This work is described later.)

The studies reviewed here arose from the discovery by Holloway and co-workers that vacuum deposition of PbTe or (Pb, Sn)Te onto heated cleaved BaF_2 crystals (12) gave layers whose quality was adequate for fabrication of high-performance thin-film PbTe photodiodes (13) and injection lasers (14). Subsequent work has been largely devoted to optimization of such devices for particular IR applications and to fabrication of photodiode arrays.

In the selection of material for the present review the earlier literature on the growth of IV–VI semiconductors on alkali halide substrates was excluded except for papers that relate directly to p–n junction formation or that provide a comparison with the technologically more promising substrates BaF_2 and SrF_2. This missing literature has been reviewed by Zemel (15). It should also be noted that there has been much excellent p–n junction work based on the epitaxial growth of IV–VI semiconductors on IV–VI substrates. Thus, molecular beam epitaxy (MBE) onto IV–VI substrates has been used by Walpole and co-workers to make injection lasers (16) [these and similar studies have been reviewed by Holloway and Walpole (17)] and liquid-phase epitaxy onto IV–VI substrates has been used to make both IR photodiodes (18) and IR injection lasers (19). Such work with IV–VI semiconductor substrates is not included here because the resulting devices are better considered in the context of bulk-crystal structures, which they closely resemble.

II. Techniques for Vacuum Deposition of IV–VI Layers

Epitaxial growth of the IV–VI semiconductors on insulating substrates has been effected by a variety of vacuum deposition techniques,* which may be broadly classified as (1) MBE (a) with near-equilibrium sources, (b) with nonequilibrium sources; (2) hot-wall methods (a) near equilibrium, (b) nonequilibrium; (3) sputtering; (4) flash evaporation. The somewhat limited evidence that is available suggests that the suitability of the epitaxial layer for p–n junction fabrication depends far less upon the deposition technique than it does upon the choice of substrate. However, the need for reproducibility of the nonstoichiometry and of the composition of pseudobinary alloys would be expected to favor methods that give well-defined and reproducible fluxes of the constituents.

1. Molecular Beam Epitaxy (MBE)

The convenient volatility ($P \sim 0.1$ Torr at 1000 K) and approximately congruent molecular sublimation of the lead and tin chalcogenides (21–24) prompt the use of effusion cells as well-controlled sources for vacuum deposition and, indeed, this arrangement has been used frequently since the early experiments with IV–VI films (10). Effusion cells are particularly suitable for growth in the presence of additional fluxes of one or the other of the elemental constituents to adjust the layer stoichiometry (25).

For deposition of pseudobinary alloys, such as (Pb, Sn)Te, the obvious choices of technique are evaporation from a source of the alloy or from separate sources of the IV–VI components. In both cases some care is needed to obtain layers with well-defined uniform compositions. Deposition of (Pb, Sn)Te from an alloy source was described by Bylander (26). However, Northrop (27) has shown that the gas phase in equilibrium with $Pb_{0.8}Sn_{0.2}Te$ is about 30% enriched in SnTe. Thus, an effusion cell will give a deposit whose composition differs from that of the source and changes with depletion of the source. Similar effects are to be expected with other IV–VI alloys. Walpole (17) has shown that this problem may be eliminated by evaporating too rapidly for the solid source to reach equilibrium with the gas phase. This leads to source configurations that resemble open boats rather than effusion cells.

---

* The present discussion is restricted to vacuum deposition techniques because these have been used exclusively for IV–VI p–n junctions on insulating substrates. In principle, other deposition methods might be used. For example, IV–VI layers have been deposited on BaF_2 substrates by pyrolysis of mixtures of Pb alkyls with H_2S, H_2Se, or Te alkyls (20).

The use of individual effusion cells for the components of a IV–VI pseudobinary alloy was described by Bis and Zemel (28), who deposited Pb(Se, Te) from sources of PbSe and PbTe. The problem with this method is the compositional inhomogeneity that arises from fluctuations in the temperatures and, hence, in the rates of effusion from the individual cells. (For a IV–VI semiconductor that is evaporated at 1000 K a fluctuation of ± 1 K gives a change in the effusion rate of about $\pm 5\%$.) This difficulty was pointed out by Holloway et al. (29), who showed that an order-of-magnitude improvement could be obtained by using pairs of cells that were thermally linked by virtue of being machined from the same block of graphite. This arrangement (Fig. 1) gives essentially constant ratios of component fluxes because the temperature fluctuations of the effusion cells occur in phase.* A detailed discussion of compositional inhomogeneities in vacuum-deposited IV–VI alloy films is given in Holloway and Walpole (17).

A different approach to alloy deposition was described by Smith and Pickhardt (30), who grew epitaxial (Pb, Sn)Te from sources of elemental Pb, Sn, and Te. The technique is relatively complex because the large differences in volatility of the components necessitate radiation-shielded sources.

Control of the concentration and type of majority carriers in MBE-grown layers of IV–VI semiconductors may be achieved by adjustment of the stoichiometry of the flux that is incident on the substrate or by the incorporation of a flux of foreign atoms. Early studies of PbTe (31) showed that n- and p-type layers with carrier concentrations in the range 10^{16}–10^{17} cm^{-3} could be deposited by use of Te- and Pb-deficient PbTe sources, respectively. Later work with PbSe and (Pb, Sn)Se (25) gave the quite different result that both Se- and metal-deficient sources yielded n-type layers that contained a second phase of metal. In this case, growth of single-phase n- and p-type layers with well-controlled carrier concentrations was obtained by use of an additional flux of the chalcogen. Similar results were obtained with PbS layers (31). The use of additional fluxes of Pb and chalcogen during the growth of PbTe and (Pb, Sn)Te has also been studied by Smith and Pickhardt (30, 32, 33). Walpole (17) has obtained n- and p-type layers of (Pb, Sn)Te on PbTe substrates by incorporation of Bi and Tl, respectively. Similar results have been obtained with BaF$_2$ substrates by Smith and Pickhardt (32, 33).

* With the use of commercially available temperature controllers (Barber-Colman model 543C), the maximum drift in the temperature of the effusion cell may be reduced to 0.2–0.3 K during the time (0.5–4 h) that is required for growth of a IV–VI layer.

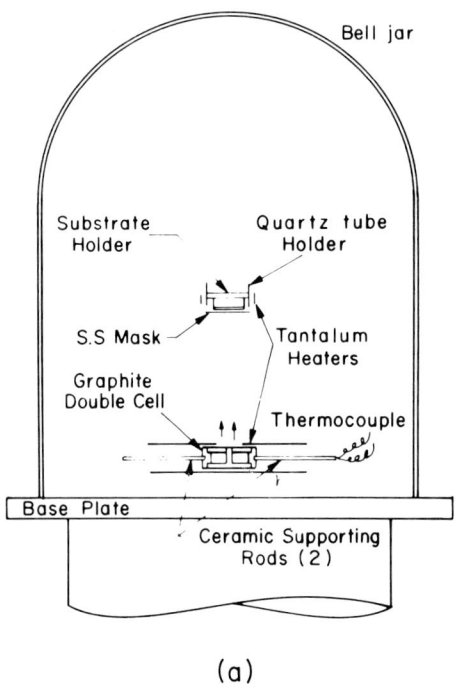
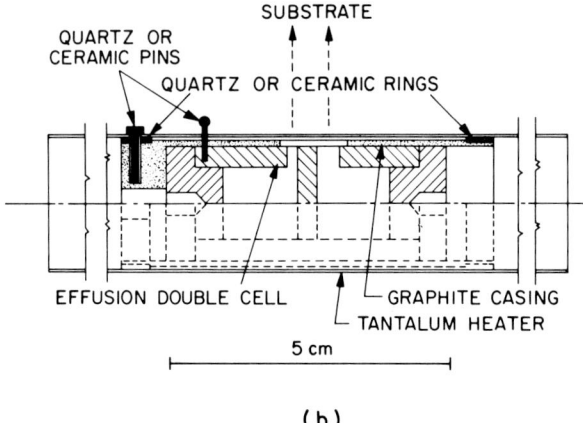

FIG. 1. Apparatus for molecular-beam evaporation of IV–VI semiconductors. (a) General arrangement, (b) details of the double-effusion cell.

2. Hot-Wall Methods

With the hot-wall technique for vacuum deposition, the evaporation source and the substrates are more or less surrounded by a heated enclosure. A typical arrangement is shown in Fig. 2. The method appears to have been invented by Koller and Coghill (*34*), who used it for deposition of ZnS, and it was later used for PbS by Hudock (*35*) and extended to the pseudobinary alloy (Pb, Sn)Te by Bis *et al.* (*36*). Subsequent work with IV-VI semiconductors (*37–48*) has been reviewed by Lopez-Otero (*44*). As with the MBE technique, the majority carrier type and concentration that is obtained with the hot-wall method may be controlled by use of a subsidiary source of the chalcogen (*38, 39, 45*).

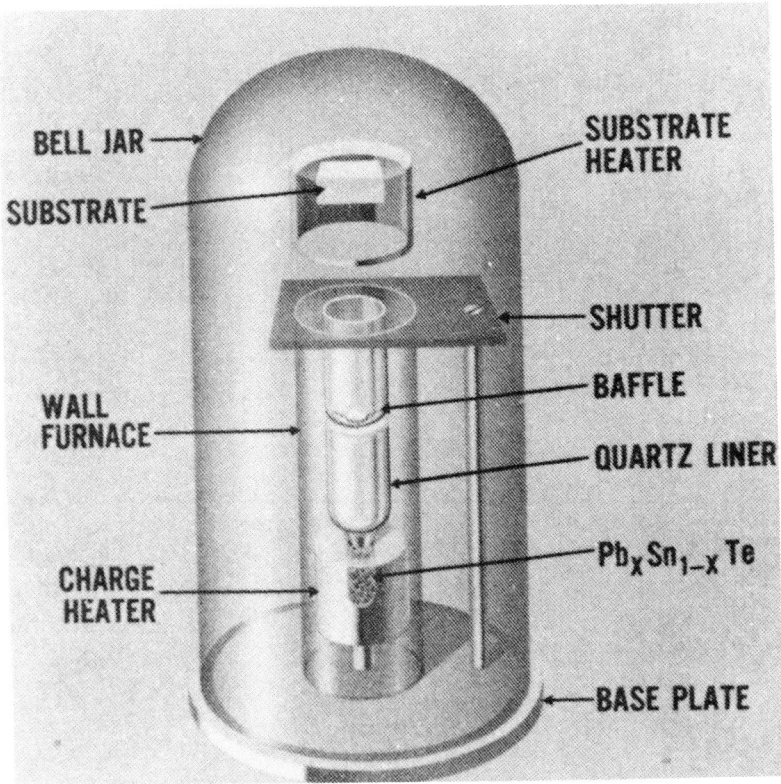

FIG. 2. A typical hot-wall evaporation system. From Bis *et al.* (*36*).

As emphasized by Hudock, the basic intent of the hot-wall method is the growth of films under conditions where the solid and gas phases are nearly in equilibrium. Qualitatively this might be expected to lead to increased crystal perfection via enhanced reevaporation of atoms and molecules from energetically unfavorable sites. However, the quantitative significance of the gas–solid interchange relative to that of atomic or molecular motion across the semiconductor surface is quite unknown. It should also be noted that adoption of the hot-wall method does not automatically ensure growth under near-equilibrium conditions. To evaluate the closeness of the approach to equilibrium conditions it is useful to consider the ratio of the growth rate to the rate of reevaporation from the semiconductor surface. For true equilibrium the growth rate would be zero, but one might still expect significant effects from the gas–solid interchange if the reevaporation rate is of the same order of magnitude as the growth rate. The existence of an effect for much smaller reevaporation rates seems hardly plausible. Figure 3 shows the calculated temperature-dependent reevaporation rate of a PbTe surface. (This assumes an evaporation coefficient of unity, which has been found for the isostructural compound SnTe (24). A smaller evaporation coefficient would give a larger departure for equilibrium.) Also plotted are the growth rates and growth temperatures for several hot-wall depositions of PbTe and of (Pb, Sn)Te (whose reevaporation rate is approximately equal to that of PbTe). It is evident that many hot-wall experiments have been performed under conditions where the growth rate greatly exceeds the reevaporation rate and the conditions are far from equilibrium. In several cases the departure from equilibrium is greater than that in typical molecular-beam depositions, which are also plotted in Fig. 3. A closer approach to equilibrium has been attained at the upper end of the range of substrate temperatures (320–500°C) that were reported by Lopez-Otero (43).

As equilibrium conditions are approached, the hot-wall method begins to resemble closed-tube transport down a temperature gradient. In this context it is of interest to note that, while closed-tube transport is usually associated with spontaneous nucleation (46) or growth on an isostructural seed (47), Pandy (48) has demonstrated use of this technique for epitaxial growth of (Pb, Sn)Te on a BaF_2 substrate.

3. FLASH EVAPORATION AND SPUTTERING

With flash evaporation a succession of small grains of evaporant are delivered to and rapidly evaporated from a hot source. The technique would be expected to eliminate the problem of incongruent evaporation of IV–VI pseudobinary alloys at the cost of increased difficulty of control of

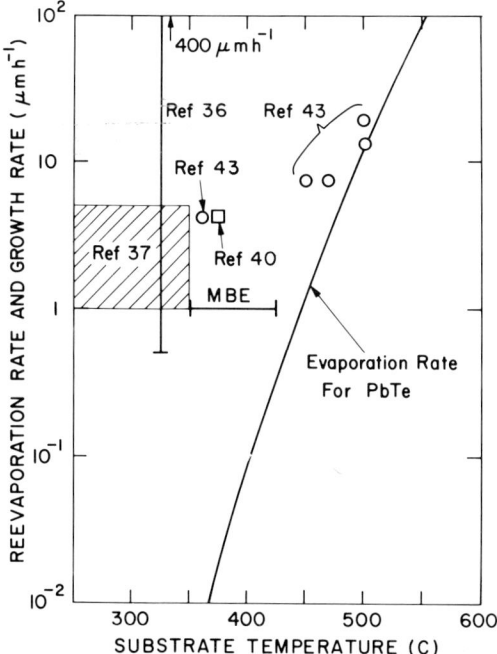

FIG. 3. The evaporation rate of PbTe as a function of temperature and the deposition conditions for several representative hot-wall experiments. Reference (36) is to growth of (Pb,Sn)Te; the other depositions are of PbTe. The line that is labeled MBE shows typical molecular-beam conditions used in the author's laboratory.

the deposition rate. This method has been used by Tao and Wang (49) to grow layers of (Pb, Sn)Se, but there is no information about its applicability to growth of device-quality material.

Sputtering has been used by Krikorian (50) and by Corsi et al. (51) to produce layers of (Pb, Sn)Te whose doping could be controlled by adding impurities to the sputtering gas or by adjustment of the layer stoichiometry. To date this technique does not appear to have been applied successfully to fabrication of p–n junction devices.

III. Properties of IV–VI Layers on Insulating Substrates

1. Layers on Alkali Halides

In an early study of epitaxial lead chalcogenides on cleaved alkali halide substrates, Zemel et al. (52) showed that such layers were well suited to measurements of optical and transport properties, despite the existence of a mosaic structure with grain size of the order of 1000 Å. Subsequent work confirmed both the mosaic structure (53) and the utility of the layers for physical measurements (15).*

The layers on alkali halides yielded much valuable information about the physical properties of the IV–VI semiconductors, but their mosaic nature made them poorly suited for p–n junction devices. Schoolar (11, 55–57) was able to demonstrate p–n junctions with the largest bandgap material (PbS), but these devices were inferior to later junctions that were obtained by the same worker with BaF_2 substrates. It is widely appreciated that bulk-crystal p–n junction technology demands a high level of structural perfection† and, consequently, the lack of significant junction device development with mosaic films is hardly surprising. However, many early and some recent papers refer to the mosaic structures as single-crystal films. This imprecise usage appears to have had a beguiling effect that diverted attention from the need to grow layers that were truly monocrystalline.

Overall the problems of characterization that arose with IV–VI semiconductors resemble those encountered initially with epitaxy of the III–V semiconductors and comments made in that context (58) apply with equal force to the IV–VI compounds. In most cases the structural characterization of IV–VI layers has been with diffraction methods whose resolution is quite inadequate for detection of the spread of orientation that occurs in a mosaic layer.

In published studies of IV–VI layers much attention has been given to the majority carrier mobilities whose resemblance to bulk-crystal values has often been noted with the implication that this provides a useful criterion for layer quality (37, 38, 59). In fact, such implications are erroneous because the mobilities are quite insensitive to departures from crystallographic perfection. The carrier mobilities of the IV–VI semiconductors tend to follow the phonon-limited values (60), for which

$$\mu \propto T^{-\alpha} \quad \text{with} \quad \alpha \approx 5/2 \qquad (1)$$

from 300 K down to temperatures at which saturation occurs. The saturation behavior of bulk-crystal Pb chalcogenides varies widely from

---

* Some use was also made of mica (54), but this substrate gives a mixture of two mutually twinned orientations.

† For example, imperfect materials are prone to reduction in their lifetimes, which arises from generation and recombination at grain boundaries. This effect degrades the junction resistance.

specimen to specimen, an effect that has been attributed to a variable concentration of compensated charged point defects (*61*). At 77 K the phonon-limited mobilities of both electrons and holes in the lead chalcogenides are about 2×10^4 cm^2 V^{-1} sec^{-1}, corresponding to scattering lengths of the order of 1000 Å for materials with carrier concentrations in the typical range 10^{16}–10^{18} cm^{-3}. Since the mobilities will be significantly affected only by imperfections that are spaced more closely than the scattering length, it is evident that mobility measurements in the range 300–77 K will not give useful information about the mosaic structure. (Even the few specimens that exhibit mobilities around 10^6 cm^2 V^{-1} sec^{-1} at temperatures near 10 K have scattering lengths that only attain a few microns.) The utility of the mobility as a criterion for crystal quality is further reduced by the observation that high-performance photodiodes have been made from bulk crystals of PbTe, whose 77 K mobilities do not attain phonon-limited values (*7*). In such cases, the low mobilities may be a consequence of compensation rather than of crystallographic imperfections (*61*).

2. Layers on BaF$_2$ and SrF$_2$ Substrates

Epitaxial growth of a IV–VI semiconductor on a fluorite-structured substrate was first reported in 1970 by Holloway *et al.* (*12*), who used the double effusion cell method to deposit (Pb, Sn)Te onto cleaved BaF$_2$.* Subsequent work showed that these and similar layers of PbTe were remarkably free from low-angle grain boundaries (*63*) and that the technique could be applied with similar results to growth of (Pb, Sn)Se (*25*), PbS (*64*), and Pb(Se, Te) (*65*).

The initial experiments with BaF$_2$ substrates were prompted by concern about the large thermal expansion mismatch between the IV–VI semiconductors and alkali halide substrates. As shown in Table I, BaF$_2$ provided a comparable lattice match and a much better thermal expansion match. There was some initial uncertainty about the attainability of epitaxy because the {111} cleavage of the fluorite-structured BaF$_2$ does not coincide with the preferred {100} growth habit of the rock-salt-

---

* Fahrinre and Zemel (*62*) had grown polycrystalline layers of (Pb, Sn)Te on CaF$_2$. Similar results have been obtained with IV–VI layers on CaF$_2$ substrates in the Ford laboratory. The significant difference from BaF$_2$ and SrF$_2$ substrates is probably the greatly increased lattice mismatch with CaF$_2$ (Table I).

TABLE I

Properties of the Lead Chalcogenides and Some Substrates

Compound	Lattice constant (Å) at 300 K	Thermal expansion coefficient (10^{-6} K^{-1}) near 300 K	Cleavage	P (Torr) at 700 K
PbS	5.94	20		
PbSe	6.12	19		
PbTe	6.46	20		
NaCl	5.64	39	(100)	4.5×10^{-7}
NaBr	5.96	42	(100)	3.7×10^{-6}
NaI	6.46	45	(100)	4.0×10^{-5}
KCl	6.29	37	(100)	2.2×10^{-6}
KBr	6.59	38	(100)	1.1×10^{-5}
KI	7.05	40	(100)	3.9×10^{-5}
CaF$_2$	5.40	19	(111)	$\sim 6 \times 10^{-20}$
SrF$_2$	5.80	18	(111)	$\sim 1 \times 10^{-20}$
BaF$_2$	6.20	18	(111)	$\sim 3 \times 10^{-17}$

structured IV–VI semiconductors.* However, {111} epitaxial growth was achieved provided that adequate care was taken to grow under clean conditions. Contaminants of unknown nature tended to give growth with a {100} texture.†

Figures 4 and 5 are electron micrographs of the surfaces of PbTe layers that were grown under identical conditions of KCl and BaF$_2$ substrates, respectively. The grainy nature of the layer on KCl is evident. In contrast, with PbTe on BaF$_2$ the most evident features are small (<1 μm diam) triangular pits that appear to arise from the intersection of dislocation

---

* Epitaxial growth of (Pb, Sn)Te with the {100} orientation has been achieved on BaF$_2$ with polished artificial {100} faces (*30*), but there is no evidence that this growth habit gives improved devices. Polishing has also been used with BaF$_2$ {111} faces (*66*). This approach eliminates the need for BaF$_2$ crystals with good cleavage, but does so at the cost of increased opportunity for contamination of the substrate surface.

† A {100} texture in (Pb, Sn)Te on BaF$_2$ {111} was also observed by Smith and Pickhardt (*30*). One reproducible cause of the {100} texture in our PbTe and (Pb, Sn)Te layers was the presence of heated freshly machined steel surfaces. The {111} orientation was only obtained after all heated steel parts had been baked *in situ* and any subsequent abrasion of their surfaces caused reversion to the {100} texture. The effect was observed with both types 304 and 316 stainless steel, but the nature of the contaminant remains unknown. Steel-induced {100} textures appear to be less common with PbSe and (Pb, Sn)Se.

FIG. 4. Parlodion replica of a layer of PbTe that had been grown epitaxially on a cleaved heated KCl substrate. The dark, approximately square, features are PbTe grains that adhered to the replica.

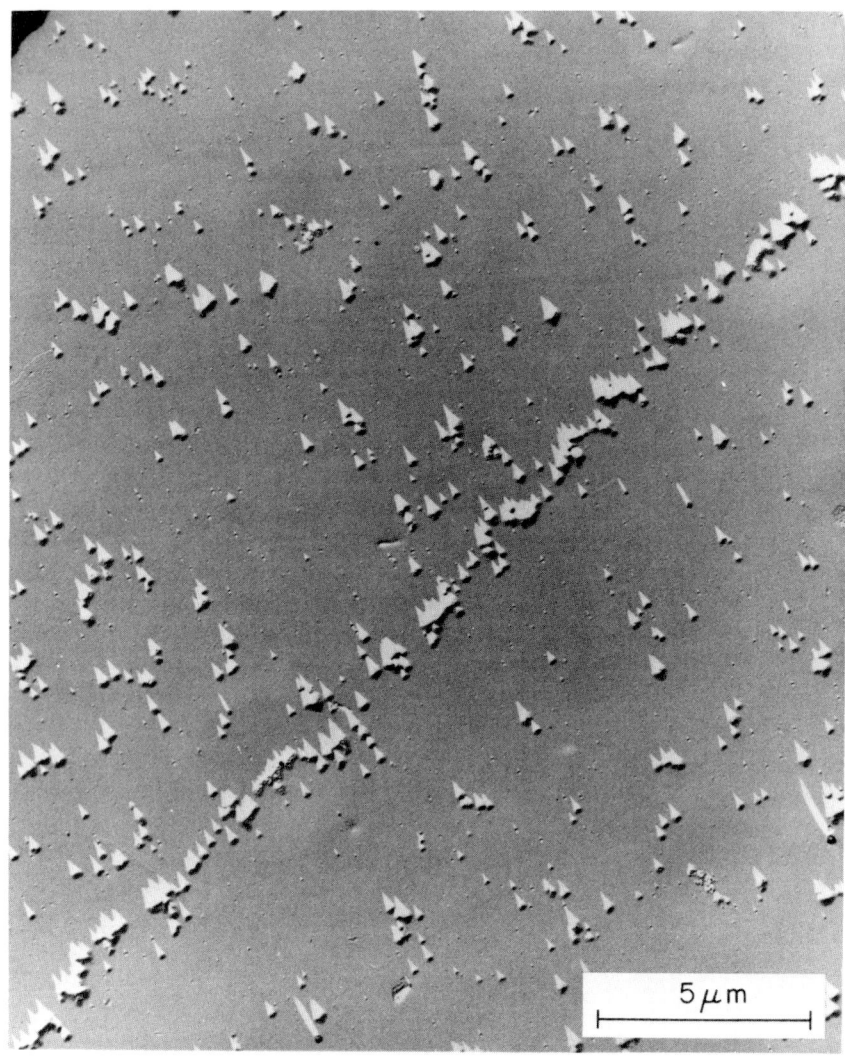

Fig. 5. Parlodion replica of a layer of PbTe that had been grown epitaxially on a cleaved heated BaF$_2$ substrate.

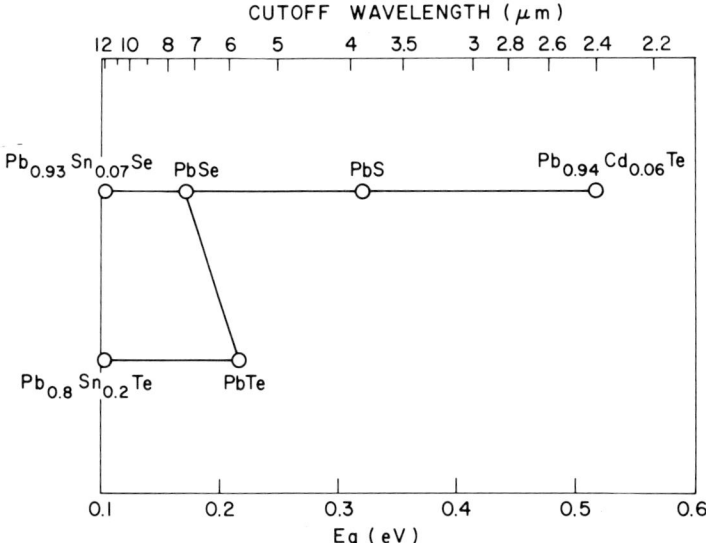

FIG. 6. The ranges of energy gaps and cut-off wavelengths at 80 K that have been used with thin-film IV–VI semiconductor photodiodes. The lines denote ranges of solid solutions.

lines with the specimen surface.* The dislocation densities so estimated range from about 10^8 cm^{-2} to substantially less than 10^6 cm^{-2}. The line of pits in Fig. 5 corresponds to a low-angle grain boundary with a misorientation of a few minutes arc. Such features occur infrequently and appear to arise from propagation of low-angle boundaries from the substrate.

Later work at the Ford laboratory and by several other groups of workers confirmed the suitability of BaF$_2$ as a substrate for PbTe and (Pb, Sn)Te (30, 40, 43, 67–70) and showed that this substrate (and to a lesser extent SrF$_2$) was also suitable for growth of a wide range of semiconductors both by MBE and by the hot-wall method. The thin-film materials that have been studied include PbSe (25), (Pb, Sn)Se (25, 42), PbS (64, 71), Pb(Se, Te) (61), Pb(S, Se) (42), and (Pb, Cd)S (71), all of which have yielded p–n junction devices. From Fig. 6 it can be seen that the

---

* Lopez-Otero (43) has reported that these features were not found with PbTe layers that were grown using the hot-wall method and suggests that their absence is a consequence of superior layer quality. However, the significance of this result is not entirely clear because the occurrence of well-defined growth pits will depend upon a large enough ratio of surface mobility to growth rate. With changes in this ratio the pits could become smaller and even more difficult to observe. This is analogous to the well-known result that not all etchants generate dislocation etch pits.

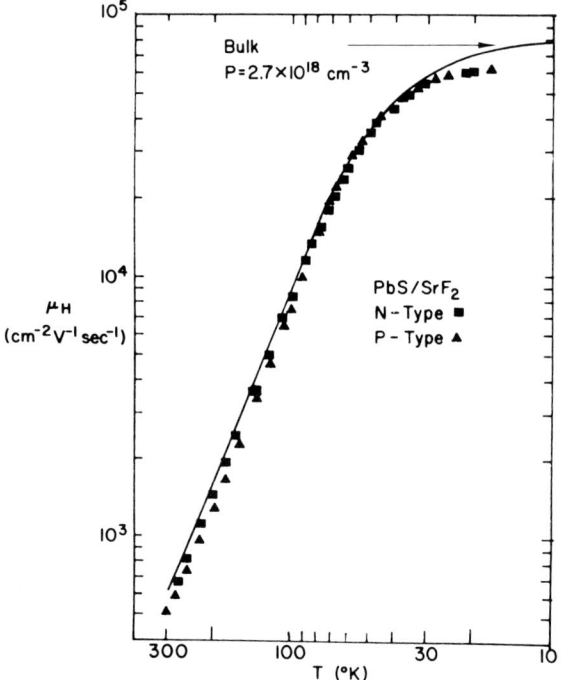

FIG. 7. Temperature dependence of the Hall mobilities of epitaxial layers of n- and p-type PbS on SrF$_2$ substrates. The curve is the result obtained by Allgaier and Scanlon (72) with a bulk crystal of PbS.

range of energy gaps covered is quite extensive and includes the values that are needed for operation in the technologically important atmospheric transmission bands at 3–5 and 8–12 μm.

As might be expected, the carrier mobilities in IV–VI layers on BaF$_2$ and SrF$_2$ were quite similar to those in good bulk crystals. Figure 7 shows the comparison between values obtained with PbS on SrF$_2$ (64) and the largest mobilities reported for bulk crystals of PbS. While the mobilities are of little consequence as a criterion for the perfection of the layers,* it is of interest that annealing at relatively low temperatures (250–350°C for 12–15 h) gave large increases in the saturation mobilities of PbTe and (Pb,

* This point is emphasized by the observation that the mobilities in PbS, PbSe, and (Pb, Sn)Se in the temperature range 300–20 K were not appreciably affected by concentrations of Pb precipitates that were sufficient to give opalescent surfaces (25, 31).

Sn)Te layers (*63*),* which were interpreted as a consequence of recombination of compensated native point defects similar to that proposed for annealed bulk crystals (*61*). However, the effects of annealing were much less marked with PbS, PbSe, and (Pb, Sn)Se layers, which tended to have larger saturation mobilities as grown. Moreover, the largest saturation mobility observed with a PbTe layer (Fig. 8) was obtained as grown.†

A curious feature of growth of the rock-salt-structured IV–VI semiconductors on fluorite-structured substrates is the occurrence of the entire deposit in an orientation that is twinned with respect to the substrate (*20, 75*),‡ i.e., the deposit is rotated by π around the $\langle 111 \rangle$ axis. This appears to be the first example of such a phenomenon with a pair of cubic materials. The result has been interpreted as a consequence of an interfacial structure (Fig. 9) that gives a stacking sequence of the type

$$\underbrace{A\beta \; \alpha B\gamma \; \beta C\alpha \; \gamma A\beta}_{BaF_2} \quad \underbrace{C \; \alpha \; B \; \gamma \; A \; \beta \; C}_{PbTe}$$

along the $\langle 111 \rangle$ growth axis. (Here the stacking symbols have their usual significance for the position of close-packed planes, with A, B, C denoting metal layers and α, β, γ denoting nonmetal layers.) Detailed analysis shows that other interfacial structures are disfavored electrostatically and sterically (*76*).

There remains to be considered the reason for the dramatic improvement in crystal quality that arises from replacement of an alkali halide substrate by BaF_2. It seems clear that the improved thermal expansion match, which prompted the initial experiments, cannot provide a satisfac-

---

* Holloway and Logothetis (*63*) describe both the annealing behavior and the evidence for monocrystallinity in PbTe and (Pb, Sn)Te layers. In subsequent papers on photodiodes we cited Holloway and Logothetis (*63*) as evidence for the crystal perfection and thereby unwittingly misled several workers into believing that the devices were made with annealed layers. This was not the case; all of the published Ford work on photodiodes was done with as-grown layers. (Some use was made of annealing for lasers.)

† Lopez-Otero (*43*) reported somewhat larger mobilities (at 13 K) in two specimens of PbTe grown on BaF_2 by the hot-wall method. The attribution of this result to an improvement in growth conditions is questionable because the specimens that are described both have $\mu_H \approx 5 \times 10^4$ cm^2 V^{-1} sec^{-1} at 77 K, which is a factor of two larger than the commonly observed phonon-limited value (*72*). Similarly anomalous mobility measurements with bulk crystals of (Pb, Sn)Te (*73*) have been interpreted as arising from inhomogeneities in the semiconductor (*74*).

‡ In an early article (*12*) we described electron diffraction evidence for the existence of two twinned orientations in the deposit. This interpretation was probably erroneous because the twinned pattern could have arisen from sampling of both the substrate and the deposit by the electron beam.

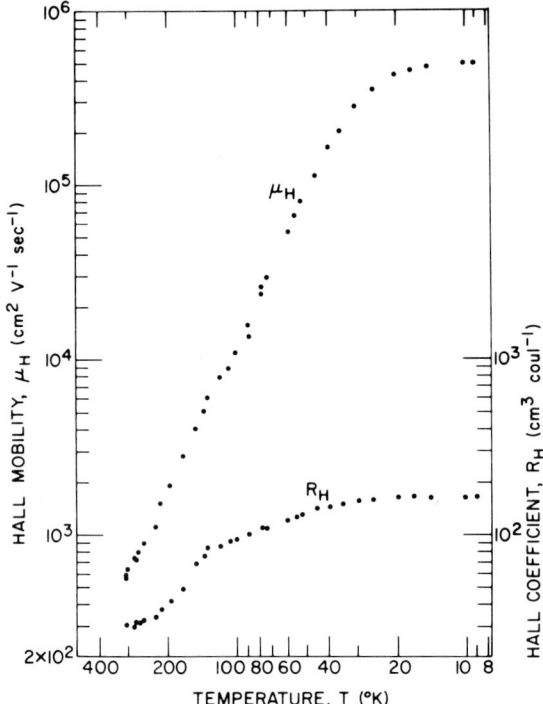

FIG. 8. Temperature dependence of the Hall mobility and Hall coefficient of an as-grown epitaxial layer of n-type PbTe on BaF_2.

tory explanation for the elimination of the mosaic structure. Qualitatively, it seems more plausible to suggest that the alkali halides are inherently unsuitable as substrates because their cleaved surfaces are hygroscopic and their substantial volatilities preclude adequate vacuum bakeout. In contrast, BaF_2 surfaces are air stable (to the extent that windows of this material do not require special care and attention) and the low vapor pressure permits routine prolonged bakeout to at least 500°C.

IV. Some Properties of IR Photodiodes

A detailed discussion of IR photodiode performance requires the use of terms and concepts that are specific to IR technology. The background for

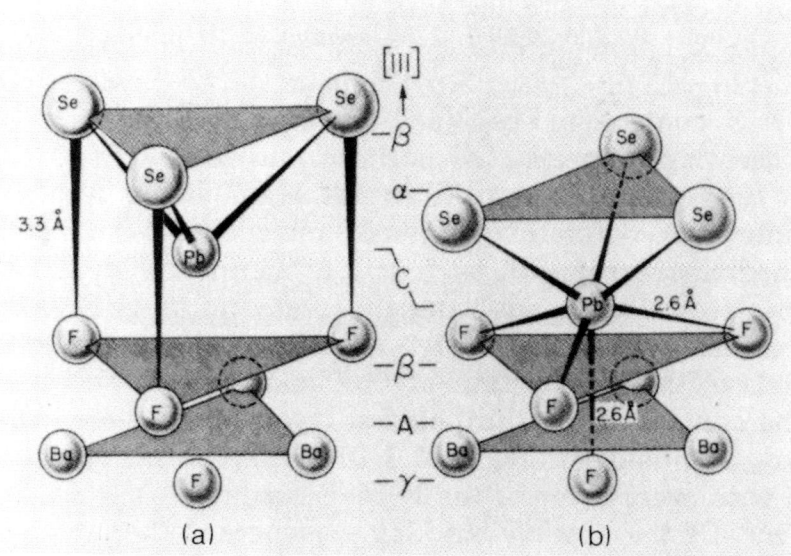

FIG. 9. Two models of the atomic structure at the epitaxial PbSe/BaF$_2$ interface. (a) Parallel orientation, (b) twinned orientation. The twinned orientation occurs exclusively.

this usage may be found in standard texts (77), but for the benefit of the nonspecialist reader a summary is given here.

The operational requirement for an IR detector is that it should give an adequately large ratio of signal to noise (S/N). This ratio depends not only upon the type and quality of the detector, but also upon the incident power, the bandwidth, and the physical dimensions of the device. For most IR detectors the signal increases linearly with the incident power and the noise increases with the square root of the bandwidth. If one defines an S/N that is normalized with respect to these quantities the usual area dependence is

$$S/N \propto A^{-1/2} \tag{2}$$

This leads to the definition of an area-independent figure of merit, the detectivity

$$D^* = SA^{1/2}/N \tag{3}$$

where again S/N is normalized with respect to incident power and bandwidth.* The parameter D^* permits comparison of the quality of IR detectors with different sizes. For practical applications, the largest S/N is obtained by choosing the detector with the largest D^* and then reducing its dimensions to the smallest values that are consistent with the properties of the associated optical system.

To obtain a theoretical estimate of the D^* of a photodiode we consider the device in the short-circuit mode illuminated with photons of energy E_λ. The signal current for unit incident power is then

$$S = \eta q/E_\lambda, \tag{4}$$

where η is the quantum efficiency and q the electronic charge. If we assume a noiseless preamplifier, the noise has contributions from fluctuations in the rate of arrival of background photons and from the Johnson noise of the junction resistance. (The series resistance is assumed to be negligible.) Thus, for unit bandwidth and a wavelength-independent quantum efficiency

$$N = (4kT/R_0 + 2\eta q^2 Q_B A)^{1/2} \tag{5}$$

where T is the operating temperature, R_0 the zero-bias junction resistance, and Q_B the flux of background photons to which the device is sensitive. Hence,

$$D^* = \frac{\eta q}{E_\lambda} \left[\frac{R_0 A}{4kT + 2\eta q^2 Q_B R_0 A} \right]^{1/2} \tag{6}$$

Useful simplifications occur when one or the other of the noise sources is dominant. Thus, for small background photon fluxes or small resistance–area products the detector becomes Johnson noise limited (JNL) and

$$D^* \text{ (JNL)} = \frac{\eta q}{E_\lambda} \left[\frac{R_0 A}{4kT} \right]^{1/2} \tag{7}$$

At the other extreme, of large photon fluxes or large resistance–area products, the detector becomes background noise limited (BNL) and

$$D^* \text{ (BNL)} = \frac{1}{E_\lambda} \left[\frac{\eta}{2Q_B} \right]^{1/2} \tag{8}$$

* In the following, D^* is defined in terms of unit incident power at a particular wavelength, which is usually that at which peak response occurs. In the literature some use is also made of a D^* that is defined in terms of unit incident power with the 500 K blackbody distribution. For an ideal photodiode (defined later) the 500 K D^* are reduced from the peak values by factors of 6 and 2.5 for cutoff wavelengths of 5 and 12 μm, respectively.

In practical terms, for a given quantum efficiency and background photon flux, the background noise limit provides a performance standard that is attained only with large enough values of the junction resistance–area product. This in turn sets a specification for the photodiode quality that is needed for optimum performance in a particular application.

Several useful results may be derived for a hypothetical ideal photodiode whose quantum efficiency is constant for $E > E_g$ and zero for $E < E_g$. In this case D^* increases linearly with wavelength up to the cutoff value, where a maximum value (peak D^*) is obtained. Now the flux of background photons to which the device is sensitive may be obtained by integrating the blackbody distribution. Substitution of Q_B into Eq. (8) then gives D^* (BNL) as a function of cutoff wavelength. A typical curve is shown in Fig. 10, where the quantum efficiency has been assumed to be 0.5, corresponding to a typical reflection-loss-limited value for a IV–VI semiconductor, and the 295 K background has been assumed to subtend a 180° field of view (FOV), corresponding to typical measurement conditions. With reduction of the FOV the background noise decreases and D^* (BNL) increases (Fig. 11). Usually detectors are used with the FOV reduced from 180°, often to values near 60° for which there is a twofold increase in D^* (BNL).

To estimate the requirements for junction resistance we may use Eq. (6) to calculate the values of R_0A that give D^* (JNL) = D^* (BNL) for 180° FOV. This will mark the transition from JNL to BNL behavior. From Eq.

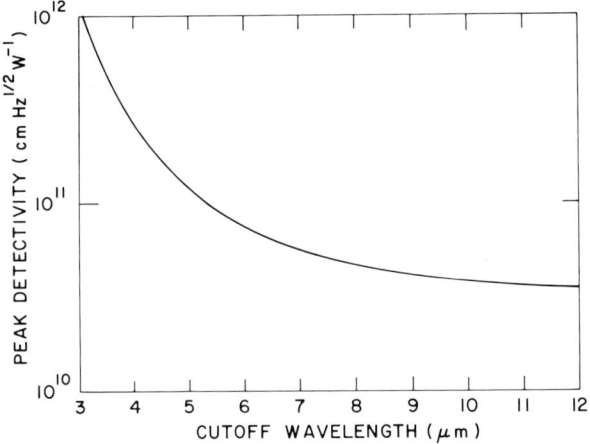

FIG. 10. Peak D^* (BNL) of an ideal photodiode that is exposed to the 295 K background at 180° FOV. The quantum efficiency is 0.5 at wavelengths below the cutoff.

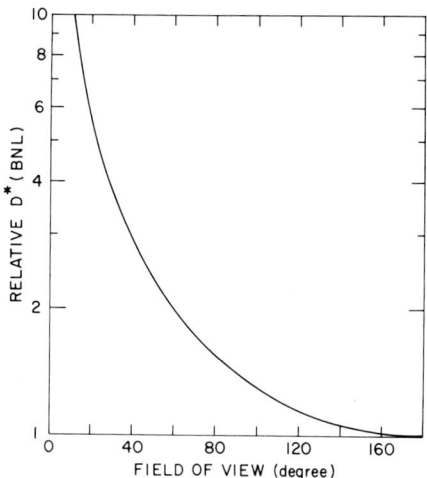

FIG. 11. Relative D^* (BNL) of an IR detector as a function of the FOV.

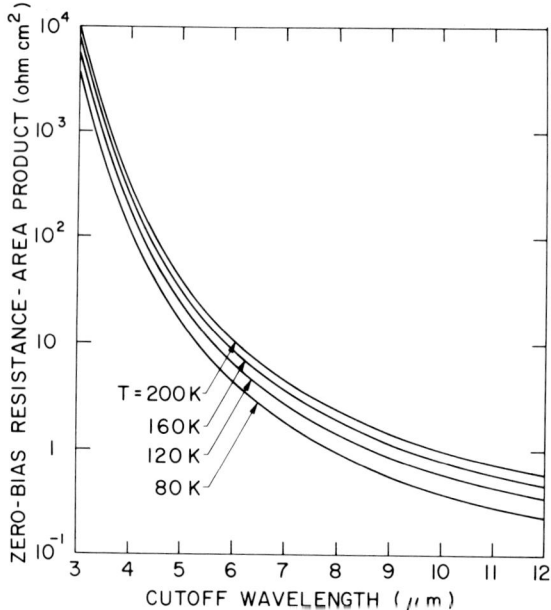

FIG. 12. Values of R_0A for which D^* (JNL) = D^* (BNL) with 180° FOV and a quantum efficiency of 0.5. The temperatures are those of the photodiode, which is exposed to 295 K background.

(6) the actual D^* under this condition will be reduced by $\sqrt{2}$ from D^* (BNL). The results of such calculations are shown in Fig. 12. Two features are of particular interest. First, the R_0A required for BNL operation decreases dramatically with increase in the cutoff wavelength. This tends to offset the increase in saturation current density and the consequent decrease in R_0A that arises with a decrease in E_g. Second, for a given cutoff wavelength there is an increase in the necessary R_0A with increase in operating temperature. Since the actual R_0A of a photodiode decreases with increase in temperature, this leads to a well-defined upper temperature for BNL operation of a particular photodiode.

The IV-VI photodiodes tend to fall into two groups that are designed for operation in atmospheric transmission bands at approximately 3-5 and 8-12 μm. The long-wavelength edges of these bands correspond to energy gaps of 0.25 and 0.10 eV, respectively. Early work with moderately cooled PbS and PbSe photoconductors was aimed at detection of warm bodies in the 3-5 μm band. Later interest in objects with temperatures near 300 K led to development of detectors with cutoffs near 12 μm to follow the resulting shift in the maximum blackbody photon flux (Fig. 13).

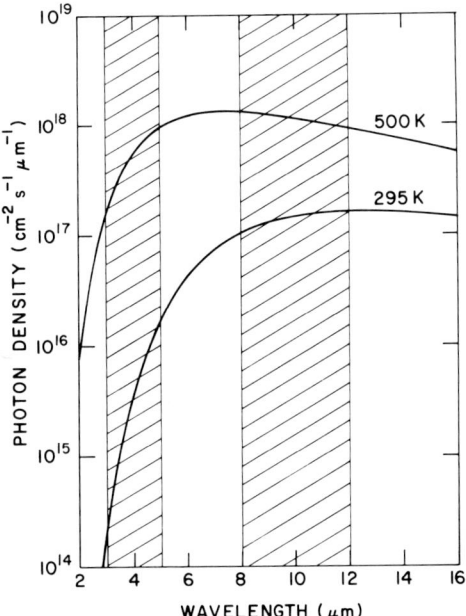

FIG. 13. Spectral photon distributions for 500 and 295 K blackbodies. The shaded regions correspond approximately to atmospheric transmission bands.

However, the 3–5 μm region has remained of interest for the detection of specific spectral features (e.g., CO_2 emissions near 4.2 μm) and because it permits the use of detectors with higher operating temperatures than those that are required for 8–12 μm operation.

V. Thin-Film IV–VI Photodiodes

1. DEVELOPMENT AND GENERAL CHARACTERISTICS

The earliest report of a IV–VI photodiode that was made by using epitaxial growth on an insulating substrate was by Schoolar (55, 56), who used PbS. A 40 μm thick layer of n-type PbS was grown on cleaved NaCl. The NaCl substrate was dissolved and a 4 μm thick layer of p-type PbS was then grown on the substrate side of the n-type PbS. The resulting devices had nonideal spectral responses that were sharply peaked near the cutoff wavelength. At 77 K the D^* at this 3.8 μm peak was about 10^9 cm $Hz^{1/2}$ W^{-1}, which is two orders of magnitude less than the 180° FOV background noise limit. In part, the small D^* appears to have been due to a small minority carrier diffusion length that reduced the collection efficiency for photogenerated carriers. This effect also accounts for the peaked response because the decreased optical absorption coefficient near the cutoff wavelength would increase the generation of carriers near the junction. However, D^* was probably also influenced by small values of R_0A. (These data were not reported, but at 77 K the back-bias resistance–area products were only 30 Ω cm².)

The first high-performance thin-film IV–VI photodiodes were reported by the Ford group (13) in 1971. These devices were made with epitaxial layers of PbTe that were grown on cleaved BaF_2 substrates using MBE. The junction fabrication technique used a Pb overlayer to invert the surface of the p-type PbTe. This method had been devised earlier by Nill et al. (78) for making injection lasers from bulk crystals of PbTe and (Pb, Sn)Te. Figure 14 shows the band structure near the Pb–PbTe interface that would be expected from the calculations of Walpole and Nill (79). The bulk-crystal devices showed photoresponse around their peripheries (80), but the Pb barrier prevented effective illumination of the junctions. With the thin-film photodiodes it became possible to illuminate the junctions from behind via the BaF_2 substrate, which is conveniently transparent for vacuum wavelengths less than 12 μm.

The early PbTe devices were made with the configuration shown in Fig. 15a. The PbTe and Pb layers were deposited as crossed stripes ($\approx$0.5 mm

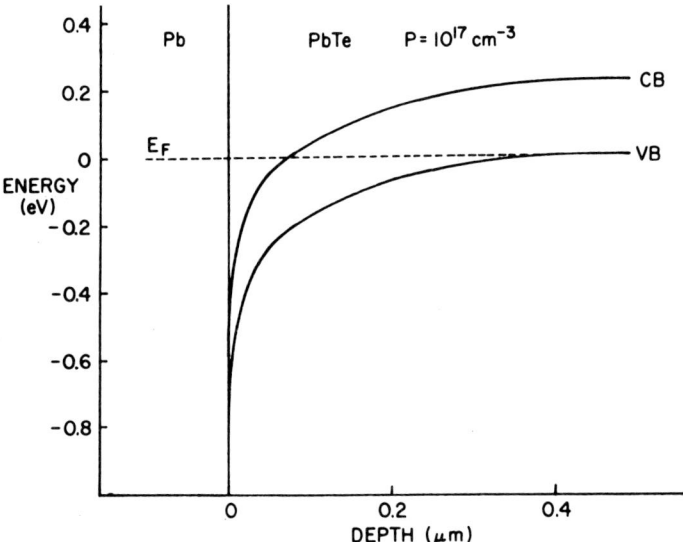

FIG. 14. Calculated energy band structure for a Pb barrier layer on p-type PbTe. After Walpole and Nill (79).

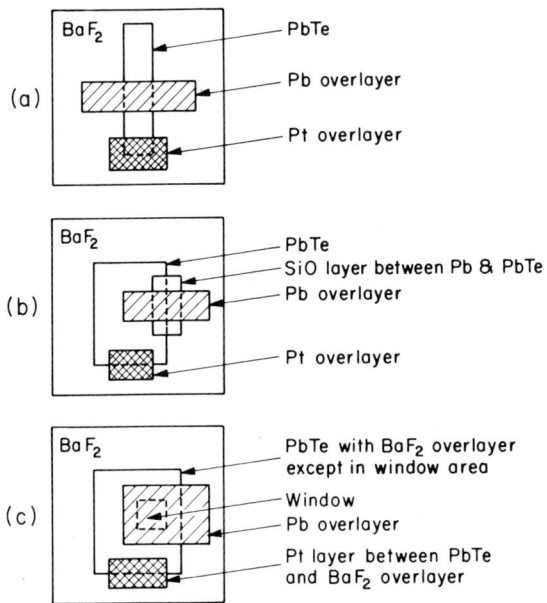

FIG. 15. Schematic diagrams of thin-film photodiode configurations. (a) Cross stripes of Pb and PbTe. (b) The PbTe edge protected with a layer of dielectric. (c) The junction area defined by a window in a layer of dielectric.

wide) that were delineated by close-spaced masks. Ohmic contact was made with sputtered Pt. At 77 K these devices gave $R_0A = 300$–2000 Ω cm^2 with quantum efficiencies in the range 0.3–0.6. Detectivities at 77 K and 33° FOV were background noise limited at 1.5–1.9×10^{11} cm Hz$^{1/2}$ W^{-1} for a peak wavelength in the range 5.35–5.5 μm. With reduction of the FOV the D^* reached a Johnson noise limit of 6×10^{11} cm Hz$^{1/2}$ W^{-1}. The same devices in forward bias were operated as injection lasers with an optical cavity that was defined by the edges of the PbTe stripe (14).

Later Pb barrier PbTe and other IV–VI thin-film detectors were made with the configurations that are shown in Fig. 15b,c. In the first of these arrangements the Pb barrier layer was isolated from the edge of the semiconductor film with an intermediate layer of SiO. This technique was introduced for IV–VI alloys, such as (Pb, Sn)Te, where the spatial separation of the component sources (e.g., of PbTe and SnTe) leads to misregistry of the mask-delineated patterns of the components. This could give edge leakage and degradation of the junction resistance when one of the components does not give a high-resistance Pb barrier diode. Removal of the misregistered edge by etching was found to be possible, but contact of the Pb barrier layer with etched edges tended to give surface leakage that was especially severe with the Se-containing materials. The final configuration adopted was that of Fig. 15c. Here the region of contact between the semiconductor and the Pb barrier layer is defined by a window in an evaporated layer of BaF$_2$. This technique was devised by Gorski (81) during the development of thin-film arrays that is described in a later section.

With the delineation technique shown in Fig. 15c and with more stringent conditions of cleanliness the Pb barrier PbTe photodiodes have given $R_0A = 3 \times 10^4$ Ω cm^2 at 80–85 K. This, together with the use of interference (described later) to increase the quantum efficiency to 0.9, corresponds to a Johnson-noise-limited $D^* = 10^{13}$ cm Hz$^{1/2}$ W^{-1} for a peak wavelength near 5 μm (82).

The demonstration of high-performance Pb barrier PbTe photodiodes was followed by the fabrication of thin-film PbTe devices with conventional p–n junctions. These junctions were made by inversion of the surface of a p-type PbTe layer, either by proton bombardment (83)* or by Sb ion implantation (86).† In each case the photodiodes had properties

* This method provides a convenient rapid method for making a p n junction, but it may not be suitable for practical applications because of a tendency for reversion of the n-type surface to p-type when heated. This annealing appears to be more rapid in (Pb, Sn)Te (84) than in PbTe (85).

† A later study of Sb-implanted PbTe diodes on BaF$_2$ substrates (87) appears to have yielded similar devices, but the detectivities that are needed for a satisfactory comparison were not reported.

TABLE II

Properties of PbTe Photodiodes at 80 K[a]

Reference	Crystal type	Junction type	R_0A (Ω cm^2)	Peak D^* under reduced background (cm Hz$^{1/2}$W^{-1})
13	thin film	Pb barrier	2×10^3	6×10^{11}
82	thin film	Pb barrier	3×10^4	1×10^{13} [b]
83	thin film	p$^+$ bombarded	5×10^2	6×10^{11}
85	bulk	p$^+$ bombarded	3×10^2	3×10^{11}
86	thin film	Sb$^+$ implanted	1.4×10^3	4×10^{11}
7	bulk	Sb$^+$ implanted	2×10^4	3×10^{12} [c]

[a] The values given for R_0A and D^* are the largest obtained. Peak response occurred near 5 μm.
[b] Based on the calculated Johnson noise.
[c] After factoring out the preamp noise, which degraded D^* by a factor of 2.

that were comparable to those of the Pb barrier thin-film devices as well as to those of bulk-crystal PbTe photodiodes that were made by proton bombardment (85) or by Sb ion implantation (7). A comparison of these devices is given in Table II. Thus, the properties of the thin-film PbTe photodiodes are a reflection of the quality of the epitaxial layer, rather than of a particular junction-forming technique.* However, the convenience of the metal-barrier method and its suitability for development of a planar array technology have led to its use for most subsequent thin-film IV–VI photodiode work.

* To date there has been little success with junction formation by successive deposition of p- and n-type layers. The only such PbTe device for which comprehensive measurements were reported (70) had D^* (5 μm) = 1.4×10^{10} cm Hz$^{1/2}$ W^{-1} at 77 K. This is an order of magnitude less than the 180° FOV background noise limit and almost three orders of magnitude less than the best Pb barrier devices under reduced FOV. This result with BaF$_2$ substrates is similar to that obtained with grown-in PbS junctions using NaCl substrates (56).

Conventional diffused junctions with mesa geometries do not appear to have been reported for PbTe films. (Probably because control of the junction depth is difficult with typical layer thicknesses around 1 μm.) Callender (67) described diffused mesa diodes made with 15 μm thick layers of Pb$_{0.82}$Sn$_{0.18}$Te on BaF$_2$. The 10 μm peak D^* at 77 K was 4×10^9 cm Hz$^{1/2}$ W^{-1}, which is an order of magnitude less than the value obtained with metal-barrier thin-film (Pb, Sn)Te devices (88).

Shortly after the announcement of Pb barrier thin-film PbTe devices Schoolar (*11, 57*) reported the use of In barriers to make PbS photodiodes. The fabrication technique partly followed the earlier PbS grown-in junction work (*56*) in that the NaCl substrate was dissolved to reveal the p-type PbS surface that was to be converted to n-type. Inversion was then achieved with a thin In layer whose IR transmittance was large enough to permit front-side illumination.* At 77 K the PbS thin-film photodiodes had R_0A up to 68 Ω cm^2 with a 3.8 μm peak D^* up to 1.5 × 10^{11} cm Hz$^{1/2}$ W^{-1}. This D^* was two orders of magnitude larger than that obtained previously with grown-in PbS junctions and was within a factor of two of the 180° FOV background noise limit. The resistance–area products were smaller than might have been expected, especially in view of the larger values that had been obtained with thin films of PbTe, which has a smaller energy gap. This result was probably a consequence of the imperfect nature of layers grown on NaCl. Schoolar *et al.* (*41*) later obtained $R_0A = 2 \times 10^4$ Ω cm^2 at 80 K with thin-film PbS photodiodes on BaF$_2$ substrates.

Further development of IV–VI thin-film photodiodes has led to metal-barrier devices for applications in both the 3–5 and 8–12 μm atmospheric transmission bands. These advances are described after a discussion of some details of the thin-film photodiode properties.

2. Quantum Efficiencies

Typically the optical absorption lengths in the IV–VI semiconductors are much smaller than the minority-carrier diffusion lengths except at wavelengths that are close to the cutoff.† Thus, the metal-barrier structure shown in Fig. 16 would be expected to have a large quantum efficiency when the semiconductor thickness is chosen to be somewhere between the diffusion and optical absorption lengths.

As an example we may consider PbTe with a moderate carrier concentration ($\leq 10^{18}$ cm^{-3}) at 80 K. Calculation of the optical absorption length using the method of Bardeen *et al.* (*92*) with Dimmock's (*93*) band-edge masses gives the results shown in Fig. 17.‡ These are in fair

---

* This barrier technique is of particular interest because it makes possible the use of bulk-crystal metal-barrier devices as photodiodes. In barriers on p-type bulk-crystal (Pb, Sn)Te have been used to make 8–12 μm photodiode arrays with excellent performance and uniformity (*89–91*).

† This condition may not have been met with some of the earlier layers of PbS on NaCl (*56*).

‡ This neglects bandtailing (*95*) and nonparabolicity. The errors are insignificant in the present context.

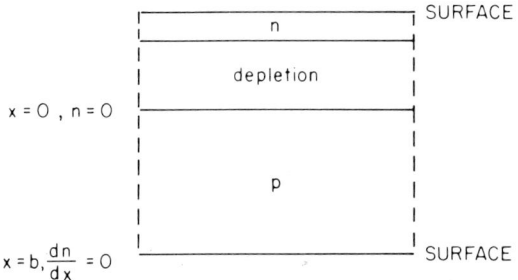

FIG. 16. Schematic cross section of a thin-film photodiode.

accord with measurements of PbTe thin films (*94*). It will be noted that, except for wavelengths close to the cutoff at 5.7 μm, the effects of bandfilling are negligible and the optical absorption length is of the order of 1 μm or less. The diffusion lengths in the IV–VI semiconductors have not been adequately characterized, but laser scans of thin-film PbTe devices (described later) show that the diffusion length in this material is of the order of 10 μm. Thus, one might expect large quantum efficiencies from PbTe devices that have thicknesses in the range 1–10 μm.

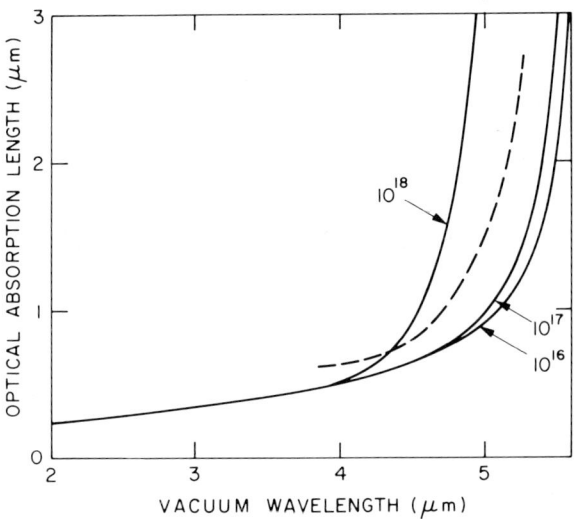

FIG. 17. Calculated wavelength dependence of the optical absorption length in PbTe at 80 K. The carrier concentrations are in cm^{-3}. The broken line is the experimental result that was obtained by Piccioli (*114*) at 85 K with a PbTe layer that had $p \approx 10^{18}$ cm^{-3}.

For a more quantitative analysis we may solve the diffusion equation

$$\frac{d^2n}{dx^2} - \frac{n}{L^2} + \frac{G}{D} = 0 \tag{9}$$

for excess minority carriers in the p region of the structure in Fig. 16. With a uniform generation rate G, the solution is

$$n = G\tau + \alpha \sinh(x/L) + \beta \cosh(x/L) \tag{10}$$

where α and β are arbitrary constants, τ the lifetime, D the diffusion coefficient, and L the diffusion length. In the short-wavelength limit the optical absorption length becomes small, so that the minority carriers are generated close to the surface. Thus, we may take $G = 0$ with the photogenerated minority carriers represented by a flux J, away from the semiconductor–BaF$_2$ interface,

$$J = D \frac{dn}{dx}\bigg|_b \tag{11}$$

Approximating the edge of the depletion region by a perfect sink, $n = 0$ at $x = 0$, whence

$$n = \frac{JL \sinh(x/L)}{D \cosh(b/L)} \tag{12}$$

Evaluating the collected flux as

$$J' = D \frac{dn}{dx}\bigg|_0 \tag{13}$$

we obtain a collection efficiency

$$\eta_c = J'/J = \text{sech}(b/L) \tag{14}$$

In the long-wavelength limit (i.e., close to cutoff) we can make the approximation that the generation rate is constant throughout the semiconductor. (This neglects the interference effects that are discussed later.) Now the boundary conditions $n = 0$ at $x = 0$ and $dn/dx = 0$ at $x = b$ give

$$n = G\tau \left(1 - \frac{\cosh[(b-x)/L]}{\cosh(b/L)}\right) \tag{15}$$

$$\eta_c = \frac{J'}{Gb} = \frac{\tanh(b/L)}{b/L} \tag{16}$$

Inspection of the collection efficiencies for the two extreme cases (Fig. 18) shows that $\eta_c \geq 0.9$ for $b/L \leq 0.5$. Thus, with PbTe thicknesses up to

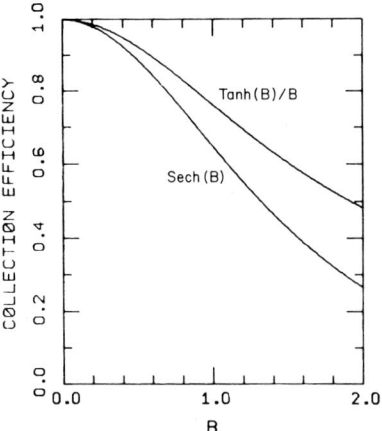

FIG. 18. Collection efficiencies of a thin-film photodiode in the long-wavelength limit [tanh(B)/B] and the short-wavelength limit [sech(B)].

about 5 μm, recombination losses are negligible and the quantum efficiency is limited only by reflection and transmission losses. For convenience this condition is designated the reflection loss limit (RLL).

Taking the refractive indices of PbTe (52) and BaF$_2$ (96) at wavelengths of interest to be approximately 6.3 and 1.5, respectively, the RLL of a thick PbTe photodiode is 0.47 for direct illumination of the PbTe surface and 0.60 for illumination via the BaF$_2$ substrate, which provides partial index matching. However, for the layer thicknesses that are commonly used the RLL may be substantially modified by interference effects.

It has long been recognized that the response of semiconductor IR detectors may be influenced by interference (97), but with bulk-crystal devices the effects are small even in extreme cases (98). The lead chalcogenide photoconductors, which are the only widely used thin-film IR detectors, also show little evidence for interference effects. This may be a consequence of the surface roughness of these precipitated layers. In contrast, the earliest work with Pb barrier thin-film PbTe devices (13) showed prominent maxima in their responses to radiation for which the PbTe optical thickness was an odd number of quarter-waves. A comprehensive analysis of the interference effects has been given by Holloway (99). The account that follows is a summary of their main features.

In evaluating the effect of interference on the RLL we must distinguish between the metal-barrier devices, which are illuminated via the BaF$_2$ substrate, and conventional p–n junctions, which may be illuminated

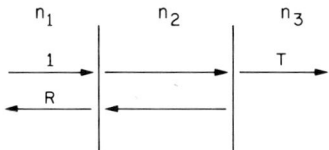

FIG. 19. Reflected and transmitted waves in a thin-film photodiode structure. The quantities 1, R, and T refer to the relative E field amplitudes.

from either side. These structures have markedly different properties. Figure 19 represents three media with refractive indices (possibly complex) n_1, n_2, n_3 that are traversed by plane waves that propagate perpendicular to the interfaces.* The boundary conditions give the well-known results (*100*) for the reflection and transmission coefficients

$$R = [(n_1 - n_2)(n_2 + n_3)e^{i\phi} + (n_1 + n_2)(n_2 - n_3)e^{-i\phi}]/\Delta \qquad (17)$$

$$T = 4n_1n_2/\Delta, \qquad (18)$$

where

$$\Delta = (n_1 + n_2)(n_2 + n_3)e^{i\phi} + (n_1 - n_2)(n_2 - n_3)e^{-i\phi} \qquad (19)$$

$$\phi = 2\pi n_2 d_2/\lambda_0 \qquad (20)$$

d_2 is the thickness of the central medium and λ_0 is the vacuum wavelength of the incident radiation.

For a conventional p–n junction n_1 and n_3 are real and the RLL quantum efficiency is

$$\eta = 1 - |R|^2 - \frac{n_3}{n_1}|T|^2 \qquad (21)$$

For such thin-film IV–VI devices on BaF_2 or SrF_2 substrates

$$n_1, n_3 < \text{Re}(n_2) \qquad (22)$$

and Eq. (21) gives maximum quantum efficiency when the semiconductor optical thickness is a multiple of half-waves. Figure 20 shows the calculated spectral quantum efficiencies of two conventional-junction PbTe devices whose thicknesses have been chosen to maximize the quantum efficiency at 5 μm. In this case the radiation is incident via the

* For most IV–VI semiconductor devices, oblique incidence may be neglected because the large refractive indices lead to almost normal propagation within the semiconductor. Grazing incidence from vacuum leads to propagation in PbTe that is only about 4° off normal.

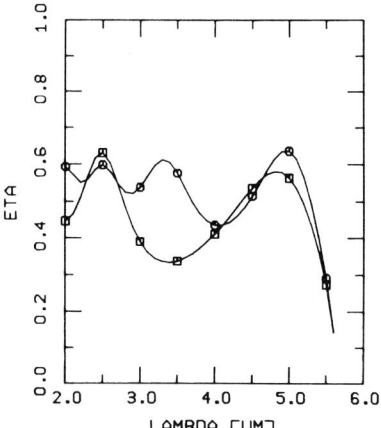

FIG. 20. Calculated RLL quantum efficiencies of two conventional-junction thin-film PbTe photodiodes at 80 K. Illumination is via the BaF$_2$ substrates and the devices are optimized for their response at 5 μm. ($\square$) PbTe thickness = 0.39 μm ≡ half-wave; ($\bigcirc$) PbTe thickness = 0.79 μm ≡ one wave.

BaF$_2$ substrate and allowance has been made for 3.5% reflection loss at the vacuum–BaF$_2$ interface. The quantum efficiencies show a small modulation, but the optimized devices only give about a 5% increase in peak quantum efficiency over the thick-film RLL ($\eta = 0.60$). The results for illumination from the PbTe side are similar, but with the modulation about a smaller thick-film RLL ($\eta = 0.47$).

The RLL quantum efficiencies of Pb barrier PbTe devices follow from Eqs. (17)–(20) if n_1, n_2, and n_3 are equated with the refractive indices of BaF$_2$, PbTe, and Pb (*101*). Transmission losses into the Pb barrier layer are 10% or less for PbTe layers with optical thicknesses of three quarter-waves or more when $\lambda_0 \approx 5$ μm. With the approximation that the Pb layer is a perfect reflector, Eqs. (17)–(20) lead to maxima in the RLL quantum efficiency for odd-quarter-wave optical thicknesses of the PbTe. As shown in Fig. 21, this result remains a good approximation when the imperfectly reflecting nature of the Pb barrier is included. Figure 21 shows that, in contrast to a conventional p–n junction, the RLL quantum efficiency of a metal-barrier device may be increased to about 0.9 for a selected peak wavelength. The experimental result in Fig. 22 shows the typical agreement with theory. In this case the structure is representative of three-quarter-wave PbTe photodiodes that have been optimized for response at 5 μm.

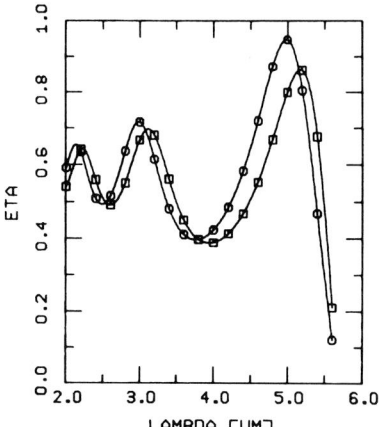

FIG. 21. Calculated RLL quantum efficiency of a metal-barrier PbTe photodiode at 80 K. The PbTe thickness is 0.59 μm ≡ three quarter-waves at λ = 5 μm. (O) Perfectly reflecting metal-barrier layer; (□) Pb barrier layer.

The RLL quantum efficiencies of 8–12 μm devices differ somewhat from those of 3–5 μm devices because band-filling effects are larger with the smaller density of states that occurs with smaller energy gaps. Figures 23 and 24 provide a comparison of the band-filling effects in PbTe and (Pb, Sn)Te devices. For the three-quarter-wave (0.59μm) PbTe device in Fig.

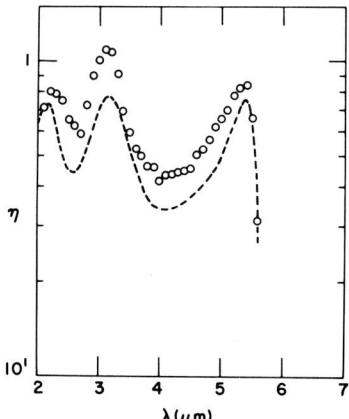

FIG. 22. Observed and calculated quantum efficiencies of a Pb barrier PbTe photodiode. The PbTe thickness is 0.64 μm.

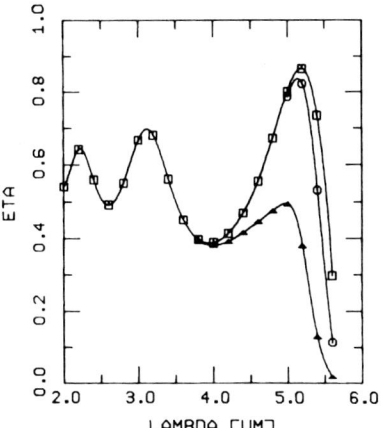

FIG. 23. Calculated effect of band-filling on the quantum efficiency of a Pb barrier PbTe photodiode at 80 K. The PbTe thickness is 0.59 μm $\equiv$ three quarter-waves at $\lambda = 5$ μm. The carrier concentrations are ($\square$) 10^{16} cm^{-3}, ($\bigcirc$) 3×10^{17} cm^{-3}, ($\triangle$) 10^{18} cm^{-3}.

23 the effects of band-filling are small for $p < 3 \times 10^{17}$ cm^{-3}, which includes the usual range of layer carrier concentrations. In contrast, the quarter-wave (0.42 μm) (Pb, Sn)Te device in Fig. 24 has a response at 11 μm that is seriously degraded by band-filling at $p = 3 \times 10^{17}$ cm^{-3}. Thus,

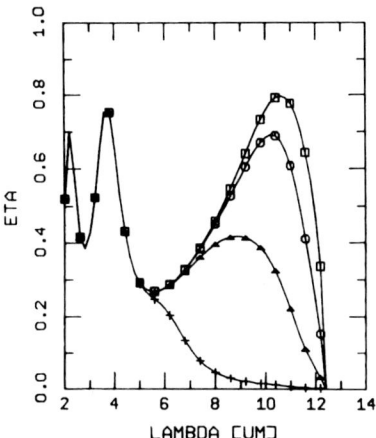

FIG. 24. Calculated effect of band-filling on the quantum efficiency of a Pb barrier (Pb, Sn)Te photodiode at 80 K. The (Pb, Sn)Te thickness is 0.42 μm $\equiv$ one quarter-wave for $\lambda = 10$ μm. The carrier concentrations are ($\square$) 10^{16} cm^{-3}, ($\bigcirc$) 10^{17} cm^{-3}, ($\triangle$) 3×10^{17} cm^{-3}, (+) 10^{18} cm^{-3}.

the 8–12 μm devices require larger semiconductor thicknesses for optimum response than do the 3–5 μm devices.

3. ESTIMATES OF RESISTANCE–AREA PRODUCTS, LIFETIMES, AND DETECTIVITIES

The preceding analysis of collection efficiency indicates that little difficulty should be encountered in making thin-film IV–VI photodiodes that have large collection efficiencies. Thus, the attainable D^* will depend almost entirely on the zero-bias resistance–area products (R_0A). Here the current level of understanding, even of bulk-crystal photodiodes, is not entirely satisfactory. This section discusses the factors that influence R_0A and provides some estimates of attainable D^* for 3–5 and 8–12 μm devices.

Ideal p–n junction devices have current–voltage characteristics of the form

$$I = I_s[\exp(qV/\beta kT) - 1] \qquad (23)$$

where typically $1 \leq \beta \leq 2$. This relationship gives

$$R_0A = \left(\frac{\partial I}{\partial V}\bigg|_{V=0}\right)^{-1} A = \frac{\beta kT}{qJ_s} \qquad (24)$$

where $J_s = I_s/A$ is the saturation current density. Thus, calculation of R_0A requires quantitative information about the possible sources of I_s. Here we may distinguish between two major mechanisms for the generation of I_s. First, the dominant mechanism may be thermal generation of electron–hole pairs outside the junction region. The saturation current then arises by diffusion to the junction, which collects the minority carriers (102). Most thin-film devices (including those with metal barriers) have junctions that are abrupt and one-sided (in the sense that I_s is dominated by the contribution from the side with the lower carrier concentration). In this case $\beta = 1$ and J_s is given by

$$J_s(\text{diffusion}) = (qn_i^2/p)(D/\tau)^{1/2} \qquad (25)$$

where p is the majority-carrier concentration, D the minority-carrier diffusion coefficient, and τ the lifetime, all on the low-carrier-concentration side. An alternative source of I_s is generation in the depletion region via states with energies in the gap (103). In this case $\beta = 2$ and

$$J_s(\text{depletion}) = qn_iw/\tau_d \qquad (26)$$

where w is the width of the depletion region and τ_d a phenomenological parameter called the effective lifetime of the depletion region. (In principle, τ_d is calculable from the properties of the states in the gap, but these are rarely known.) These two sources of I_s have been discussed by Sze (*104*). In addition, the junction resistance may be degraded by surface leakage and, especially with the smaller-gap materials, by tunneling. Excluding, for the moment, the possibility of tunneling, the most general temperature dependence of R_0A is shown in Fig. 25. The occurrence of n_i^2 and n_i in Eqs. (25) and (26) gives Arrhenius plots [$\ln(R_0A)$ vs. $1/T$] with slopes of approximately E_g/k and $E_g/2k$, respectively, for the two models. This leads to diffusion current limitation at higher temperatures. At lower temperatures depletion current limitation may occur and at the lowest temperatures R_0A may saturate due to the existence of a weakly temperature-dependent shunt resistance.

If we postulate that states that permit generation within the depletion region are not an inherent property of the semiconductor (i.e., that they arise from impurities or imperfections), it follows that, in principle, J_s(depletion) may be made insignificantly small. In this case the attainable R_0A will depend upon J_s(diffusion) and hence upon the lifetime. The lifetime might be degraded by recombination centers, but its ultimately attainable value will depend upon fundamental physical processes in the

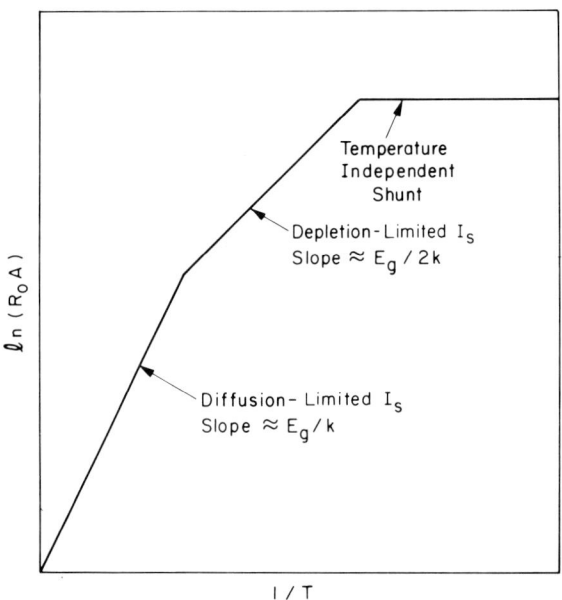

FIG. 25. Generalized Arrhenius plot of R_0A.

semiconductor. Thus, we may tentatively estimate the upper limit of R_0A from Eq. (25), which leads to

$$(R_0A)_{\text{diffusion}} = \frac{p}{q^{3/2}n_i^2}\left(\frac{kT\tau}{\mu}\right)^{1/2} \tag{27}$$

where μ is the minority carrier mobility and the lifetime τ is limited by an as yet unspecified recombination process.

Equation (27) has been derived for a semiinfinite bulk crystal. If the boundary conditions are changed to those appropriate to a thin-film device the diffusion-limited saturation current may be reduced because it is collected from a smaller volume. In the absence of recombination at the semiconductor/substrate interface this leads to the result (*105*)

$$(R_0A)_{\text{diffusion, film}} = (R_0A)_{\text{diffusion, bulk}} \coth(b/L) \tag{28}$$

where b is the semiconductor thickness and $L = (D\tau)^{1/2}$ is the minority-carrier diffusion length. If there is significant recombination at the semiconductor–substrate interface Eq. (28) is modified to give

$$(R_0A)_{\text{diffusion, film}} = (R_0A)_{\text{diffusion, bulk}} \left[\frac{\cosh(b/L) + \gamma \sinh(b/L)}{\sinh(b/L) + \gamma \cosh(b/L)}\right] \tag{29}$$

where $\gamma = sL/D$ and s is the surface recombination velocity.

Lifetimes in the IV–VI semiconductors are still rather poorly understood. Early analysis (*6, 106*) considered radiative recombination (*107*) and recombination via Shockley–Read centers (*108*). More recently a theoretical analysis by Emtage (*109*) has shown that, contrary to earlier belief (*110*), significant recombination may occur by the Auger process, in which the recombination energy is transferred to a majority carrier rather than a photon, as in the radiative recombination process.

As an example of a material for 3–5 μm photodiodes we may consider PbTe, for which Fig. 26 shows the calculated radiative and Auger lifetimes at temperatures in the range 80–300 K.* The radiative lifetimes

---

* In an earlier comparison of these lifetimes (*111*) the radiative values are about an order of magnitude larger than those given here. The larger values arose from the assumption that the band-to-band transition had unity oscillator strength. In fact, the oscillator strength is inversely proportional to the effective mass (*112, 113*) and this introduces a large correction for small-bandgap materials. Here we follow Melngailis and Harman (*6*) in taking the oscillator strength to be the reciprocal of the conductivity effective mass. The oscillator strength enters the radiative lifetime calculation via the optical absorption coefficient, which may also be calculated to check the validity of the method. The optical absorption coefficients that are obtained (*105*) do agree quite well with experimental values (*94, 114*) (Fig. 17).

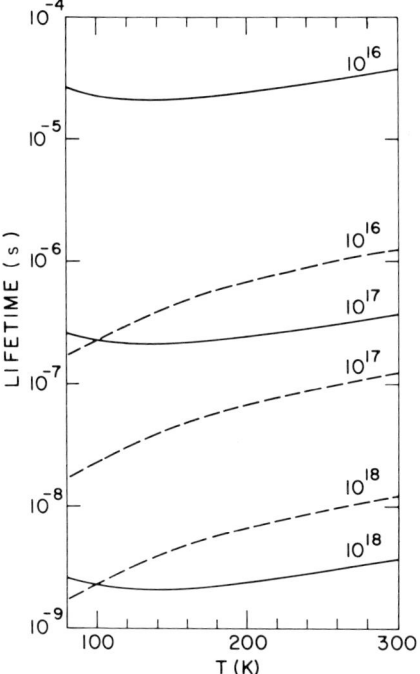

FIG. 26. Calculated lifetimes for PbTe. The solid curves are Auger lifetimes and the broken curves are radiative lifetimes. The majority carrier concentrations are in cm^{-3}.

have been calculated by the method of van Roosbroeck and Shockley (*107*) using Dimmock's (*93*) band-edge masses. (This neglects nonparabolicity, but the approximation has little effect on the optical absorption coefficient (*113*) and even less effect on the radiative recombination rate.) The effects of degenerate carrier statistics are negligible for PbTe with majority carrier concentrations up to 10^{18} cm^{-3} at 80 K and above. Also, the effective masses in the alloy system (Pb, Sn)Te are approximately proportional to the energy gap and, to this level of approximation, the radiative lifetimes follow

$$\tau_{\text{rad}} \propto T^{3/2}/p \qquad (30)$$

with no dependence on the energy gap. (Here p is the majority carrier concentration.) The Auger lifetimes calculated by Emtage's method (*109*) (which also assumes nondegenerate statistics) follow

$$\tau_{\text{Aug}} \propto 1/p^2 \tag{31}$$

with only a weak temperature dependence for PbTe in the temperature range 80–300 K. Thus, the Auger process would be expected to be dominant at higher temperatures and larger carrier concentrations. However, inspection of Fig. 25 indicates that, even at 300 K, the Auger process is not significant in PbTe with carrier concentrations less than about 3×10^{17} cm^{-3}.

At this stage there are few lifetime measurements that may be compared with the calculations for PbTe. Lischka and Huber (*111*) measured photoconductive decay in n-type PbTe and obtained two time constants the smaller of which was assumed to be the band-to-band recombination lifetime. In the temperature range 50–200 K this time was approximately 2×10^{-7}–10^{-6} sec for $n = 6.5 \times 10^{16}$ cm^{-3} and 3×10^{-8}–10^{-7} sec for $n = 2.9 \times 10^{17}$ cm^{-3}. These values are in fair accord with the radiative lifetimes in

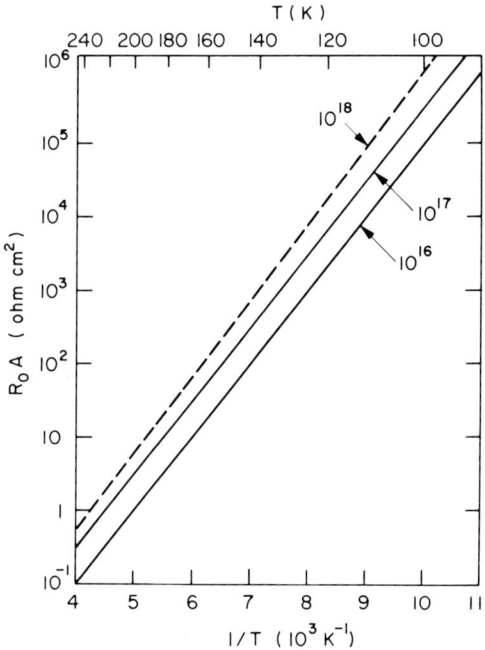

FIG. 27. Calculated R_0A for one-sided abrupt junctions in PbTe, assuming diffusion-limited saturation current. The solid lines indicate limitation by the radiative lifetime and the broken line indicates limitation by the Auger lifetime. The carrier concentrations are in cm^{-3}.

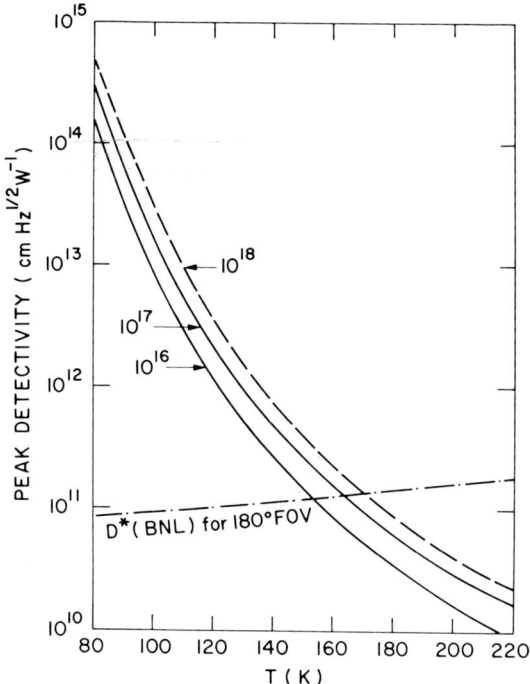

FIG. 28. Calculated peak D^* for ideal PbTe photodiodes with quantum efficiencies of 0.5. The dot–dash line is D^* (BNL) for the 295 K background at 180° FOV. The other curves show D^* (JNL) based on values of R_0A from Fig. 27. The broken curve indicates limitation by the Auger lifetime rather than the radiative lifetime. The carrier concentrations are in cm^{-3}.

Fig. 25. Lifetimes in the range 10^{-8}–10^{-7} sec have also been inferred by Heinrich *et al.* (*115*) from the transient behavior of thin-film PbTe diodes, although in this case neither the carrier concentrations nor the temperatures of the measurements were reported.

Figure 27 shows the calculated temperature-dependent diffusion-limited R_0A for an abrupt one-sided PbTe junction that is obtained from the lifetimes in Fig. 25. These calculations neglect the possible enhancement of R_0A by the reduction in generation volume that occurs with thin-film devices. The mobilities have been assumed to have phonon-limited values ($\mu_n = 1.56 \times 10^9 T^{-5/2}$ cm^2 V^{-1} sec^{-1}). The mobilities in device quality PbTe at 80 K are often lower than the phonon-limited value by about a factor of two. This would lead to a 40% increase in R_0A at 80 K. Figure 28 shows the corresponding estimated upper limits of D^* (JNL) for PbTe devices. Here the quantum efficiencies have been chosen to have the bulk-crystal RLL value of 0.5 (this neglects possible enhancement by interference effects) and again the possible thin-film enhancement of R_0A has been neglected. The curves for $p = 10^{16}$ cm^{-3} and $p = 10^{17}$ cm^{-3} correspond to the radiative lifetime, while that for $p = 10^{18}$ cm^{-3} corresponds to the Auger lifetime for temperatures above 100 K. However, the reduction in D^* due to Auger processes in the latter case is hardly significant.

The fundamental limits on the performance of bulk-crystal IV–VI photodiodes for 8–12 μm operation have received considerably more attention than the 3–5 μm case. Detailed discussions of (Pb, Sn)Te devices have been given by DeVaux et al. (116) and by Johnson et al. (117).* Emtage (109) has pointed out the significance of Auger processes in (Pb, Sn)Te and Preier (118) has provided a comparative analysis of (Pb, Sn)Se and (Pb, Sn)Te.

Figure 29 shows calculated radiative and Auger lifetimes of (Pb, Sn)Te with $E_g = 0.10$ eV at 80 K. The radiative lifetime calculation follows that for PbTe except that here one must include the effects of degenerate statistics on the majority carriers. For a condition of complete degeneracy with the effective masses proportional to E_g, Eq. (30) is replaced by

$$\tau_{\text{rad}} \propto 1/E_g \tag{32}$$

with no significant dependence on either the majority carrier concentration or the temperature. The calculations show that at 80 K there are significant departures from nondegenerate behavior when the majority carrier concentration is greater than 10^{17} cm^{-3}. The Auger lifetimes have been derived directly from Emtage's result (109) for this material,

$$\tau_{\text{Aug}} = 1/\gamma p^2 \tag{33}$$

with $\gamma = 5 \times 10^{-26}$ cm^6 sec^{-1}. In this case no allowance has been made for degenerate carrier statistics. The results suggest that recombination will be dominated by the Auger process for carrier concentrations greater than 10^{16} cm^{-3}.

Melngailis and Harman (6, 106) measured photoconductive lifetimes in (Pb, Sn)Te with $E_g = 0.10$–0.11 and obtained values of 1.5×10^{-8} sec for both n- and p-type material with majority carrier concentrations of 6–8×10^{15} cm^{-3} at 77 K. These values are more than an order of magnitude smaller than the radiative lifetimes that would be expected to dominate at

* Johnson et al. (117) also includes a discussion of graded-junction 3–5 μm devices.

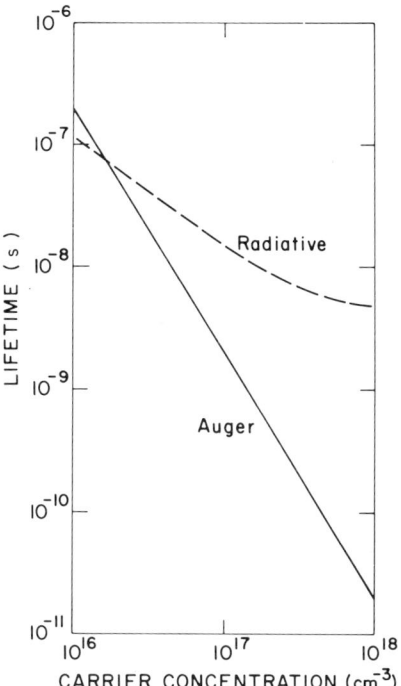

FIG. 29. Calculated lifetimes for (Pb, Sn)Te with $E_g = 0.10$ eV at 80 K. The broken curve shows the radiative lifetime. Here the reduced slope at larger carrier concentrations is a consequence of the onset of degenerate carrier statistics. The solid line is the Auger lifetime. This is calculated, following Emtage (*109*), without allowance for degeneracy. Both lifetimes are not much affected by small changes (~10%) in E_g.

these carrier concentrations. More recently Hoai and Herrmann (*119*) have reported photoconductive lifetimes at 100 K of (Pb, Sn)Te specimens with carrier concentrations in the range 4×10^{14}–10^{18} cm^{-3}. The largest lifetimes that were observed were in excellent agreement with the calculated Auger lifetimes for carrier concentrations in the range 10^{16}–10^{18} cm^{-3}.* Several specimens gave lifetimes that were much smaller than the Auger values. This suggests the frequent occurrence of a Shockley–Read recombination mechanism.

Figure 30 shows the calculated values of R_0A for (Pb, Sn)Te at 80 K with $E_g = 0.10$–0.12 eV and carrier concentrations in the range 10^{16}–10^{18}

* These values were obtained with $E_g = 0.125$ eV, but this small change in E_g does not lead to a significant change in τ_{Aug} from Emtage's calculation.

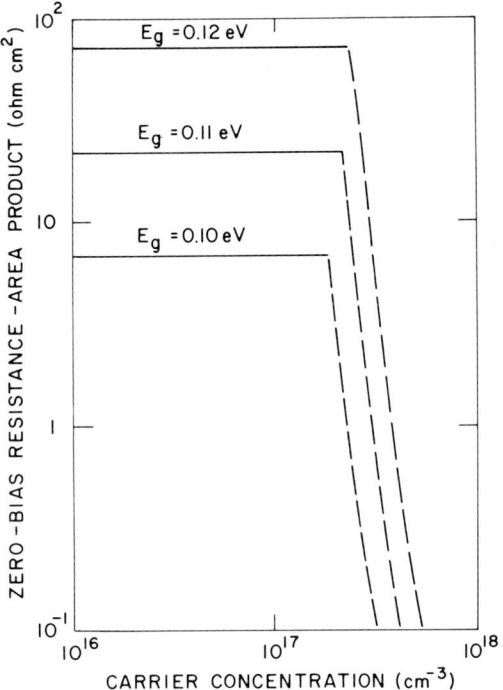

FIG. 30. Calculated R_0A for one-sided abrupt junctions in (Pb, Sn)Te. The solid lines assume a diffusion-limited saturation current that is determined by the Auger lifetime. The broken curves denote limitation by tunneling following Preier's calculation (118).

cm^{-3}, which is of most interest for junction fabrication. The Auger-limited values have been calculated from Emtage's result (109). For carrier concentrations larger than $2-3 \times 10^{17} cm^{-3}$, tunneling dominates the values of R_0A. The calculations of tunneling-limited R_0A are tentative because the values are very sensitive to changes in the model that is used. However, the steep decrease of R_0A with increase in carrier concentration gives an onset of limitation by tunneling at a carrier concentration that is not very sensitive to the choice of model. The calculations here follow Preier (118); the results of Johnson et al. (117) give the onset of tunneling limitation at a somewhat larger carrier concentration. The corresponding values of peak D^* (assuming a quantum efficiency of 0.5) are shown in Fig. 31. In the Auger-limited region D^* (JNL) is independent of the carrier concentration and significantly exceeds D^* (BNL) for 180° FOV. For carrier concentrations greater than $2-3 \times 10^{17}$ cm^{-3}, the

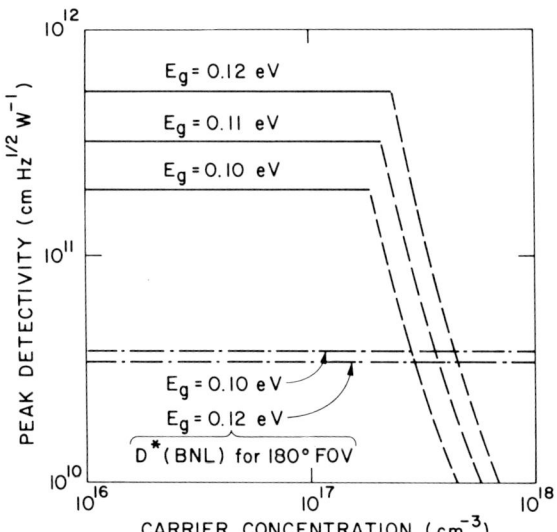

FIG. 31. Calculated peak D^* for ideal (Pb, Sn)Te photodiodes with quantum efficiencies of 0.5. The dot–dash lines show D^* (BNL) for the 295 K background seen at 180° FOV. The other curves show D^* (JNL) based on values of R_0A from Fig. 30. Solid lines indicate limitation by the Auger lifetime and the broken curves indicate limitation by tunneling.

attainable D^* (JNL) is significantly degraded by tunneling. The results for the alternative 8–12 μm material (Pb, Sn)Se are generally similar except that, as pointed out by Preier (118), a reduction in the anisotropy of the effective masses leads to an increase of the Auger lifetime that would be expected to increase D^* (JNL) by about a factor of two.

Comparison of the calculated performance of (Pb, Sn)Te photodiodes with that achieved in practice with bulk-crystal devices ($116, 117$) shows that diffusion-limited values of R_0A may be obtained at temperatures as low as 80 K, although many specimens show evidence for depletion-limited behavior. Under reduced FOV the D^* of specimens (116) with $E_g = 0.10$ eV have exceeded 10^{11} cm Hz$^{1/2}$ W^{-1} and thus approach the theoretical limits that are set by Emtage's values of the Auger lifetime.

4. Detector Noise

The preceding analysis of detectivity has assumed that the noise characteristics of the photodiodes are ideal. Mostly this assumption is valid, but the conditions for its validity require some discussion. It is often found (120) that IV–VI semiconductor photodiodes exhibit excess noise with approximately $1/f$ power spectrum when they are operated at other than zero bias. The source of the excess noise is undetermined, but one may speculate that it is associated with surface leakage. The phenomenon is not unique to IV–VI semiconductors, having also been observed in early work with InSb photodiodes (121). An example of the bias-dependent noise of a thin-film (Pb, Sn)Se photodiode is shown in Fig. 32. It will be noted that the position of minimum noise is offset slightly from zero bias. The offset corresponds to the voltage drop that arises from the background photocurrent flowing through the series resistance of the thin-film device. Thus, minimum noise occurs with zero bias across the depletion region, rather than across the external connections.

At zero bias most of the thin-film devices, like those made from bulk crystals, exhibit noise that is in excellent agreement with that calculated as the sum of contributions from the Johnson noise, the background photon noise, and the preamplifier. An example of the noise behavior of a thin-film PbTe photodiode is shown in Table III. Occasionally a significant contribution from $1/f$ noise has been observed at the optimum, near-zero bias. Such behavior, of which Fig. 33a shows an example from a (Pb, Sn)Se device, tends to be associated with abnormally large resistance at the nominally ohmic Pt–semiconductor contact. With the more typical noise characteristic shown in Fig. 33b, the $1/f$ noise contribution is negligible at frequencies above 40 Hz.

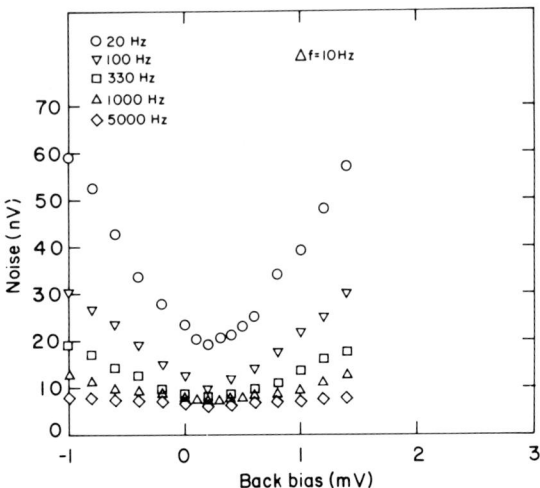

FIG. 32. Bias and frequency dependences of the noise of a thin-film (Pb, Sn)Se photodiode at 80 K and 77° FOV.

TABLE III

Temperature-Dependent Noise Properties of a Thin-Film PbTe Photodiode[a]

T (K)	Junction resistance (Ω)	Background current (μA)	Background noise current (pA)		Noise voltage (nV)			
					Calculated			Observed
			Calc.	Obs.	Background	Johnson	Total	
80	1.3×10^5	2.8	3.4	3.0	—	—	—	—
150	1900	1.6	—	—	4.3	12.5	13.4	13
170	510	0.6	—	—	0.7	6.9	7.3	7.8
190	185	0.6	—	—	0.3	4.4	5.0	5.8

[a] This specimen was made with a 4.1 μm thick PbTe film. The junction area was 3.2×10^{-3} cm². The measurements were made at 330 Hz with 10 Hz bandwidth. At 80 K the current mode preamplifier that was used contributed a negligible amount of noise. At higher temperatures the voltage mode preamplifier contributed 2.4 nV of noise and this is included in the calculated total noise.

In a few cases bulk-crystal (Pb, Sn)Te devices have been found to be operable in backbias without significant $1/f$ noise. Under such conditions, with R_0A limited by diffusion, the detector noise is theoretically capable of reduction by a factor of $\sqrt{2}$ from the Johnson noise of the junction resistance. Such noise reductions have been observed under reduced background conditions by DeVaux et al. (*116*). A reduction in the backbias $1/f$ noise has also been achieved with thin-film PbTe photodiodes (*122*), although in this case there are no reduced background measurements that might reveal a reduction in detector noise from that at zero bias. For these devices, the junction areas were delineated with windows in an insulating layer of BaF_2 and the windows were opened by a photolithographic lift-off technique. (This is described in a section that deals with array fabrication.) The reduction in backbias $1/f$ noise was achieved by prolonged (1–5 min) washing with pure water at the stage between delineation of the photoresist and deposition of the BaF_2 insulator.* At 80 K the resulting devices were capable of operation with significant backbias at 180° FOV without appreciable deviation from

---

* The original motivation for this procedure was concern that the semiconductor surface might be contaminated with alkali ions from the developer used for the positive resist (Shipley AZ 1350J). However, attempts to restore the backbias $1/f$ noise by deliberate contamination with dilute developer after the wash step were unsuccessful. Thus, washing appears to passivate the surface, but the mechanism remains obscure.

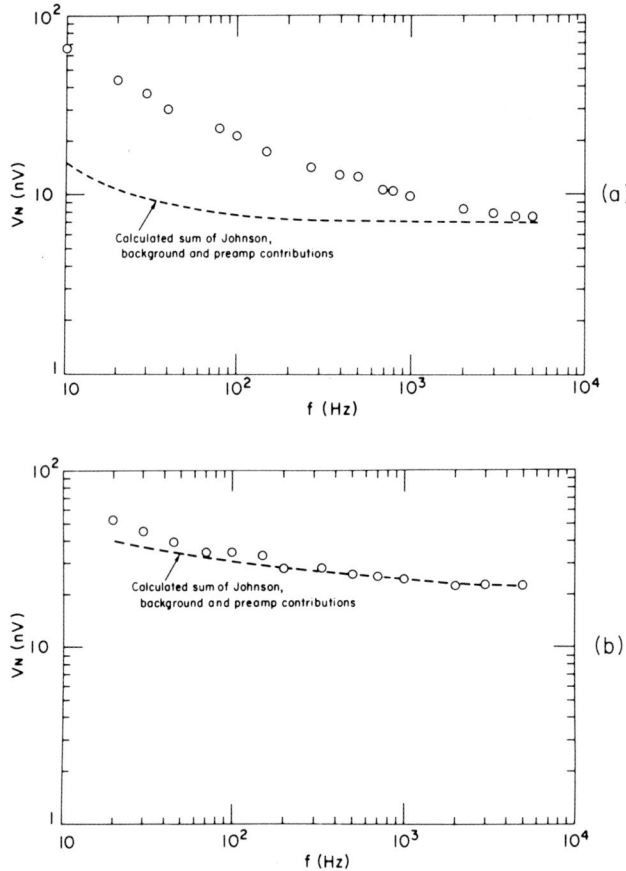

FIG. 33. Frequency dependence of the noise of thin-film (Pb, Sn)Se photodiodes at optimum (near-zero) bias. (a) With large $1/f$ noise, (b) with typical $1/f$ noise.

background-limited behavior. However, the amount of backbias that could be applied without increased noise was quite variable, even with devices whose other characteristics were almost identical. This variability is illustrated by the array of PbTe photodiodes that is described in Table IV and Fig. 34. At temperatures above 100 K such PbTe devices still gave significant $1/f$ noise with small backbiases.

The reduction in backbias $1/f$ noise permitted operation of PbTe photodiodes in the low-capacitance pinched-off mode that is described later. The washing technique also led to improvements in the thermal

TABLE IV

Properties of a Thin-Film PbTe Photodiode Array at 80 K[a]

Diode no.:	a	b	c	d	e
R_0A (10^4 Ω cm^2)	3.1	3.3	3.5	2.6	2.5
Background current (μA)	1.08	1.10	1.12	1.06	1.06
Calc. background noise (pA)	1.86	1.87	1.89	1.84	1.84
Measured noise (pA)	1.7	1.7	1.8	1.8	1.8
$\Re_I$ (AW^{-1})	0.57	0.57	0.58	0.56	0.56
D^* (4.8 μm) (10^{11} cm Hz$^{1/2}$ W^{-1})	1.8	1.8	1.7	1.6	1.6

[a] This specimen was made from a PbTe layer with thickness 0.54 ± 0.03 μm and the devices had area 6.0 × 10^{-4} cm^2. The properties shown were measured at zero bias and show typically good uniformity. The backbias noise characteristics shown in Fig. 34 are much less uniform. The measurements were made at 990 Hz with 10 Hz bandwidth and 180° FOV. $\Re_I$ is the 500 K blackbody current responsivity.

stability of the IV–VI metal-barrier devices. This effect is discussed in the next section.

5. Surface Effects and the Nature of the Metal-Barrier Devices

The surfaces of the IV–VI compounds, like those of other semiconductors, would be expected to have a significant influence on the properties of p–n junction devices. Such an influence is evident in the dependence of $1/f$ noise on surface treatment. At this stage the knowledge of surface effects, even with bulk-crystal IV–VI devices, is quite fragmentary. In particular, the lack of knowledge about the metal–semiconductor interface in these materials leads to uncertainty about the nature of the p–n junction in metal-barrier photodiodes.

The most widely investigated IV–VI semiconductor surface effect is the interaction with gaseous oxygen. In an early study Egerton and Juhasz (123) found that thin epitaxial layers of PbTe on mica substrates exhibited changes in the Hall coefficient when they were exposed to oxygen. These changes were in the direction of increased acceptor concentrations. Subsequent Hall measurements of oxygen-exposed Pb chalcogenides have been interpreted in terms of a p$^+$ skin (124). Recent work by Sun et al. (125, 126) has shown that the effect of oxygen on Pb$_{0.8}$Sn$_{0.2}$Te is quite different from that on PbTe. With layers of PbTe on BaF$_2$ substrates the oxidation tends to saturate at about monolayer coverage. In contrast, the

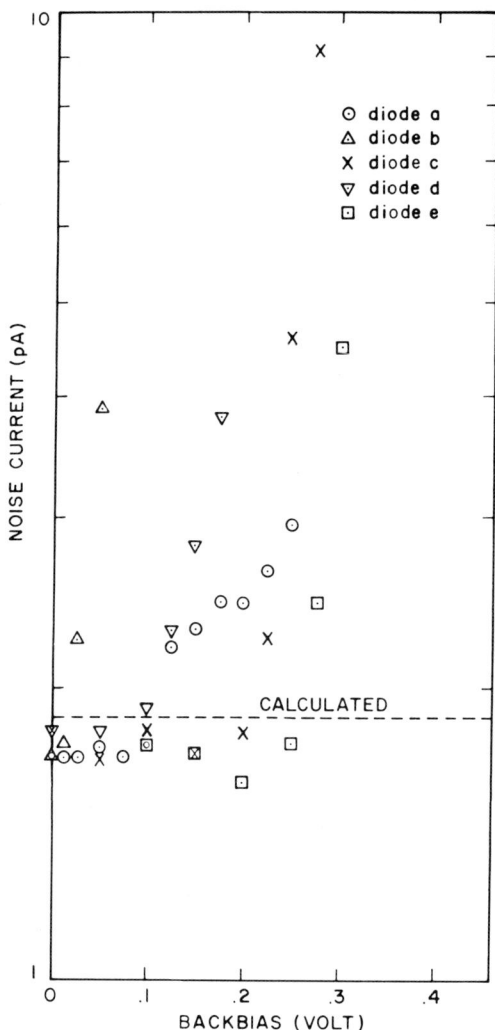

FIG. 34. Backbias dependence of the noise at 80 K of five Pb barrier thin-film PbTe photodiodes that were made on the same chip. The measurement frequency was 990 Hz and the bandwidth was 10 Hz. The calculated noise was derived from the DC background current.

oxidation of (Pb, Sn)Te does not saturate and the continuing oxidation is accompanied by outdiffusion of Sn to the surface. Further experiments by Buchner *et al.* (*127*) showed that deposition of Pb onto a $Pb_{0.8}Sn_{0.2}Te$ surface gave a rectifying barrier only if the semiconductor had been previously exposed to the atmosphere. The result was interpreted as a consequence of rapid diffusion of Sn from the clean (Pb, Sn)Te surface into the Pb barrier layer. With PbTe layers, rectifying contacts to Pb were obtained with both clean and air-exposed semiconductor surfaces.

The original work by Nill *et al.* (*78*) on Pb barriers used bulk-crystal PbTe and (Pb, Sn)Te specimens that were cooled (nominally to 77 K) during Pb deposition, and the resulting devices seem to have had limited thermal stability. The early Pb barrier thin-film PbTe devices that were made at the Ford Laboratories (*13*) used air-exposed PbTe surfaces that were nominally at room temperature during the Pb deposition. Subsequent studies showed that the PbTe surfaces had, in fact, been heated radiatively from the Pb evaporation source and with reduction in such radiative heating the PbTe photodiodes, as made, tended to exhibit reduced photocurrent and abnormal I–V characteristics.*

Normal I–V characteristics and IR responses were obtained with mild heat-treatment (typically 10–30 min at 120–150°C) of the as-made PbTe devices. The need for heat treatment to obtain normal IR response was correlated with air exposure of the PbTe surface. This is illustrated by Fig. 35, which shows the laser spot-scan of an as-made Pb barrier PbTe device in which part of the Pb barrier was deposited on the as-grown PbTe surface without air exposure. Although normal photoresponse may be obtained without heat-treatment by using devices that are made by deposition of Pb onto clean PbTe surfaces, practical delineation techniques have not yet been developed to take advantage of this possibility. Thus, most of the Pb barrier photodiodes that have been described appear to have been heated, either inadvertently during the Pb deposition or deliberately afterward.† A likely explanation for the transition to normal behavior on heating is that a thin layer of interfacial oxide dissolves into the Pb barrier layer. However, at this stage, we cannot exclude the possibility that the Pb layer changes the stoichiometry of the IV–VI layer

---

* The abnormal I–V characteristics were of two kinds. The first kind had a small zero-bias resistance together with a negative second derivative near the origin and little or no response to IR radiation. The second kind showed an abnormally large resistance for a range on either side of zero bias. Response to IR, if present, occurred only in backbias. Neither abnormality has been satisfactorily analyzed.

† Devices that were made with (Pb, Sn)Se and Pb(Se, Te) layers tended to require lower temperatures and shorter heating times than PbTe devices.

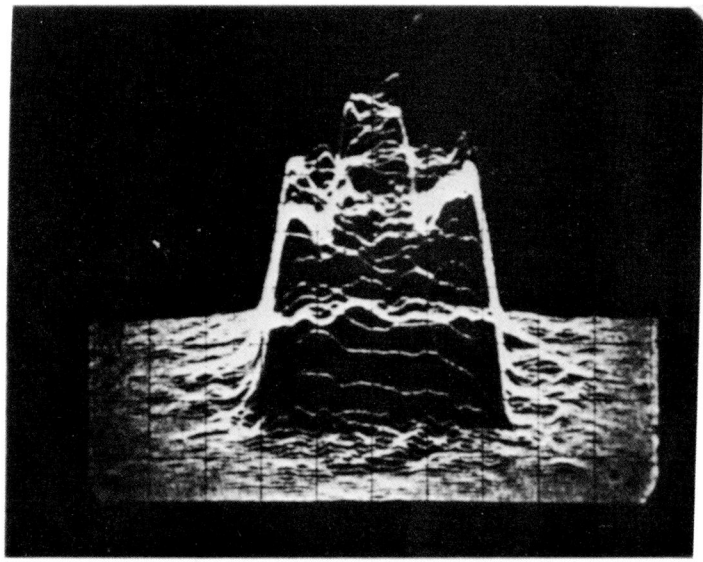

FIG. 35. Laser scan of an as-made 225 μm square Pb barrier PbTe photodiode at 80 K. The central 75 μm square region was made by depositing Pb onto the clean PbTe surface without breaking the vacuum. The surrounding part of the Pb layer was deposited after atmospheric exposure of the PbTe surface. The vertical displacement is proportional to the photocurrent.

surface to give a shallow diffused n^+–p junction rather than the electrostatically inverted structure described earlier.*

Our early experiments with Pb barrier PbTe devices revealed some problems with thermal stability. Thus, baking *in vacuo* at 100–110°C for periods in excess of 1 h tended to give decreases in R_0A that significantly degraded the intermediate-temperature (170–200 K) performance. This was a matter for concern because such baking was essential for the outgasing step of effective vacuum packaging. However, the washing step that was used to reduce backbias $1/f$ noise also gave a dramatic increase in thermal stability to the point where the PbTe devices would withstand 150°C for periods of up to 12 h. The 80 K reverse-bias resistance of a PbTe device after various periods of 150°C baking is shown in Fig. 36. The only significant change with baking is a hardening of the reverse characteristic.

The influence of the washing step on the $1/f$ noise and the thermal

---

* Similar heat-treatment of bulk-crystal In barrier devices has been interpreted as a diffusion of In to give a doped surface layer (*91*).

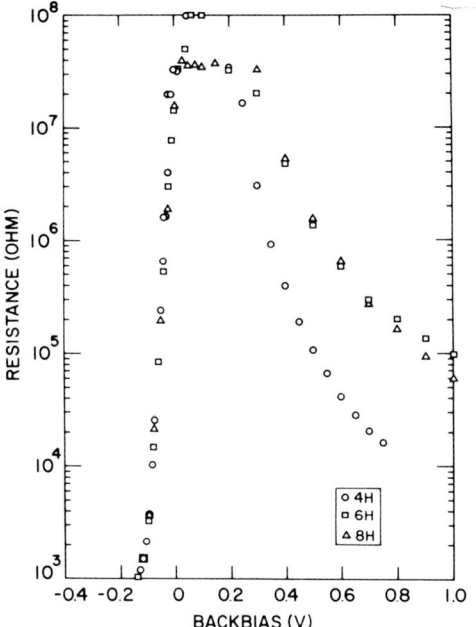

FIG. 36. Bias dependence of the junction resistance of a Pb barrier thin-film PbTe photodiode with area 10^{-3} cm^2. The different curves were obtained after varying periods of heating at 150°C. (○) 4 h; (□) 6 h; (△) 8 h.

stability remains unclear. It is worth noting that the BaF$_2$ window technique that is used exposes the semiconductor outside the junction region to a photoresist developer that lightly etches PbTe. Thus, the effects might be related to observations (128, 129) that etched (Pb, Sn)Te surfaces bear a layer of elemental Te that appears to reduce surface leakage.

VI. Thin-Film Photodiodes for 3–5 μm Operation

Photodiodes that are sensitive in the 3–5 μm atmospheric transmission band have widespread practical applications. Some of these applications make use of IR emissions that are concentrated in the 3–5 μm band, for example, emissions from CO$_2$ near 4.2 μm or from relatively hot (700–1200 K) bodies with peak photon flux in the range 3–5 μm. A less obvious application is to thermal imaging, where the peak photon flux from near-300 K backgrounds occurs at about 12 μm. In this case, the reduced signal level in the 3–5 μm band is only partially offset by the increased D^* (BNL) of 3–5 μm devices. However, the increase in E_g of the 3–5 μm devices over that of the 8–12 μm devices permits operation in the intermediate temperature range ($\approx$170–200 K) that is attainable with thermoelectric cooling. This feature is attractive for portable thermal imaging systems where the ultimate in performance may be traded for convenience and reduced weight.

The temperature-dependent performance of thin-film PbTe devices was discussed in an earlier section. Here we consider the observed performance in the light of these theoretical estimates. In an early study of bulk-crystal PbTe diffused p–n junctions, Butler (*130*) found that R_0A was limited by saturation current that was generated in the depletion region. This behavior was shown by the Arrhenius plot to occur over the whole of the temperature range 80–300 K (although some specimens appeared to exhibit surface leakage below 100 K) and was confirmed at 80 K by analysis of the forward-bias I–V characteristic. At 170 K, the peak wavelength was 4.8 μm and with $R_0A = 0.4$ Ω cm^2 a peak $D^* = 9 \times 10^9$ cm Hz$^{1/2}$ W^{-1} was obtained. Thus, the depletion-limited R_0A led to an order of magnitude reduction in performance from that predicted in Fig. 28.

Studies of the temperature-dependent properties of thin-film Pb barrier PbTe photodiodes (*131*) showed that background-limited detectivities could be maintained at temperatures close to those predicted by Fig. 28. This result is exemplified by Fig. 37, which shows the temperature dependence of the spectral detectivity of a Pb barrier PbTe device. It will be noted that the peak D^* actually increases as the operating temperature is increased from 84 to 130 K. This is a consequence of the increase in the energy gap of PbTe, which reduces the background current and hence the background noise of the device. Above 130 K, the decreasing junction resistance reaches a value where the device becomes Johnson noise limited and further increase in temperature leads to a decrease in D^*. In the early example shown in Fig. 37, the transition from BNL to JNL behavior occurs at about 140 K rather than near 170 K as expected. (For $p = 3 \times 10^{17}$ cm^{-3}.) Similar results were obtained subsequently by McMahon (*69*).

Further investigation of thin-film PbTe devices showed that there was a considerable variability of R_0A in the intermediate-temperature range. Arrhenius plots of R_0A tended to be grouped into two main classes. In the first class the slopes $d \log(R_0A)/d(1/T)$ were approximately 1000 K, in good agreement with the slope of 980 K that is calculated for diffusion-

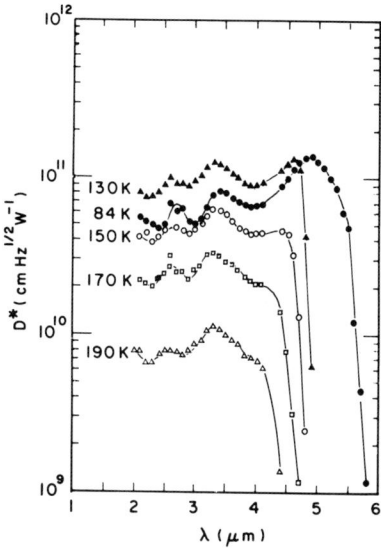

FIG. 37. Temperature dependence of the spectral D^* of a Pb barrier PbTe photodiode.

limited behavior in PbTe.* Thus, these specimens may be tentatively classified as showing diffusion-limited behavior. An example is shown in Fig. 38. As the temperature was decreased from 170 to 80 K, most specimens showed a tendency for R_0A to saturate. This behavior is consistent with the existence of surface leakage, although it may also be due in part to the increased significance of the depletion-limited saturation current at these lower temperatures. The diffusion-limited specimens had $R_0A = 10$–15 Ω cm^2 at 170 K. These values are a factor of 2–3 less than those calculated for the carrier concentrations ($p = 1$–3×10^{17} cm^{-3}) that were obtained with Hall measurements of these specimens. (The agreement is significantly closer if allowance is made for an overestimate of the carrier concentration because of an oxygen-induced p$^+$ skin.) With interference enhancement of the quantum efficiency near 4 μm to 0.9, these values of R_0A correspond to peak values of D^* (JNL) $= 9.5 \times 10^{10}$–1.2×10^{11} cm Hz$^{1/2}$ W^{-1} at 170 K.

The diffusion-limited resistances of the thin-film PbTe photodiodes did not give evidence for values that were increased from those of bulk-

---

* This includes the temperature dependence of the preexponential factor. The approximation $d \ln(R_0A)/d(1/T) = E_g/k$, where E_g is the zero-temperature energy gap, gives a value that is about 7% smaller.

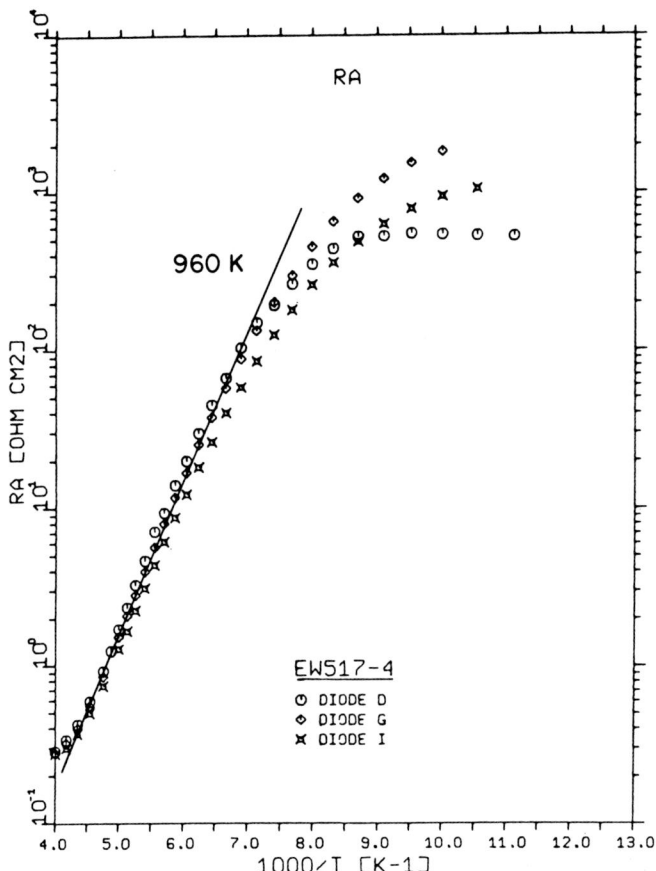

FIG. 38. Arrhenius plot of R_0A of a set of Pb barrier PbTe photodiodes on the same chip showing diffusion-limited behavior.

crystal devices, although such an effect might be expected from the analysis given earlier. To some extent the lack of evidence for a thickness effect may be a consequence of lack of data, since most of the PbTe studies were with three-quarter-wave structures that had layer thicknesses in the range 0.5–0.6 μm. However, the absence of an effect is suggested strongly by the closeness of the values of R_0A to those calculated for bulk-crystal devices. The predicted thin-film effect has been observed by Schoolar (132) with Pb barrier PbS devices on BaF_2 substrates and by Lanir et al. (133) with 3–5 μm (Hg, Cd)Te photodiodes on CdTe substrates.

The second class of PbTe specimens showed Arrhenius slopes of about 500 K (Fig. 39) and considerably reduced values of $R_0A \approx 1-4\ \Omega$ cm² at 170 K. (A few specimens displayed intermediate behavior.) These devices appeared to be showing depletion-limited behavior like that reported for bulk crystals of PbTe by Butler.

Confirmation of the interpretation of the two Arrhenius slopes in terms of two saturation-current-generating mechanisms is provided by results obtained with the lateral-collection photodiodes that are discussed in detail in a later section. These studies show that p–n junctions made on the same PbTe layer may show either of the Arrhenius slopes, with the choice of slope being determined by the junction area. The effect arises because, with small enough junction areas, further reduction of the

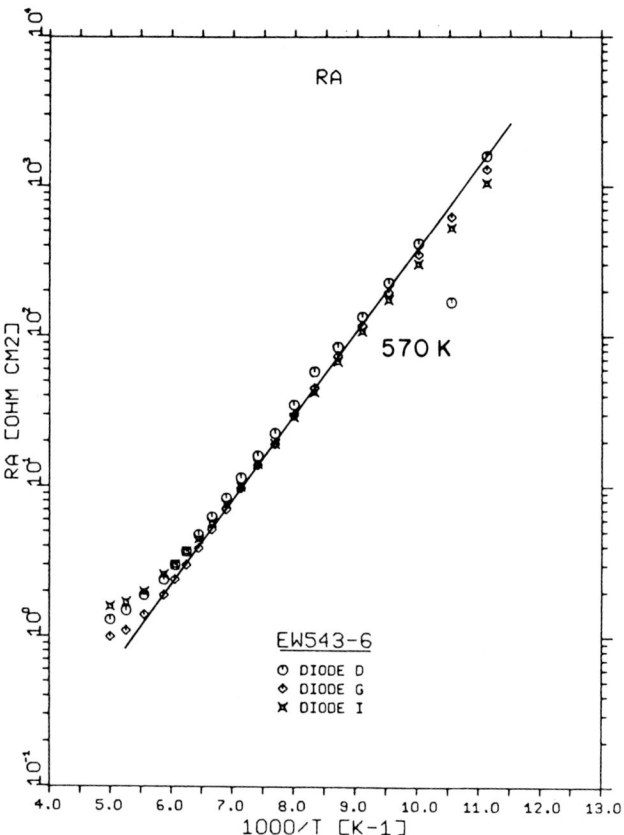

FIG. 39. Arrhenius plot of R_0A of a set of Pb barrier PbTe photodiodes on the same chip. The slope suggests depletion-limited behavior.

junction area reduces the depletion-region-generated saturation current without having much effect upon the diffusion current, which may be collected laterally from outside the junction area. Thus, with decrease of the junction area there may be a transition from depletion-limited to diffusion-limited R_0A.

The variability of R_0A and of the dominant mechanism for generation of the saturation current in thin-film PbTe devices suggests the presence of a variable concentration of generation–recombination centers. Some weight is given to this interpretation by the observation that the device characteristics were fairly consistent among specimens that had been made from PbTe layers that originated in the same film growth run. However, neither the defects nor the growth conditions that determine their presence or absence have been identified.

For further discussion of 3–5 μm photodiodes, it is necessary to make a brief digression to consider the IR-transmitting properties of the atmosphere. Examination of the 3–5 μm transmission band reveals that it is divided by a strong CO_2 absorption at about 4.2 μm. The transmission in both of the resulting subbands is reduced by atmospheric water vapor, but this effect is much larger at the longer wavelengths (*134*). Thus, with increased range or humidity, the available 3–5 μm signal from the near-300 K scene tends to be concentrated in the shorter wavelength subband. This leads to a system trade-off because the additional signal that is obtained by including the longer wavelength subband may be offset by increased noise, either from the increase in the background photon flux that is collected, or from an increase in the saturation current that would be expected to accompany the necessary reduction in energy gap. Some details of such trade-offs are discussed by Lloyd (*135*). Here it is sufficient to note that, for specific applications, 3–5 μm photodiodes may be required to have IR responses that extend either to 4 or to 5 μm.

Operation of PbTe photodiodes near 170 K does not permit full use of the longer wavelength subband because at these temperatures the energy gap of 0.26 eV gives a cutoff at about 4.7 μm.* In practice, the reduced optical absorption near the cutoff wavelength would be expected to displace the peak D^* to somewhat smaller wavelengths. This displacement is particularly significant when the PbTe layer is made thin enough to give an interference-enhanced quantum efficiency. In this case, detailed calculations show that the optimum thickness is three quarter-waves at the wavelength for peak response (this corresponds to a

---

* Unlike the tetrahedral semiconductors, most of the IV–VI semiconductors have energy gaps that increase with increased temperature. The positive temperature coefficients are approximately 4.5×10^{-4} eV K^{-1}.

thickness of about 0.5 μm) and that this interference peak cannot be profitably located at wavelengths longer than about 4.2 μm when the device is operated at 170 K. Thus, the thin-film PbTe devices are well suited to intermediate-temperature applications that use only the shorter-wavelength subband of the 3–5 μm atmospheric transmission. For use of both subbands a different material is desirable.

The response of the thin-film Pb barrier devices was extended to somewhat longer wavelengths by replacement of the PbTe with PbSe, also on BaF$_2$ substrates (*136*). At 77 K, the PbSe photodiodes had a cutoff wavelength of 7.2 μm and a peak quantum efficiency of 0.7 at 6.1 μm. With an $f/0.6$ aperture, the peak D^* was background noise limited at 6×10^{10} cm Hz$^{1/2}$W^{-1}. Reduction of the field of view increased the D^* to the JNL of 5×10^{11} cm Hz$^{1/2}$ W^{-1}. The spectral properties of such a device are shown in Fig. 40. The values of R_0A that were obtained with the PbSe junctions tended to saturate at lower temperatures, with the best device giving 90 Ω cm^2 at 77 K. In the range 120–200 K, Arrhenius plots indicated that R_0A was depletion limited with the best specimen having $R_0A \approx 0.5$ Ω cm^2 at 170 K. These results are quite similar to those obtained with the early PbTe thin-film photodiodes.

At 170 K the IR response of PbSe extends to a somewhat longer wavelength (≈ 5.8 μm) than is required for 3–5 μm applications. However, the PbSe devices provided a starting point for development of the pseudobinary alloy Pb(Se, Te). Here the existence of strains in films with

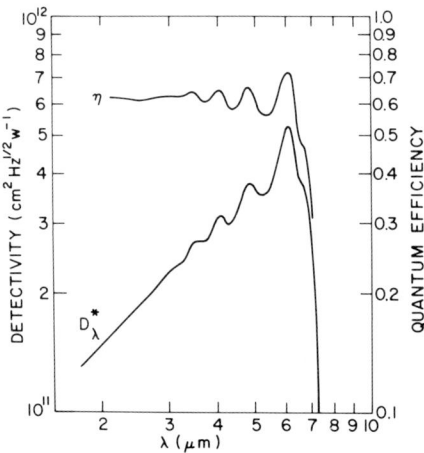

FIG. 40. Spectral detectivity and quantum efficiency of a Pb barrier PbSe photodiode at 80 K.

compositions near the middle of the range led to instability of the layers on BaF$_2$ substrates (65). Stable layers were achieved with the composition of PbSe$_{0.8}$Te$_{0.2}$, which gave a cutoff wavelength of 5.4 μm at 170 K. This cutoff is about optimal for location of a three-quarter-wave response peak at 5 μm and, consequently, the Pb(Se, Te) devices are suitable for applications that use both of the 3–5 μm subbands.

Apart from the difference in cutoff wavelength, the Pb barrier Pb(Se, Te) devices were found to be similar to the PbTe photodiodes. The values of R_0A were quite variable and it is possible that the mechanism for generation of the saturation current varies like that in PbTe, although there are insufficient data for certainty on this point. The temperature-dependent D^* of a typical Pb(Se, Te) device (137) is shown in Fig. 41. As with PbTe (Fig. 37), the D^* shows an initial increase as the temperature is increased from 80 K. The transition from BNL to JNL behavior occurs at about 160 K, and at 170 K the peak $D^* = 6.4 \times 10^{10}$ cm Hz$^{1/2}$ W^{-1} occurs at 5 μm. Improved processing techniques developed by Asch and co-

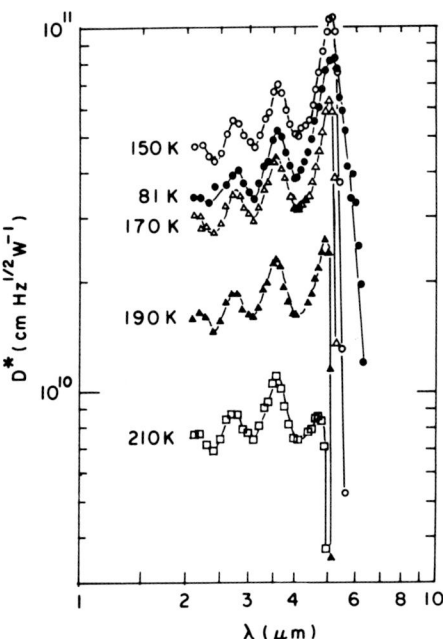

FIG. 41. Temperature dependence of the spectral D^* of a Pb barrier Pb(Se, Te) photodiode.

workers (138) have given D^* (5 μm) $\approx 10^{11}$ cm Hz$^{1/2}$ W^{-1} with 180° FOV at 170 K. Under these conditions, the background and Johnson noise contributions are approximately equal and D^* (JNL) is about a factor of $\sqrt{2}$ larger than the D^* that is measured with 180° FOV. Some details of a Pb(Se, Te) device are given in Table V.

Extensive studies of Pb barrier thin-film devices made with other IV–VI alloys on BaF$_2$ substrates have been reported by Schoolar et al. (71), who examined Pb(S, Se) over the whole of its composition range and (Pb,Cd)S with up to 5.6 mole-% CdS. At 77 K these alloys have cutoff wavelengths that cover the range 2–7 μm. A particularly interesting feature of this work was a demonstration of room temperature operation of the (Pb, Cd)S devices with D^* (1–2 μm) = 5 × 10^9 cm Hz$^{1/2}$ W^{-1}. Application of Schoolar's Pb(S, Se) devices to narrow band IR detection is described later.

TABLE V

INTERMEDIATE-TEMPERATURE PERFORMANCE OF PbSe$_{0.8}$Te$_{0.2}$ PHOTODIODES (138)[a]

T (K)	Diode no.	R_0A (Ω cm^2)	$\mathfrak{R}_1$ (A W^{-1})	D^* (4.5 μm) (10^{10} cm Hz$^{1/2}$ W^{-1})
170[b]	1	10.6	0.36	9.9
	2	10.0	0.35	9.6
	3	8.7	0.36	8.8
190	1	4.7	0.30	6.8
	2	4.7	0.30	5.7
	3	4.1	0.31	5.7
210	1	1.7	0.26	3.9
	2	1.6	0.27	3.5
	3	1.4	0.27	3.2

[a] These devices were made on the same chip. The geometric area of the junctions was 6.2 × 10^{-5} cm^2. Lateral collection of minority carriers that were generated outside the junction region increased the active area by approximately a factor of two. The performance figures given here have been corrected downward to allow for this increased area. The values of R_0A are based on the geometric area. The parameter $\mathfrak{R}_1$ is the 500 K blackbody current responsivity. The measurements were made at 1 kHz.

[b] With the approximately 180° FOV that was used for these measurements at 170 K, the background and Johnson noise contributions were approximately equal. For example, the calculated background and Johnson noise contributions for diode no. 2 were 97 and 103 nV, respectively, with 7 Hz bandwidth. This corresponds to a total noise of 142 nV. The measured noise was 140 nV.

VII. Thin-Film Photodiodes for 8–12 μm Operation

Bulk-crystal studies of IV–VI semiconductor photodiodes for 8–12 μm operation have been devoted almost exclusively to the alloy (Pb, Sn)Te with about 20 mole-% SnTe. Early work with thin films of (Pb, Sn)Te on BaF$_2$ substrates showed that the layers were truly single crystalline (63) and measurements of photoconductive lifetimes (139) gave values that were comparable to those obtained with bulk crystals (6). However, attempts to extend the thin-film Pb barrier devices from PbTe to (Pb, Sn)Te were unsuccessful for reasons that remain unclear. The Pb barrier (Pb, Sn)Te devices showed poor photocurrents and abnormal I–V characteristics.

The alternative thin-film semiconductor (Pb, Sn)Se with about 7 mole-% SnSe did give Pb barrier devices with nearly ideal characteristics (140). Devices that were made using layers with $p = 2$–6×10^{17} cm^{-3} and $E_g = 0.10$–0.11 eV gave values of R_0A at 80 K that were typically near 0.5 Ω cm^2 and reached 2 Ω cm^2 with a few specimens. These values are comparable to the R_0A that is calculated for bulk-crystal (Pb, Sn)Te (Fig. 30), although the uncertainty in the tunneling calculation precludes firm conclusions about the agreement with theoretical values.

Typical thin-film (Pb, Sn)Se devices had peak D^* in the range 10–11.5 μm that were close to the calculated background limit with 180° FOV and that saturated at JNL values in the range 2–5×10^{10} cm Hz$^{1/2}$ W^{-1} when the FOV was reduced. Representative spectral D^* are shown in Fig. 42. The best specimen gave D^* (10.1 μm) $= 8 \times 10^{10}$ cm Hz$^{1/2}$ W^{-1} under reduced FOV; details of this device are given in Table VI. The performance obtained with thin-film (Pb, Sn)Se is comparable to that obtained with bulk-crystal (Pb, Sn)Te devices. (120).*

It is of interest that Schoolar et al. (71) have obtained thin-film (Pb, Sn)Se photodiodes with properties that are very similar to those described above. In this case, growth on BaF$_2$ substrates was effected with the hot-wall method rather than the MBE technique that had been used previously by the Ford group. The similarity of the results obtained with the two growth methods lends support to the thesis that, for growth of device quality layers, the choice of substrate is important, but the choice of deposition technique is inconsequential.

---

* There has not been enough work with bulk-crystal (Pb, Sn)Se photodiodes to permit a valid comparison with the thin-film devices. In an early bulk-crystal study (6) $D^* \sim 3 \times 10^9$ cm Hz$^{1/2}$ W^{-1} was obtained near 10 μm with a device that was operated at 77 K.

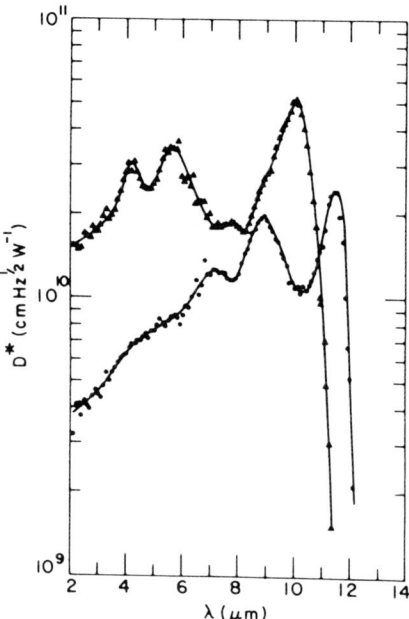

FIG. 42. Spectral detectivities of two Pb barrier (Pb, Sn)Se photodiodes at 80 K.

TABLE VI

Properties of a Thin-Film (Pb, Sn)Se Photodiode at 80 K[a]

Field of view (deg):	86	47	20
Calculated noise (nV)			
Johnson[b]	12.4	11.4	11.3
background	21.0	10.3	5.8
preamp	16.5	16.5	16.5
total	29.4	22.5	20.8
Measured noise (nV)	31	22	18
D^* (10.1 μm) (10^{10} cm $Hz^{1/2}$ W^{-1})			
measured	3.7	4.3	5.2
corrected	5.0	5.9	7.3[c]

[a] This device with area = 6.2×10^{-4} cm^2 had $R_0A = 1.6$ Ω cm^2 and was measured at 1 kHz with 10 Hz bandwidth. The corrected value of D^* was obtained by correcting the measured noise for the noise of the preamplifier.

[b] The variation in the calculated Johnson noise reflects a small variation in the zero-bias resistance. This was probably a result of a small change in the temperature that occurred during the measurements.

[c] The calculated Johnson noise corresponds to D^* (10.1 μm) = 8.3×10^{10} cm $Hz^{1/2}$ W^{-1}.

Thin-film photodiodes have been made with (Pb, Sn)Te by the use of evaporated In barriers. Mathur (*88*) described such devices that had been made by growth of (Pb, Sn)Te on polished artificial BaF_2 {100} faces.* These In barrier devices gave $R_0A = 0.1-3$ Ω cm² at 77 K and approached the BNL condition with D^* (10.0–10.1 μm) = $1.2-3.5 \times 10^{10}$ cm Hz$^{1/2}$ W^{-1} at 180° FOV.

VIII. Unconventional Thin-Film Devices

The thin-film devices that have been described so far are conventional in the sense that, apart from interference modulation of their quantum efficiencies, they closely resemble the corresponding bulk-crystal IV–VI semiconductor photodiodes. This section describes several devices that exploit thin-film structures to obtain properties that differ from those of bulk-crystal IR detectors. Two of these unconventional devices give reduced junction capacitance and their description is preceded by a discussion of the significance of this device parameter.

1. The Junction Capacitance of Photodiodes

The relatively large static dielectric constants of the IV–VI semiconductors give rise to zero-bias junction capacitances that are of the order of 1 μF cm^{-2} for typical carrier concentrations in the range $10^{17}-10^{18}$ cm^{-3}. For a one-sided abrupt junction the differential junction capacitance for unit area is

$$C \equiv \frac{dQ}{dV} = \left(\frac{Nq\epsilon}{2(V + V_{bi})}\right)^{1/2} \quad (34)$$

where Q is the depletion layer charge, N the dopant concentration on the low-carrier-concentration side, ϵ the permittivity, V the bias, and $V_{bi} \approx E_g/q$ the built-in voltage. As shown in Table VII, the capacitances of junctions in the IV–VI semiconductors are larger than those in the III–V compounds by a factor of 3–5. The junction capacitance imposes a limitation on the operating frequency of an IR system. The origin of this

* Callender *et al.* (*141*) also reported In barrier devices on thin-film (Pb, Sn)Te. In this case, the carrier concentrations in the layers were adjusted with an annealing step at 550°C before deposition of the In barrier layer. Work damage that was introduced during the annealing process was found to greatly reduce the stability of the devices when they were cycled between room temperature and the operating temperature of 77 K.

TABLE VII

Junction Capacitances of III–V and IV–VI Photodiodes[a]

Semiconductor	E_g (eV) at 77 K	Dielectric constant	Junction capacitance (μF cm^{-2})
InAs	0.38	15	0.17
InSb	0.25	18	0.23
PbS	0.32	205	0.67
PbSe	0.20	280	1.0
PbTe	0.22	400[b]	1.1

[a] The capacitances have been calculated for a one-sided abrupt junction at zero bias with doping to 10^{17} cm^{-3}.

[b] This is the value that is appropriate for 300 K or for large polarizations. Bate et al. (142) have shown that PbTe is paraelectric and that at 77 K the small-field dielectric constant is larger by about a factor of 2.

limitation may be understood by referring to the equivalent circuit shown in Fig. 43. Here the signal current flowing through the junction impedence gives rise to a signal voltage that is degraded at high frequencies by the junction capacitance. However, with an ideal preamplifier S/N is unaffected because the noise voltage is degraded in the same way as the signal voltage. The frequency limitation arises because real preamplifiers have a voltage noise that is not degraded by the input capacitance. Thus, above a critical frequency S/N becomes limited by the preamplifier at a frequency-dependent value. A useful approximation to the frequency dependence of S/N and D^* may be obtained with the following simplifying assumptions:

(1) The preamplifier input capacitance is negligible.
(2) The preamplifier current and voltage noises are independent of frequency and have negligible effect at low frequencies.
(3) The photodiode has negligible $1/f$ noise.

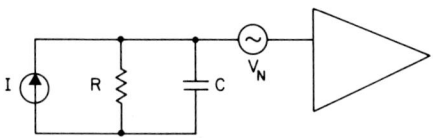

Fig. 43. Equivalent circuit of a photodiode/preamplifier combination. I represents both the signal current and the noise current of the photodiode and V_N is the preamplifier voltage noise.

This leads to a detectivity for the detector/preamplifier combination that is no longer independent of the frequency and the detector area

$$D^*(f) = \frac{\eta q}{E_\lambda[(4kT/\rho) + 2q^2\eta Q_B + (2\pi f \sigma V_N)^2 A]^{1/2}} \quad (35)$$

where $\rho = R_0 A$ is the zero-bias resistance-area product, σ the junction capacitance for unit area, and V_N the preamplifier voltage noise in unit bandwidth. At low frequencies $D^*(f)$ reduces to the usual expression for D^*,

$$D^*(0) = \frac{\eta q}{E_\lambda[(4kT/\rho) + 2q^2\eta Q_B]^{1/2}} \quad (36)$$

and at high frequencies it becomes inversely proportional to the frequency,

$$D^*(f \to \infty) = \frac{\eta q}{E_\lambda 2\pi f V_N A^{1/2}} \quad (37)$$

The 3 dB point for the rolloff of $D^*(f)$ is then given by the frequency at which $D^*_\infty(f) = D^*_0$. Some typical results are given in Fig. 44, which shows

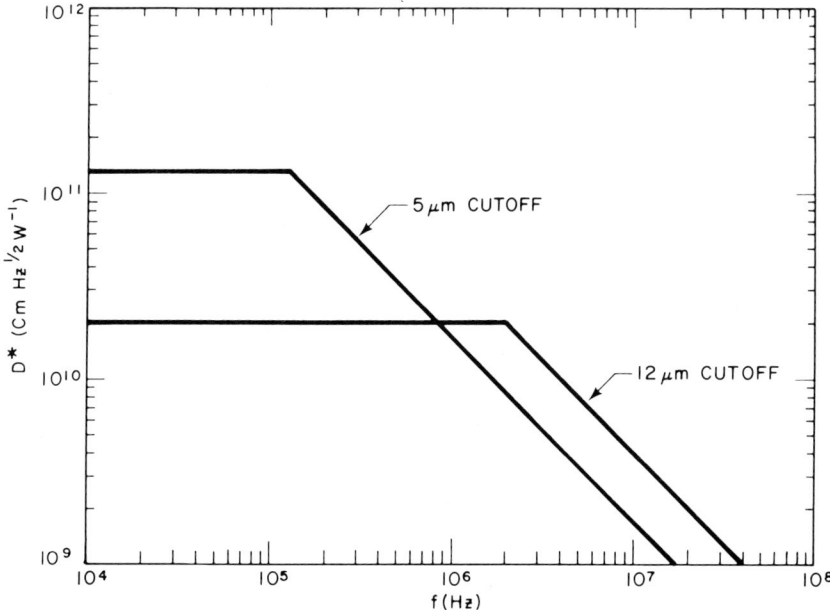

FIG. 44. Frequency-dependent detectivities of typical IV–VI semiconductor photodiode–preamplifier combinations. The photodiodes are 50 μm square.

calculations for 50 μm square photodiodes that have 5 and 12 μm cutoffs with low-frequency peak detectivities of 1.3×10^{11} and 2.0×10^{10} cm Hz$^{1/2}$ W^{-1}, respectively.* The junction capacitances have been assumed to be 1 μF cm^{-2}, the quantum efficiencies have been assumed to have typical RLL values of 0.5, and V_N has been chosen to have a value of 4 nV Hz$^{-1/2}$ that is typical of an FET.

The most striking feature of Fig. 44 is the increase in severity of the bandwidth limitation that occurs when the cutoff wavelength is changed from 12 to 5 μm. In the example chosen, the 3 dB point shifts from 2 to 0.1 MHz. The numerical results change somewhat with changes in the initial assumptions, but the distinction between devices for the two spectral regions remains. This indicates that attempts to reduce the junction capacitance of IV–VI semiconductor photodiodes are especially important for 3–5 μm devices. Thus, in the absence of techniques for capacitance reduction, high-frequency 3–5 μm applications will tend to favor the III–V semiconductors over the IV–VI compounds. (The performance of 3–5 μm III–V photodiodes compares quite well with that of the IV–VI devices; e.g., at 80 K, InSb photodiodes (143) are quite similar to PbTe photodiodes.)

It should be noted that techniques for capacitance reduction are not exclusive to the thin-film devices. From Eq. (34) it can be seen that the junction capacitance may be reduced by reducing the dopant concentration, although the square root dependence leads to the requirement of a large reduction in the doping.† Essentially this method was used by Andrews et al. (18), who used liquid-phase epitaxy on PbTe substrates to obtain graded (Pb, Sn)Te junctions with average dopant concentrations of 10^{14}–10^{15} cm^{-3}. Operation of these devices at 1.5 V backbias gave junction capacitances as small as 0.07 μF cm^{-2}. However, the quantum efficiencies were only 0.05–0.10, which is a factor of 5–10 less than the RLL. An alternative approach by Andrews et al. (144) used total internal reflection within a PbTe tapered mesa structure to obtain a tenfold concentration of the optical signal at a PbTe/(Pb, Sn)Te heterojunction. Relative to the optical collection area this gave an order of magnitude reduction in junction capacitance. Finally, Noreika et al. (145) have reduced the capacitance of bulk-crystal (Pb, Sn)Te photodiodes by the use of lateral

---

* These values are typical of those obtained with IV–VI semiconductor photodiodes at 80 K and 180° FOV. The value for a 12 μm cutoff includes a significant. Johnson noise contribution. The 3 dB points for other values of the low-frequency detectivity may be obtained by extrapolating $D^*(f \to \infty)$ to the new values of $D^*(0)$.

† This assumes a one-sided abrupt junction. However, other types of junction require a similar degree of control over the dopant concentration.

collection of photogenerated carriers. These last devices are discussed in conjunction with the corresponding thin-film devices.

2. Pinched-off Photodiodes

The principle of the pinched-off photodiode is illustrated in Fig. 45. Application of increased backbias to a conventional p–n junction (Fig. 45a) widens the depletion region and thereby causes a change in the depletion layer charge that gives rise to a differential capacitance. The pinched-off device (Fig. 45b) is configured to have a depletion region that extends from the n^+ region through the semiconductor to an insulating substrate. In this case the change in the depletion layer width occurs only around the periphery of the diode. Thus, the change in the depletion layer charge and the junction capacitance are reduced.

Pinched-off operation may be achieved with semiconductor layers that are thin enough to be pinched off at zero bias or with somewhat thicker layers by application of sufficient backbias. The range of semiconductor thicknesses that is applicable to pinched-off devices is such that interference effects are substantial. Consequently, optimization of the spectral quantum efficiency for a particular application leads to a discrete set of semiconductor thicknesses. For example, with PbTe photodiodes that are to be operated at 80 K with peak response at 5 μm one obtains the set of odd-quarter-wave thicknesses that are shown in Table VIII. It will be noted that PbTe layers with optical thicknesses in the range $\lambda/4$–$7\lambda/4$ are all calculated to give good peak quantum efficiencies with little variation in the 500 K blackbody current responsivity. However, examination of

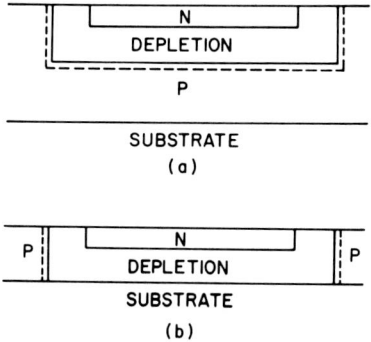

Fig. 45. The concept of the pinched-off photodiode. (a) A conventional device. (b) A pinched-off photodiode. In each case the broken line indicates the position of the edge of the depletion region after application of a small additional backbias.

TABLE VIII

PINCHED-OFF VOLTAGES FOR PbTe PHOTODIODES AT 80 K[a]

m	d (μm)	Pinched-off voltage (V)		η	$\mathfrak{R}_1$ (A W^{-1})
		$\epsilon = 400\epsilon_0$	$\epsilon = 800\epsilon_0$		
1	0.19	−0.14	−0.18	0.60	0.47
3	0.56	0.48	0.13	0.87	0.46
5	0.93	1.8	0.76	0.82	0.44
7	1.33	3.8	1.8	0.78	0.43

[a] The calculations are for PbTe layers with $N_A = 10^{17}$ cm^{-3} and optical thicknesses $nd = m\lambda/4$ (m odd), for $\lambda = 5$ μm. The quantum efficiencies at the peak (η) and the 500 K blackbody current responsivities ($\mathfrak{R}_1$) are calculated from the RLL. Positive voltages correspond to backbias. Much of the depletion region is in the low-field condition and the low-field dielectric constant ($\epsilon = 800\epsilon_0$) might be expected to give a closer approximation to the pinched-off voltage than the high-field value ($\epsilon = 400\ \epsilon_0$).

the calculated pinch-off voltages for a typical acceptor concentration of 10^{17} cm^{-3} shows that layers with optical thicknesses greater than $3\lambda/4$ require backbiases that exceed those for which low-noise operation of PbTe diodes has been achieved. Thus, the choice of optical thicknesses for practical applications is restricted to $\lambda/4$ and $3\lambda/4$.

Studies of pinched-off Pb barrier PbTe photodiodes by Holloway and Yeung (122) showed that, as predicted, the quarter-wave structures were pinched-off with reduced capacitance at zero bias. However, the performance of these devices was relatively poor because abnormally high-resistance Pt contacts gave substantial $1/f$ noise that degraded D^* at 1 kHz by a factor of 2 or 3 from D^* (BNL). An even more serious problem with these devices was a tendency for degradation of the photocurrent after heat-treatment at 100–150°C. Laser scans of the degraded devices showed that the photoresponse had become confined to the peripheries of the junctions.* The results are consistent with rapid diffusion of an n region through the PbTe to its BaF$_2$ substrate. This might be a consequence of reduced crystal perfection in the very thin (<2000 Å) PbTe layers that were used for the quarter-wave devices.

Useful pinched-off devices were obtained with three-quarter-wave PbTe structures. Pinch-off occurred in the expected range of backbias and the only significant change that was observed with 150°C heat-treatment was an initial tendency for the pinch-off voltage to decrease. An example

---

* These laser scans were kindly furnished by R. E. Callender and D. R. Kaplan of the U.S. Army Night Vision Laboratory.

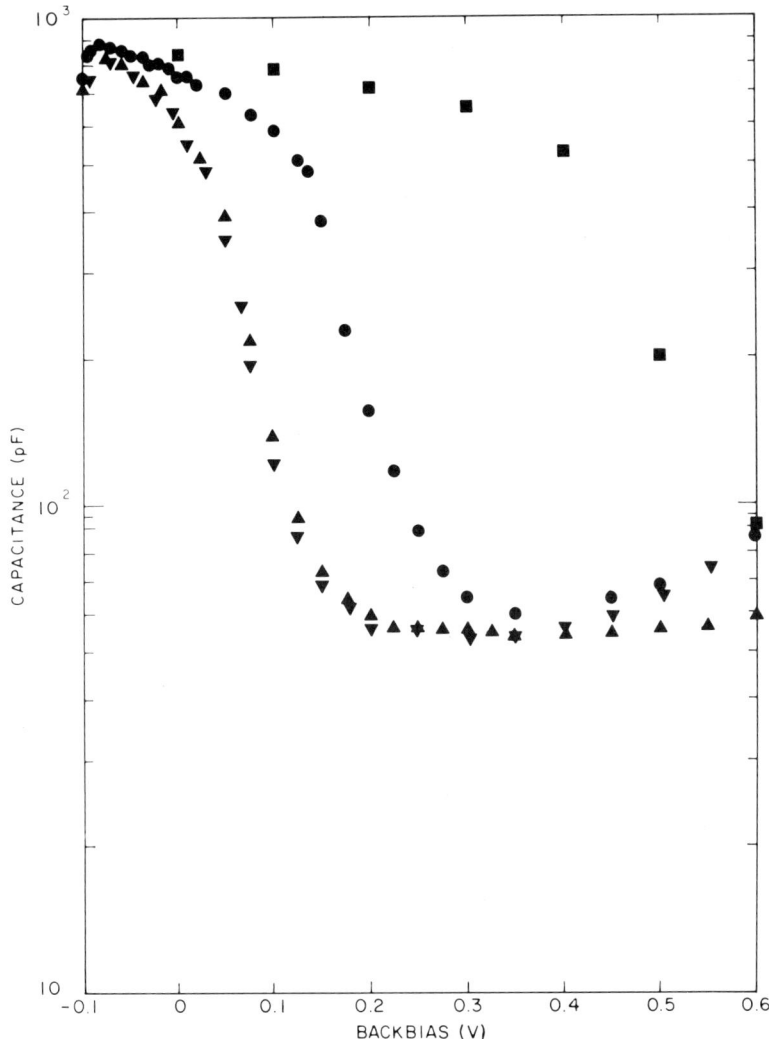

FIG. 46. C/V characteristic at 80 K of a pinched-off PbTe photodiode after successive heat-treatments at 150°C. ■, 20 min; ●, 1 h, 50 min; ▲, 2 h, 20 min; ▼, 4 h, 50 min.

of this change is shown in Fig. 46. (This figure also shows an unexplained reduction of capacitance in forward bias that was observed quite commonly.) The specimen used for Fig. 46 is the same as those whose resistance is shown in Fig. 36.

At first sight the changes in the C/V characteristic that occur on

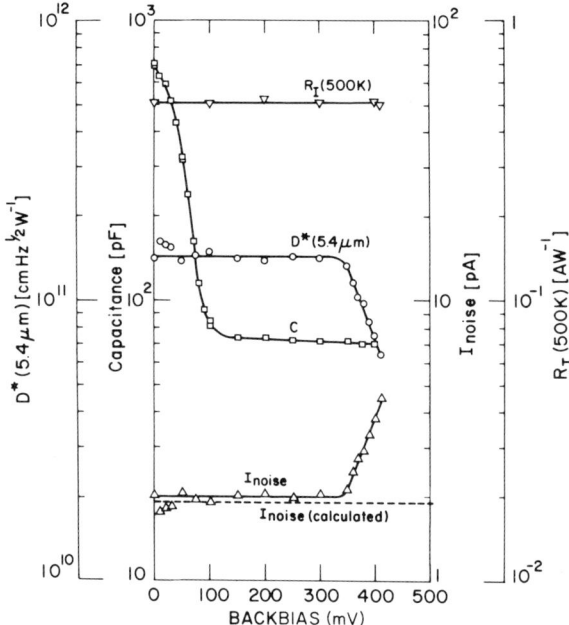

FIG. 47. Bias-dependent properties at 80 K of a three-quarter-wave pinched-off PbTe photodiode. The noise was measured at 1 kHz and 10 Hz bandwidth and the calculated value was derived from the DC background current. The photodiode area is 10^{-3} cm².

baking suggest that diffusion of the n region is occurring in the same way as with the quarter-wave structures. However, the eventual stabilization of the C/V characteristic is inconsistent with a junction diffusion effect. A possible explanation of the change in C/V is in terms of removal of oxygen acceptors that are introduced from prior atmospheric exposure of the PbTe surface. Reaction of these oxygen acceptors with the Pb barrier layer would widen the zero-bias depletion region and thereby reduce the pinch-off voltage.*

Figure 47 shows the bias dependence of the properties of a typical three-quarter-wave PbTe device at 80 K and 180° FOV. At zero bias the

* Capacitance measurements of thicker Pb barrier PbTe diodes at 80 K gave linear plots of $1/C^2$ against V, as might be expected with a one-sided abrupt junction. However, while the slopes of such plots were in reasonable accord with Eq. (34), their intercepts often gave anomalously large values of V_{bi} that were in the range 0.4–0.6 V instead of near 0.2 V as expected. These results are consistent with an inhomogeneous acceptor concentration that could arise by removal of oxygen acceptors from a region close to the Pb/PbTe interface. [Walpole and Nill (79) have shown that an increase in V_{bi} is to be expected with Pb barrier devices, but the predicted effect is much smaller than that described here.]

capacitance has a value of 690 pF that corresponds to 1.15 μF cm^{-2} for this device with area 6 × 10^{-4} cm^2. Application of backbias gives a rapid decrease in the capacitance, which levels out at about 70 pF beyond 0.1 V backbias. This almost constant value of capacitance contains a large contribution from the lead-out capacitance. The 500 K blackbody current responsivity $\Re_I$ has a value of 0.51 A W^{-1} that agrees quite well with the value of 0.46 A W^{-1} calculated from the RLL quantum efficiency. The value of $\Re_I$ is constant, within experimental uncertainty, for backbias up to 0.4 V. This shows that the quantum efficiency is unaffected by pinch-off. The noise at 1 kHz remains close to that calculated for the background contribution with backbiases up to 0.34 V. (Larger backbiases give a significant contribution from $1/f$ noise.) Thus, the BNL condition is maintained beyond pinch-off and the photodiode has D^* (5.4 μm) = 1.7 × 10^{11} cm Hz$^{1/2}$ W^{-1} both at zero bias and over a range of backbias that gives an order of magnitude reduction in its junction capacitance.

Table IX shows the properties of several quarter-wave and three-quarter-wave pinched-off PbTe photodiodes. Of particular interest is a 700 μm square device that showed almost two orders of magnitude reduction in its junction capacitance. In this case about half of the 35 pF of backbiased capacitance was estimated to arise from lead-out capacitance. Overall, the backbiased results are consistent with a simple model

TABLE IX

PROPERTIES OF PINCHED-OFF PbTe PHOTODIODES AT 80 K

PbTe thickness^a (μm)	Area (cm^2)	R_0A^b (Ω cm^2)	λ_{peak}^b (μm)	Back-bias (V)	Capacitance (pF)	$\Re_I$ (500 K) (A W^{-1})	$D^*(\lambda_{\text{peak}}$, 1 kHz) (10^{11} cm Hz$^{1/2}$ W^{-1})
0.16	6.0 × 10^{-4}	7.1 × 10^3	4.6	0	37	0.45	0.63^c
0.18	5.9 × 10^{-4}	4.4 × 10^3	4.8	0	96	0.30	1.0^c
0.62	6.0 × 10^{-4}	1.7 × 10^3	5.4	0	690	0.51	1.7
				0.15	73	0.52	1.7
0.56	5.4 × 10^{-4}	3.0 × 10^3		0	600	0.62	1.8
				0.175	88	0.62	1.7
0.54	6.0 × 10^{-4}	2.5 × 10^4	4.8	0	1220	0.54	1.6
				0.25	64	0.57	1.7
0.47	5.0 × 10^{-3}	3.3 × 10^3	4.9	0	2800	0.60	1.9
				0.50	35	0.60	1.8

^{a} With an uncertainty of ±0.03 μm.
^{b} Measured at zero bias.
^{c} These specimens exhibited $1/f$ noise at zero bias.

in which the ratio (junction capacitance)/(peripheral area) has values of the order of 1 μF cm^{-2}, like those obtained for the ratio (junction capacitance)/(junction area) of conventional devices. This model would predict a pinched-off capacitance (excluding lead-out capacitance) near 50 pF for the device whose properties are shown in Fig. 47.

The reduction in capacitance that has been demonstrated with the pinched-off devices is sufficient to significantly relax the constraints on operating frequency that were described earlier. There remains unanswered the question of the ultimate speed of these devices. This speed would be limited by the transit times of photogenerated carriers for which there is at present no satisfactory analysis.

3. Lateral-Collection Photodiodes

In the lateral-collection photodiode (LCP) the junction capacitance is reduced by making the junction area smaller than the optical collection area. The concept is illustrated in Fig. 48. In a typical conventional photodiode (Fig. 48a) the junction depth is smaller than the minority-

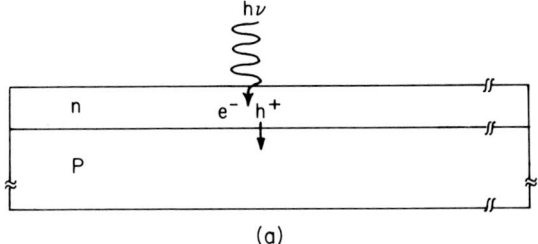

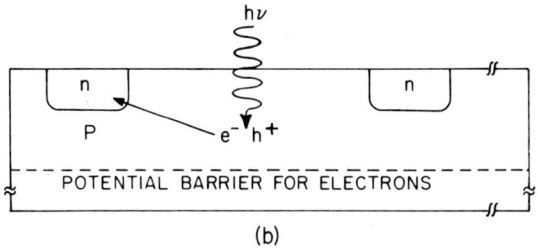

Fig. 48. The concept of the lateral-collection photodiode. (a) A conventional device. (b) A lateral-collection device.

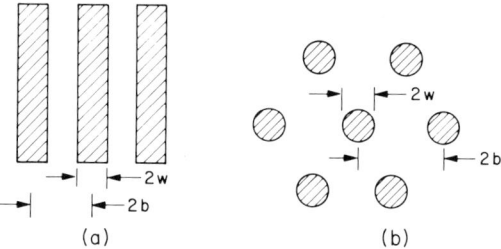

FIG. 49. Collector geometries for lateral-collection photodiodes. (a) An array of stripe collectors. (b) A matrix of circular collectors. The junction areas are hatched.

carrier diffusion length and larger than the optical absorption length. Under these conditions most of the photons are absorbed on the surface side of the junction, which acts as a nearly perfect sink so that the quantum efficiency approaches the RLL. In the LCP (Fig. 48b) the n region is replaced by a set of smaller n regions that collect photogenerated minority electrons from the intervening p region. A possible problem with this configuration is that photogenerated minority carriers may diffuse away from the surface and recombine, rather than be collected by the junctions. (The quantitative significance of this loss mechanism is discussed later.) For this reason a potential barrier has been introduced to confine the photogenerated minority carriers to a skin near the surface. If this skin has a thickness that is much less than the minority-carrier diffusion length, then the minority-carrier concentration will be almost uniform across the skin and significant diffusion will only occur laterally toward the collector p–n junctions.

In principle, the potential barrier may be introduced into an LCP in a variety of ways. Thus, a potential barrier with height greater than $2\ kT$ may be achieved by increasing the doping level of the underlying material, or by increasing its energy gap. A simpler scheme that has been used employs an epitaxial layer of the semiconductor on an insulating substrate.

An analysis of the properties of thin-film LCPs has been given by Holloway (*105*). This work shows that illumination of an array of stripe p–n junctions with the geometry of Fig. 49a gives a quantum efficiency

$$\eta = \eta_c \eta_{\text{RLL}} \tag{38}$$

where η_{RLL} is the RLL quantum efficiency,* η_c a collection efficiency, and

---

* In general, η_{RLL} will be different for the junction and nonjunction regions. For simplicity this difference is neglected here. The error that is so introduced is negligible with typical LCPs whose junction areas are only a small fraction of the optical collection area.

$$\eta_c = \frac{W}{B} + \frac{\tanh(B-W)}{B} \qquad (39)$$

where the dimensions $W = w/L$ and $B = b/L$ are expressed in units of the minority-carrier diffusion length L for the nonjunction region.

Similarly, for the matrix of circular collectors that is shown in Fig. 49b the collection efficiency is

$$\eta_c = \frac{W^2}{B^2} + \frac{2W}{B^2}\left(\frac{I_1(B)K_1(W) - K_1(B)I_1(W)}{I_1(B)K_0(W) + K_1(B)I_0(W)}\right) \qquad (40)$$

where $I_n(x)$ and $K_n(x)$ are modified Bessel functions of the first and second kinds. The plots of the two collection efficiencies that are shown in Figs. 50 and 51 indicate that large reductions in the ratio (junction area)/(total area) may be obtained with quite modest reductions in the quantum efficiency from that of a conventional device. Some details of the trade-off of collection efficiency for reduced capacitance are shown in Fig. 52. For example, with a matrix of circular collectors whose radius is 0.1 diffusion length, an order of magnitude reduction in capacitance is achievable with less than 10% sacrifice in quantum efficiency. Assuming a diffusion length near 10 μm this would require a collector diameter of the order of 2 μm, which is within the capability of standard photolithographic techniques.

An interesting feature of the LCP is that, under some conditions, the decrease in capacitance from that of a conventional device may be accompanied by an increase in D^* (JNL). Unfortunately, detailed studies of 3–5 μm photodiodes have tended to show that this feature is less useful than was originally believed.

The values of D^* (JNL) that are attainable with LCPs were found to depend on the source of the saturation current (I_s). With diffusion-limited I_s the thermally generated minority carriers are collected laterally in the same way as signal-generated carriers and R_0A of the small-area p–n junctions is degraded by lateral collection. This leads to the relationship

$$D^* \text{ (JNL)} \propto \eta_c^{1/2} \qquad (41)$$

and capacitance reduction requires a trade-off with detectivity of the same kind as that described above for quantum efficiency. In contrast, generation of I_s in the depletion region leads to a value of R_0A that is not degraded with small-area p–n junctions. In this case R_0A for the combination of junction and nonjunction areas increases as the proportion of junction area decreases. Thus, for constant collector diameter, the effect of increasing the collector separation is an increase in R_0A (for the device as a whole) and a decrease in quantum efficiency. This leads to a broad maximum in D^* (JNL) at a collector separation of about $2L$. For

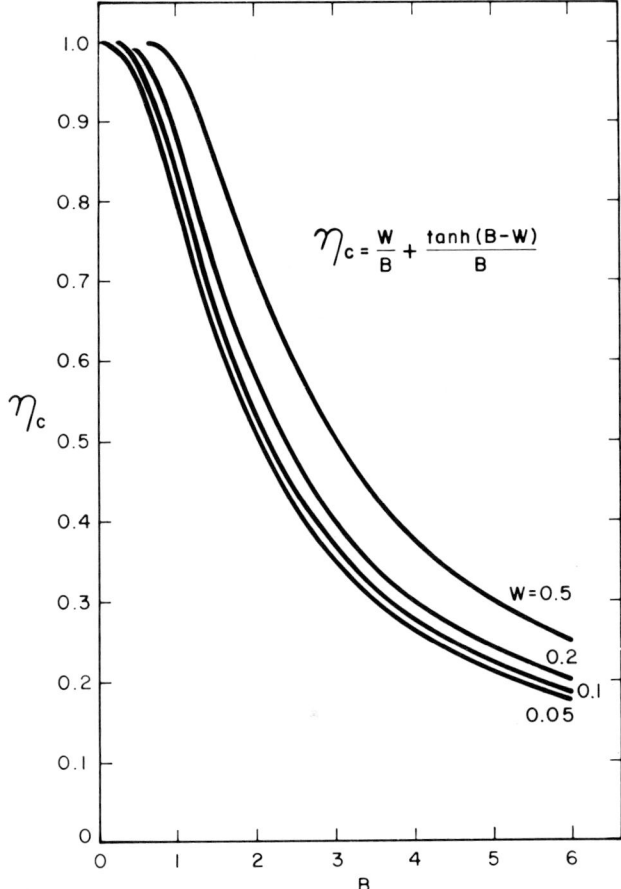

FIG. 50. Collection efficiency of an array of stripe collectors. The dimensions W and B are in units of the diffusion length.

example, a collector radius of $0.1L$ gives a sixfold increase in D^* (JNL) over that of a conventional device.

Thin-film LCPs made with layers of PbTe and Pb(Se, Te) on BaF_2 substrates have been described by Holloway et al. (146). The existence of large lateral-collection effects was shown directly with laser scans, of which Fig. 53 is an example. Detailed measurements showed that the junction capacitance could be reduced by more than an order of magnitude and that in some cases there was an increase in D^* (JNL). An example of these effects is given in Table X.

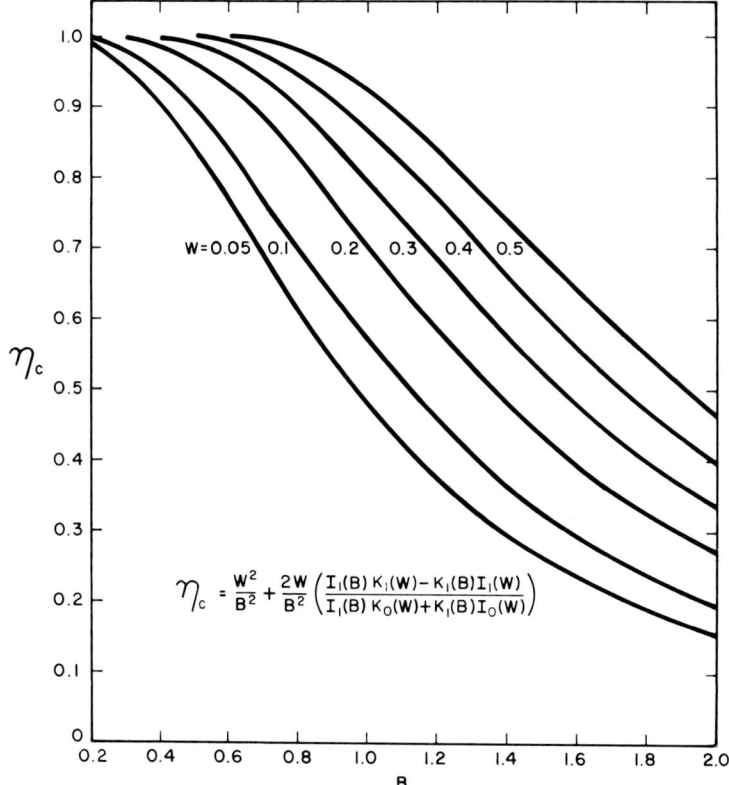

FIG. 51. Collection efficiency of a matrix of circular collectors. The dimensions W and B are in units of the diffusion length.

Further studies of thin-film LCPs for intermediate-temperature operation (77) showed that, while capacitance reduction was commonly achieved, the increase in D^* (JNL) was only obtained with epitaxial layers that gave relatively poor conventional photodiodes.* This result is related to the variability of the dominant mechanism for generation of I_s that was described earlier. Arrhenius plots of R_0A of the poorer conventional devices showed that, in such cases, I_s was depletion generated and

---

* Attempts to make 8–12 μm LCPs with thin films of (Pb, Sn)Se or (Pb, Sn)Te were unsuccessful. These devices had small quantum efficiencies and laser scans showed that the minority-carrier diffusion lengths were much smaller than in PbTe or Pb(Se, Te) layers. It is possible that this result is a consequence of surface recombination.

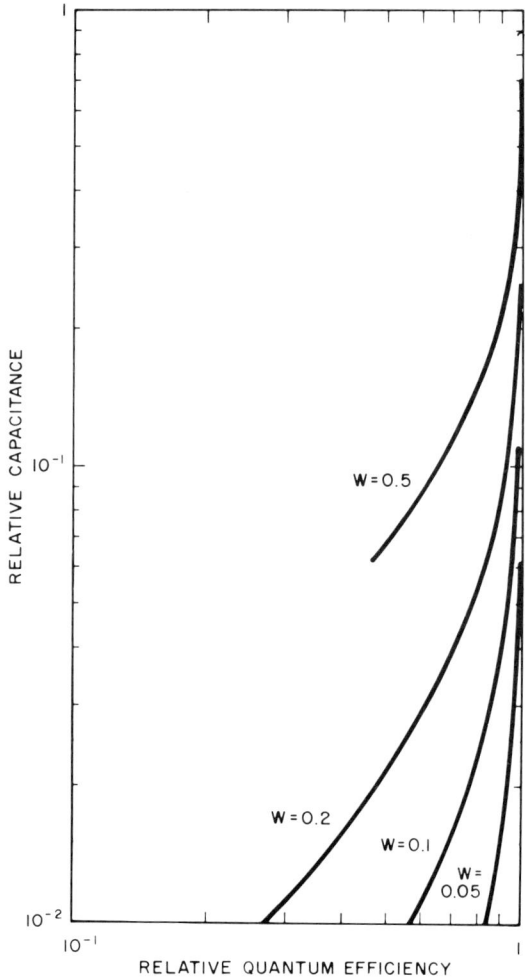

FIG. 52. Collection efficiency as a function of relative capacitance for a matrix of circular collectors.

hence capable of reduction with the lateral-collection structures. The better conventional devices had I_s that was diffusion limited and thereby capable of lateral collection. Support for this interpretation was provided by specimens (Fig. 54) whose Arrhenius plots indicated depletion-generated I_s for conventional devices, but showed a transition to diffusion-limited behavior when I_s (depletion) was reduced with the lateral-collection structure.

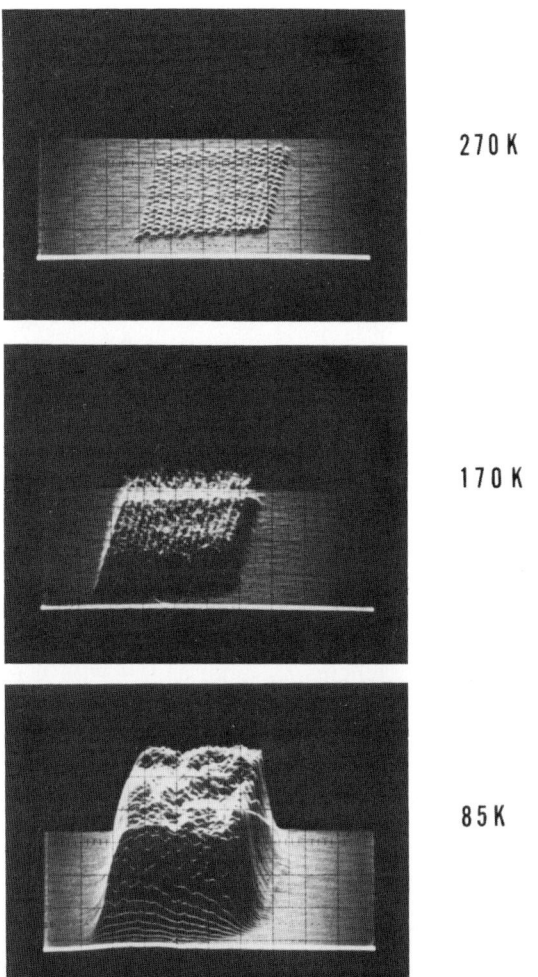

FIG. 53. Laser scans of a Pb barrier PbTe lateral-collection photodiode at 270, 170, and 85 K (top to bottom). The 300 μm-square device had a matrix of 5 μm diam collectors on 20 μm centers. The vertical displacement is proportional to the photocurrent and the resolution is about 10 μm.

It is of considerable interest that efficient LCPs in bulk crystals of (Pb, Sn)Te have been demonstrated by Noreika *et al.* (*145*). In this case the photogenerated minority carriers are subject to loss by diffusion away from the surface. Despite the possibility of such losses the bulk-crystal

TABLE X

PROPERTIES OF Pb(Se, Te) LCPs AT 170 K[a]

	Device type			
	Conventional	LCP	LCP	LCP
Collector spacing (μm)	—	10	15	25
Capacitance (pF)[b]	1050	260	108	53
$R_0 A$[b] (Ω cm^2)[c]	6.5	13	29	100
Peak wavelength (μm)	3.7	4.1	4.1	4.1
Peak quantum efficiency	0.85	0.85	0.73	0.51
Peak D^* (10^{10} cm Hz$^{1/2}$ W^{-1})[d]	4.8	8.0	10.0[e]	11.6[e]

[a] These devices were made on an 0.43 μm thick layer with $p = 10^{17}$ cm^{-3}. The matrices of 5 μm diam collectors were 320 μm square and had total areas of 10^{-3} cm^2.
[b] Measured at zero bias and 80 K.
[c] At zero bias. A is the active area.
[d] At zero bias and 1 kHz.
[e] Under these conditions the background and Johnson noises are approximately equal. Thus, D^* (JNL) is approximately a factor of $\sqrt{2}$ larger than the measured value that is tabulated.

LCPs gave quantum efficiencies as large as 0.4 with 6 μm wide stripe junctions on 31 μm centers. This led to a threefold reduction of capacitance relative to a conventional device on the same chip. (A fivefold reduction of capacitance was obtained with $\eta = 0.19$.)

An extension of the analysis of collection efficiency from a thin film to a bulk crystal (147) shows that Eq. (39) for an array of stripes is replaced by

$$\eta_c = \frac{W}{B} + \frac{1 - \exp[-2(B - W)]}{2B} \quad (42)$$

provided that $W \gtrsim 0.3$. Comparison of Eqs. (39) and (42) shows that the collection efficiency of the bulk-crystal LCP is always at least one-half that of the thin-film device.* Thus, while losses to the interior of the bulk-crystal LCP do occur, their influence on the collection efficiency is surprisingly small. The collection efficiency that was obtained by Noreika et al. suggests that the minority-carrier diffusion length was of the order of

---

* The lateral photocurrent into an isolated thin-film stripe collector is equivalent to complete collection for a range of one diffusion length on either side. With a bulk-crystal device the range is reduced to one-half diffusion length.

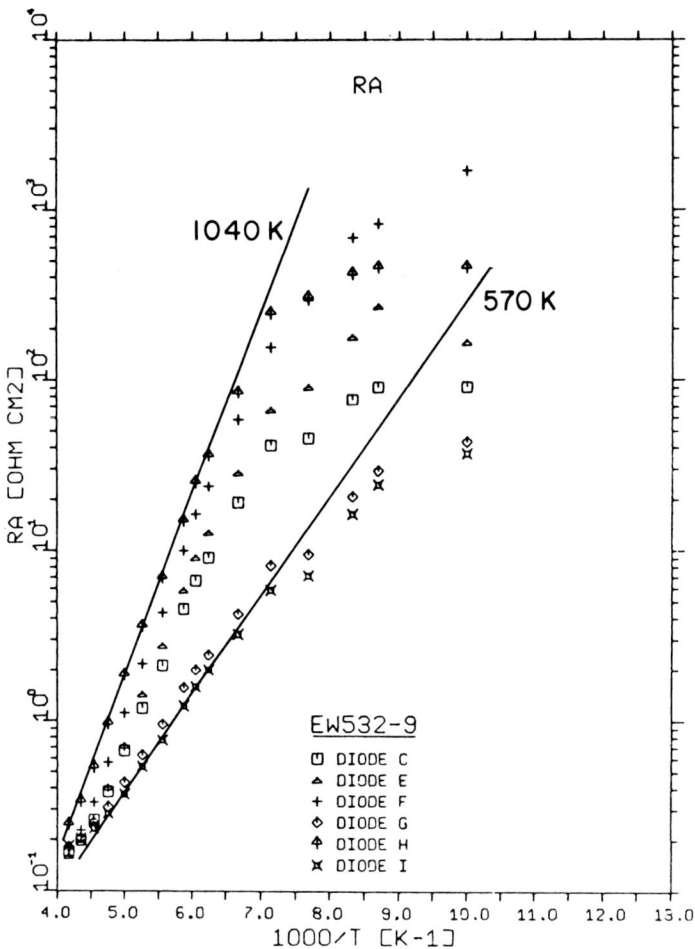

FIG. 54. Arrhenius plots of R_0A of conventional and lateral-collection PbTe photodiodes on the same chip. Diodes g and i (with slope ≈ 570 K) are conventional. The other devices (with slope ≈ 1000 K) are matrices of 5 μm diam collectors. The collector spacings are c, 10 μm; e, 15 μm; f, 20 μm; h, 25 μm.

20 μm in their specimen of (Pb, Sn)Te at 80 K. This, together with a mobility of about 2×10^4 cm² V⁻¹ sec⁻¹, would imply a lifetime of about 6 nsec. This appears to be significantly larger than the value of 0.5–2 nsec that would be expected from Emtage's calculation with carrier concentrations in the range $1-2 \times 10^{17}$ cm⁻³ that were used by Noreika et al.

4. Self-Filtered Narrow-Band Photodiodes

Schoolar *et al.* (*41*) have demonstrated thin-film IV–VI photodiodes whose response is concentrated in a narrow band of wavelengths. The arrangement and the principle of these devices are shown in Fig. 55. Layers of IV–VI semiconductors with slightly different energy gaps are deposited on either side of a cleaved BaF_2 substrate. The layer with the smaller energy gap is used to make a metal-barrier photodiode and the layer with the larger energy gap acts as a cooled absorption filter. With appropriate choices of layer compositions the photoresponse occurs as a spike that is located between the cut-on of the filter and the cutoff of the photodiode.

Examples of the spectral response of narrow-band Pb(S, Se) photodiodes (*41*) are shown in Fig. 56. These devices gave peak widths of 0.1–0.2 μm at half-maximum and peak quantum efficiencies of 0.4–0.5. In

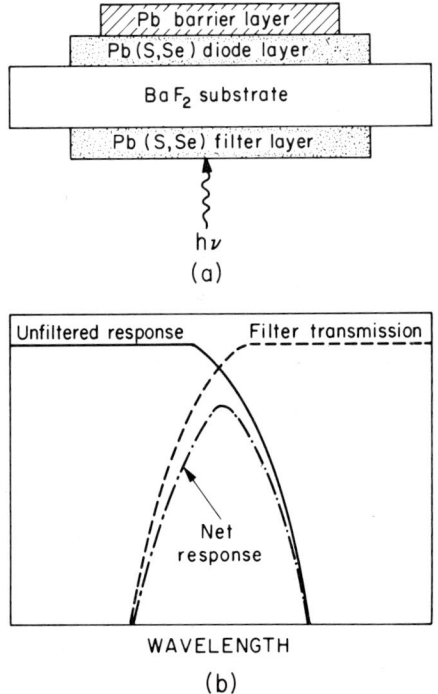

FIG. 55. Combination of a metal-barrier photodiode with an integral absorption filter after Schoolar and Jensen (*42*) (a) Schematic diagram of the structure. (b) Spectral properties of the photodiode and the filter.

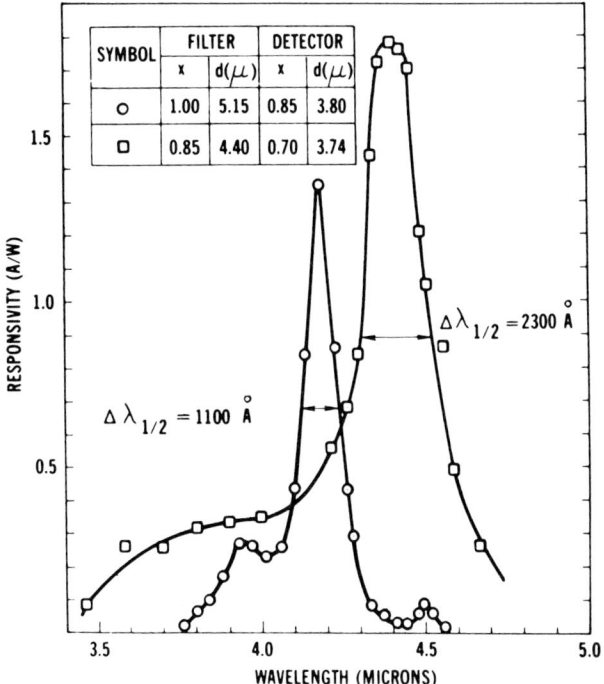

FIG. 56. Spectral responses of thin-film Pb(S, Se) photodiodes with integral absorption filters. Schoolar et al. (41).

addition to the demonstration of narrow-band response, this work showed that high-performance thin-film photodiodes could be made over the whole range of compositions from PbS to PbSe. At 77 K the values of R_0A ranged from 30 Ω cm² with PbSe to 2×10^4 Ω cm² with PbS. Measurements of the detectivity of unfiltered photodiodes under reduced (20°) FOV gave $D^*(3.7\ \mu\text{m}) = 1.0 \times 10^{13}$ cm Hz$^{1/2}$ W^{-1} with PbS and $D^*(6.9\ \mu\text{m}) = 2.7 \times 10^{11}$ cm Hz$^{1/2}$ W^{-1} with PbSe.

Applicability of the self-filtered devices to the 10 μm region was demonstrated by Schoolar and Jensen (42) with combinations of photodiodes and filter layers that had been prepared on different BaF$_2$ substrates. For this work some use was also made of interference to enhance the peak quantum efficiency, but the layers that were used were too thick for large interference effects. With a combination of (Pb, Sn)Se layers a 0.6 μm wide response peak was obtained at a wavelength of 10.6 μm, where the quantum efficiency was 0.32. With a 20° FOV the photodiode was Johnson noise limited with $D^*(10.6\ \mu\text{m}) = 7 \times 10^{10}$ cm Hz$^{1/2}$ W^{-1}.

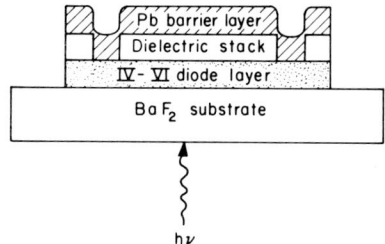

FIG. 57. Schematic diagram of a lateral-collection photodiode with an integral interference filter.

An alternative narrow-band device whose spectral properties depend entirely on interference affects within the device has been proposed by Holloway (99). In this case (Fig. 57) a backside-illuminated photodiode on a BaF_2 substrate is covered with a combination of a dielectric stack and a metal reflector. The photodiode may be either a conventional p–n junction or an LCP in which most of the optical absorption takes place in a region of the semiconductor that is not in direct contact with the metal barrier layer. Detailed calculations show that narrow response peaks are obtained when the optical thickness of the semiconductor is an odd multiple of quarter-waves. For minimum response at wavelengths on either side of the peak the optimum PbTe optical thickness is one quarter-

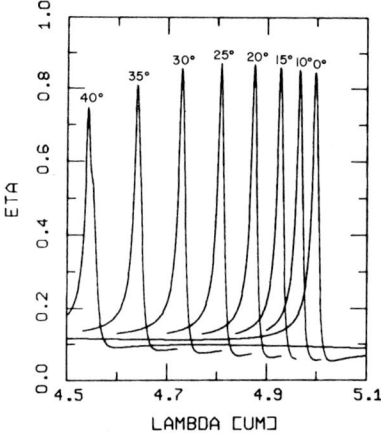

FIG. 58. Calculated spectral quantum efficiency of a thin-film photodiode with an integral interference filter. The curves show the tuning of the response peak with change in the angle of incidence.

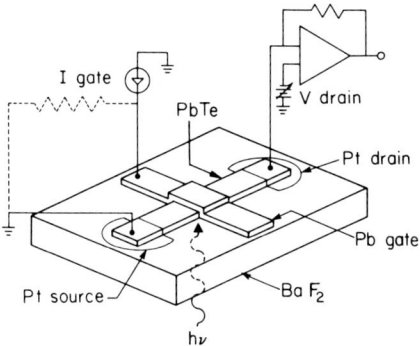

FIG. 59. Schematic diagram of a thin-film PbTe phototransistor and its associated circuitry.

wave. In its most rudimentary form the dielectric stack is a one-wave-thick layer of vacuum-deposited BaF_2. This structure has been calculated to have a spectral response that is similar to that obtained by Schoolar *et al.* with absorption-filtered Pb(S, Se).

While the interference-filtered devices have not yet been demonstrated it is of interest that the more complicated structure

BaF_2	PbTe	BaF_2	Te	BaF_2	Pb
$\lambda/4$	$\lambda/4$	$\lambda/4$	$\lambda/2$		

has been calculated to give the narrow response peaks that are shown in Fig. 58. In this case the peak position may be tuned by changing the angle of incidence and the structure is capable of operation as a simple spectrometer.

5. Phototransistors

A preliminary report has been given of a PbTe phototransistor (*148, 149*) with the structure shown in Fig. 59. This appears to be the first 3–5 μm photovoltaic device that gives amplification of the photocurrent.* Unlike conventional phototransistors, which are bipolar devices, the

---

* The term phototransistor has also been applied to a PbS metal–insulator–semiconductor (MIS) structure (*150*). However, this device was not a photocurrent amplifier. The depletion mode was used to enhance S/N of a PbS photoconductor by reducing the concentration of mobile carriers. MIS structures have also been used to make field effect transistors from thin films of PbTe (*151, 152*).

PbTe structure is a junction field effect transistor (JFET). Early studies of thin-film Pb barrier PbTe structures had shown JFET characteristics (*64, 153*), but the excessively noisy nature of these devices had discouraged further work. A reexamination of the PbTe JFETs, and particularly a study of their properties as phototransistors, was prompted by the development of techniques for making PbTe junctions that were capable of low-noise backbias operation.

During operation of the PbTe phototransistor, photons that are incident under the gate cause a change in the junction bias and thereby change the current that flows to the drain. With typical fields of view the long-wavelength response of the PbTe junction gives a significant background current that would forward-bias a floating gate. Thus, for control of the gate voltage, the circuit shown in Fig. 59 includes a source of constant gate current. The gate may be operated in the forward, zero, or back-biased condition by adjusting the gate current to be less than, equal to, or greater than the background current.

For the phototransistor we may define a photocurrent gain

$$\alpha \equiv \frac{\partial I_D}{\partial I_G} = \frac{\partial I_D}{\partial V_G} \frac{\partial V_G}{\partial I_G} \tag{43}$$

Now, making use of the definition of transconductance

$$g_m \equiv \partial I_D / \partial V_G \tag{44}$$

we obtain

$$\alpha = g_m X_{jn} \tag{45}$$

where X_{jn} is the dynamic impedance of the junction.

With PbTe at 80 K the frequency dependence of g_m is insignificant when compared to that of X_{jn}. Thus, one would expect the roll-off of α to be determined by the *RC* time constant of the p–n junction. This leads to an estimated 3 dB point near 1 kHz for a zero-biased gate. To obtain a flat response (with smaller α) to higher frequencies the junction resistance may be artificially degraded by adding a shunt resistance between source and gate, as shown in Fig. 59. In this case the BNL condition will be maintained if the shunt resistance is significantly larger than the junction resistance that is needed for BNL operation.

The frequency-dependent photocurrent gain that was actually obtained with a PbTe JFET is shown in Fig. 60. It will be noted that α is very large at low frequencies, where a single absorbed photon corresponds to a change of more than 10^4 electrons flowing into the drain. However, the frequency dependence of α extends down to at least 1 Hz rather than, as

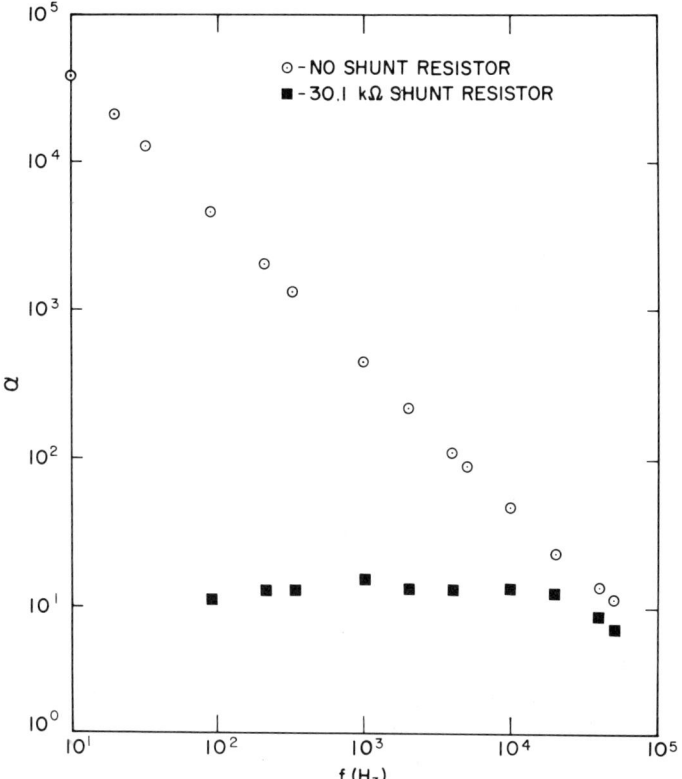

FIG. 60. Photocurrent gain of a thin-film PbTe phototransistor at 80 K. The squares and the circles show the gain with and without a shunt resistor.

expected, to 1 kHz.* With the addition of a shunt resistor a photocurrent gain of 10 was maintained to 40 kHz. It is of interest that the phototransistor is capable of relatively low-noise operation. Thus, in the absence of the shunt resistance, the device whose gain is shown in Fig. 60 had a detectivity that was within 30% of the BNL at frequencies in the range 20–600 Hz.

IX. Thin-Film Photodiode Arrays

Arrays of up to 100 thin-film IV–VI semiconductor photodiodes have been developed by Asch and co-workers (*138*). These devices have been

* Subsequent work showed that the abnormal frequency dependence of the gain was a consequence of a large source-to-drain resistance in the thin-film structure. With improved geometry this resistance was reduced and the rolloff in gain and D^* was shifted to higher frequencies (*149*).

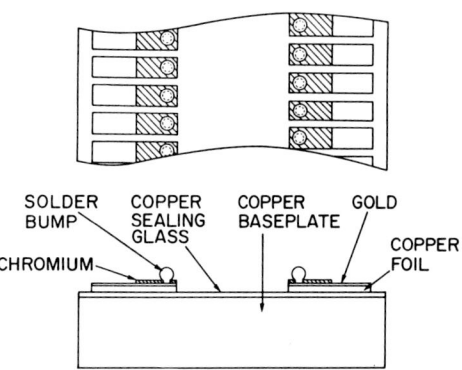

FIG. 61. Schematic diagram of a header for flip-chip mounting of thin-film IV–VI semiconductor photodiode arrays. Gorski et al. (156).

made with Pb barriers on Pb(Se, Te) (154, 155), PbTe, and (Pb, Sn)Te to permit operation in both the 3–5 and 8–12 μm atmospheric transmission bands. In general, the properties of the photodiodes in the arrays are similar to those of discrete devices. However, the transition from laboratory devices to practically packaged arrays has required the solution of several novel problems.

A problem that required early solution, even with discrete devices, arose from the lack of a practical scheme for mounting and connecting to the photodiodes.* Initial attempts at mounting the photodiodes had shown that a good thermal expansion match with the header was needed to avoid cracking the BaF_2 substrates. Further, the backside illumination of the Pb-barrier devices imposed the requirement of a lead-out pattern between the devices and the header. These needs were met by a header that was devised by Gorski et al. (156). The basic structure was a Cu/glass/Cu sandwich (Fig. 61) in which one of the Cu layers was a thin foil that was delineated to give a lead-out pattern that was insulated from the thicker Cu base layer. Further thin metalizations of Cr and Au were used to control the contact area of an electrodeposited Pb/Sn solder bump. Connection between the thin-film array and the solder bumps was made by a flip-chip process that was found to be better than 98% effective for 100 element arrays.

The second major development by Asch and co-workers (155) was of a

---

* For laboratory measurements these difficulties had been bypassed with the crude expedients of thermal compound contact to a varnished copper header and connection with a hand-painted silver lead-out pattern.

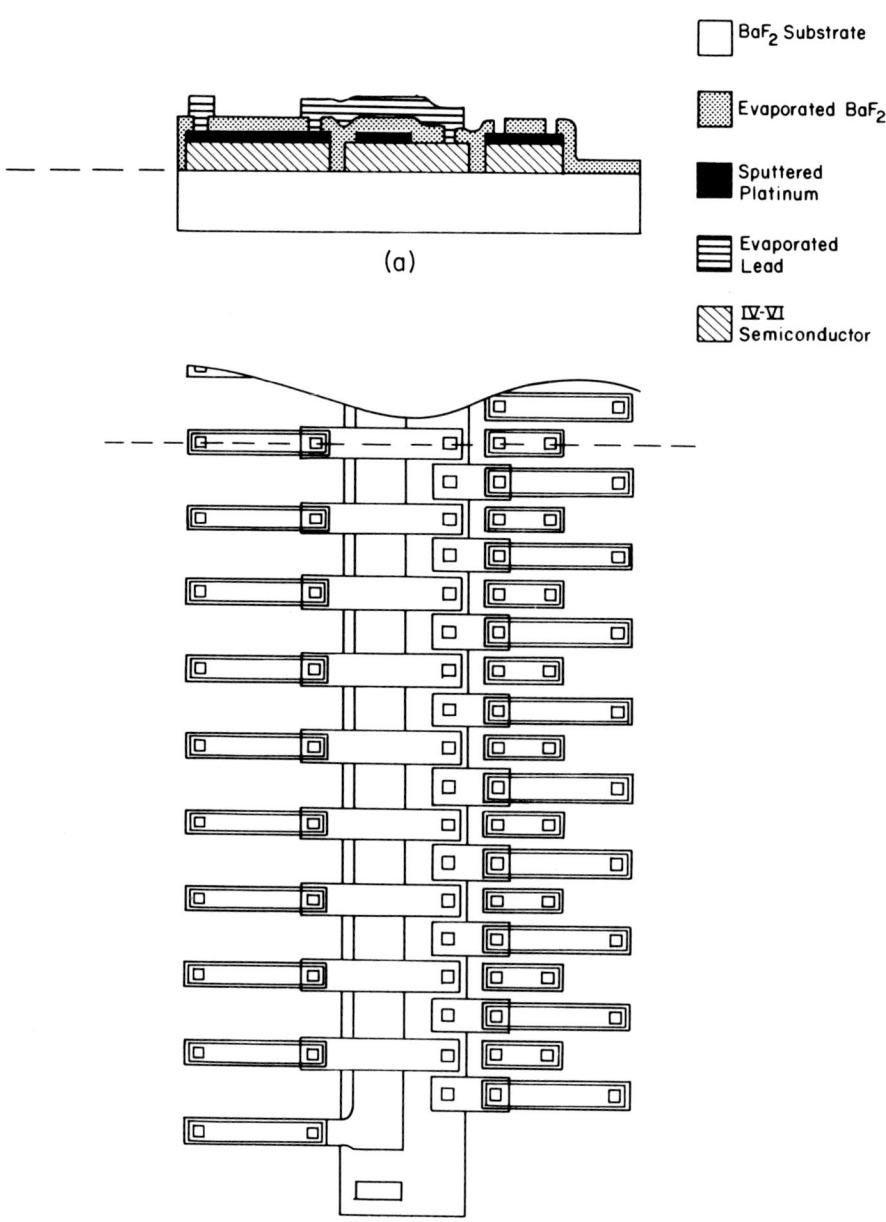

FIG. 62. Schematic diagram of a thin-film IV–VI semiconductor photodiode array. (a) Cross section. (b) Plan. Asch (*155*).

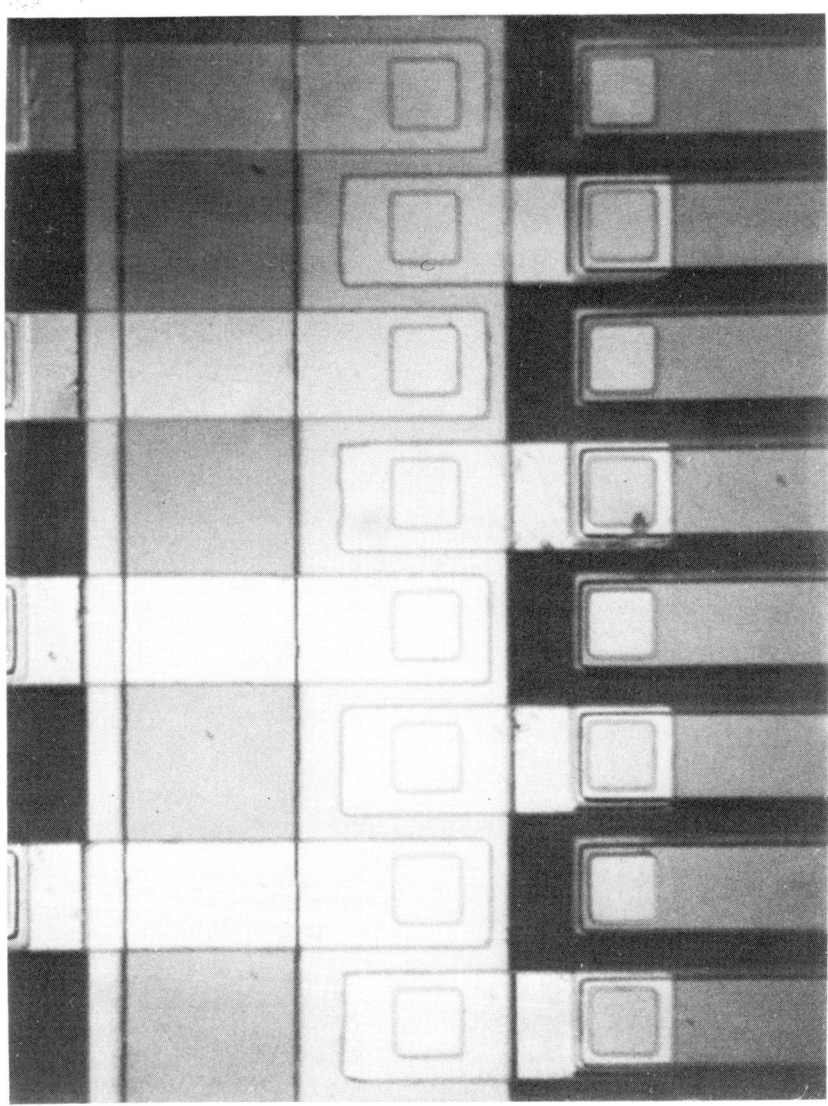

FIG. 63. Micrograph of part of a Pb(Se, Te) photodiode array made in the configuration of Fig. 62. The diodes are 100 μm square. Asch (*138*).

FIG. 64. A 48 element thin-film PbTe photodiode array with a BaF_2 immersion lens. Asch (*138*).

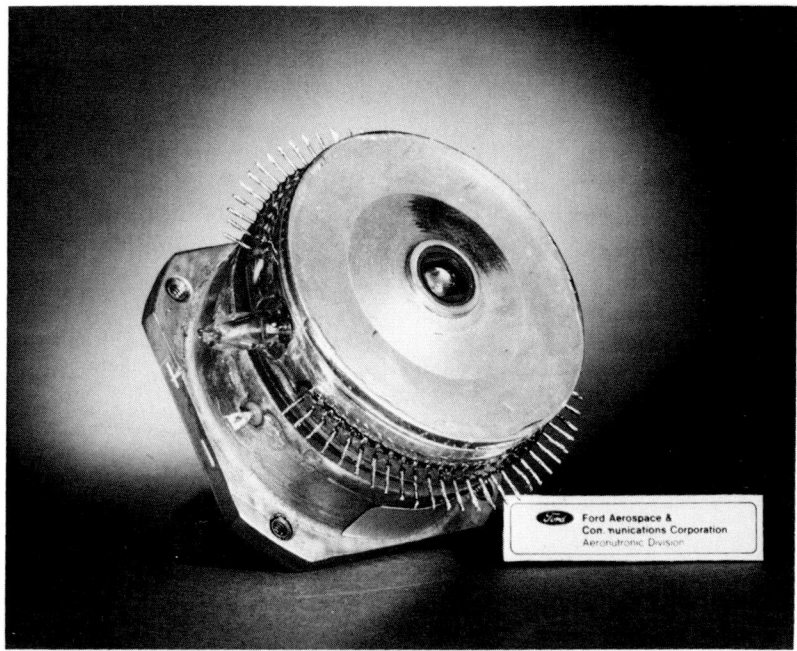

Fig. 65. The array of Fig. 64 mounted on a vacuum-packaged six-stage thermoelectric cooler for 170–195 K operation. Asch (*138*).

set of fabrication techniques for small closely spaced thin-film photodiodes. These photolithographic methods included the BaF_2 window technique (*81*) that was described earlier together with delineation techniques for the semiconductor and the Pb and Pt metallizations.* A schematic diagram of such a thin-film photodiode array is shown in Fig. 62 and a micrograph of part of a typical array is shown in Fig. 63. The process has yielded vacuum-packaged arrays with up to 100 elements and

* The IV–VI semiconductor may be delineated by protecting the wanted regions with photoresist (Shipley AZ 1350J) and removing the remainder by etching with a solution of bromine in hydrobromic acid. The metallizations and the BaF_2 insulator layer may be delineated with a lift-off technique in which the layer is deposited over a developed photoresist pattern. Dissolution of the resist pattern then removes the unwanted portions of the overlayer. An alternative delineation technique for the Pb metallization uses etching of unprotected regions with a dilute aqueous solution of acetic acid and hydrogen peroxide. This reagent etches the IV–VI semiconductors much more slowly than it does the Pb metallization.

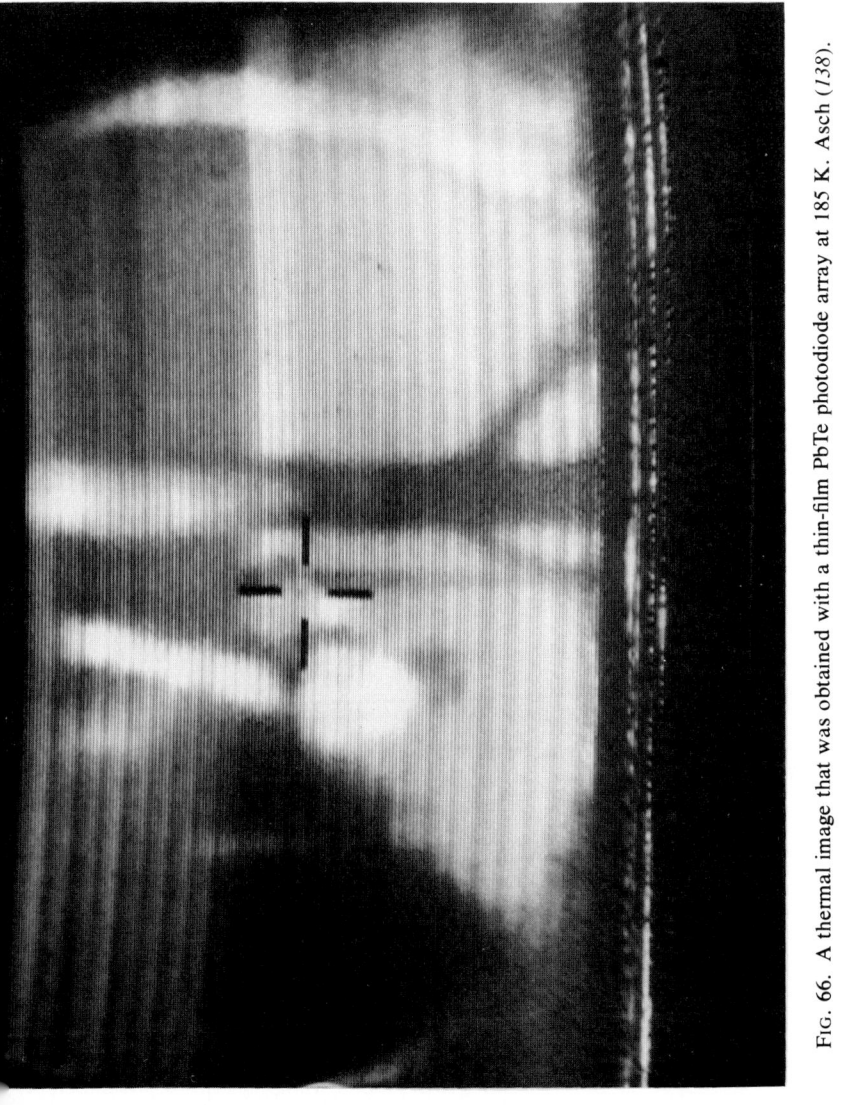

FIG. 66. A thermal image that was obtained with a thin-film PbTe photodiode array at 185 K. Asch (138).

it has been used to make array elements with dimensions down to about 20 µm.*

Figures 64 and 65 show a 48 element thin-film PbTe photodiode array before and after mounting on a vacuum-packaged six-stage thermoelectric cooler for 170–195 K operation. In this example a BaF_2 lens that was cemented to the BaF_2 substrate was used to provide immersion optics. An example of a thermal image that was obtained with a thin-film PbTe photodiode array at 185 K is shown in Fig. 66.

At this stage of the development of thin-film photodiode arrays there exists a set of proven techniques that permit exploitation of the high quality of layers of the IV–VI semiconductors on BaF_2 substrates. However, recent work by Beck *et al.* (*157*) suggests that future development might be in a radically different direction. In these studies the sticking coefficient for (Pb, Sn)Te on a heated BaF_2 substrate was drastically reduced by electron beam pyrolysis of residual gases to give a thin layer of carbon. Scanning techniques permitted deposition of carbon patterns and, thereby, direct deposition of (Pb, Sn)Te patterns. At present there is no information about the suitability of such *in situ*-delineated semiconductors for photodiode fabrication. However, if the IV–VI semiconductor layers are of high enough quality it is likely that similar in situ delineation techniques could be adopted for other layers that are used in photodiode fabrication.† This provides the intriguing prospect of photodiode arrays that are made with a single pumpdown. Such a fabrication technique would be cheap and relatively immune to problems with surface contamination.

X. Conclusions

Thin-film photodiodes on BaF_2 substrates have now been demonstrated with epitaxial layers of a wide range of IV–VI semiconductors. These devices are suitable for operation in both the 3–5 and the 8–12 µm spectral regions and their performance has been demonstrated by several groups of workers to be quite competitive with that of bulk-crystal p–n junctions.

* The methods are suitable for substantially smaller dimensions. The BaF_2 window technique was used to make the small collector p–n junctions in the lateral-collection photodiodes that were described earlier. Such circular junctions have been made with diameters down to 2 µm.

† Green and Albrecht (*158*) have shown that Pb patterns may be delineated in vacuum via selective photooxidation of PbI_2. Deutsch *et al.* (*159*) have shown that micron resolution may be obtained by selective photolysis of metal alkyls with a UV laser.

Any doubts about the practical applicability of the thin-film photodiodes appear to be eliminated by the successful fabrication and vacuum packaging of thin-film arrays.

There remain unanswered several questions, not all of which are unique to the thin-film nature of the devices that have been reviewed here. Thus, practically nothing is known about the processes that generate saturation current within the depletion region of a IV–VI semiconductor p–n junction. The limited evidence that is available suggests that generation occurs via centers whose concentration varies from specimen to specimen. This would imply that we do not yet know how to specify material that will yield the best devices. We also remain rather poorly informed about the nature of the IV–VI semiconductor surfaces. More information is required here before we will gain an understanding of the variability of the $1/f$ noise, which is common to both bulk-crystal and thin-film devices, or of the nature of the Pb barriers that have been employed for most of the thin-film photodiodes.

Acknowledgments

I thank M. H. Francombe for his very constructive criticisms of an early draft of the manuscript and G. Jesion for his patience in finding many bugs in the present version. Any errors that remain are, of course, my own. Thanks are also due to the workers whose private communications are cited and whose figures are reproduced. Preprints of Buchner et al. (127) and Beck et al. (157) were kindly supplied by N. E. Byer.

The present review makes frequent use of a learning experience that was shared with many co-workers. In this respect I am immeasurably indebted to A. E. Asch, D. K. Hohnke, E. M. Logothetis, A. J. Varga, and K. F. Yeung. I have also benefited greatly from much encouragement by and discussion with R. E. Callender, L. M. Cameron, E. T. Hutcheson, and D. R. Kaplan of the U.S. Army Night Vision Laboratory.

References

1. H. Hintenberger, Z. Phys. **119**, 1 (1942).
2. L. Sosnowski, J. Starkiewicz, and O. Simpson, Nature (London) **159**, 818 (1947).
3. F. Kicinski, Chem. Ind. (London) **17**, 54 (1948).
4. D. Bode, in "Physics of Thin Films" (G. Hass and R. E. Thun, eds.), Vol. 3, p. 275 Academic Press, New York, 1966.
5. J. O. Dimmock, I. Melngailis, and A. J. Strauss, Phys. Rev. Lett. **16**, 1193 (1966).
6. I. Melngailis and T. C. Harman, in "Semiconductors and Semimetals" (R. K. Willardson and A. C. Beer, eds.), Vol. 5, Chapter 4. Academic Press, New York, 1970.
7. J. P. Donnelly, T. C. Harman, A. G. Foyt, and W. T. Lindley, Appl. Phys. Lett. **20**, 279 (1972).
8. R. Dalven, Infrared Phys. **9**, 141 (1969).

9. A. J. Elleman and H. Wilman, *Proc. Phys. Soc., London* **61**, 164 (1948).
10. R. B. Schoolar and J. N. Zemel, *J. Appl. Phys.* **35**, 1848 (1964).
11. R. B. Schoolar, *Nav. Ordnance Lab. Rep.* **NOLTR-71-223** (1971).
12. H. Holloway, E. M. Logothetis, and E. Wilkes, *J. Appl. Phys.* **41**, 3453 (1970).
13. E. M. Logothetis, H. Holloway, A. J. Varga, and E. Wilkes, *Appl. Phys. Lett.* **19**, 318 (1971).
14. H. Holloway, W. H. Weber, E. M. Logothetis, A. J. Varga, and K. F. Yeung, *Appl. Phys. Lett.* **21**, 5 (1972).
15. J. N. Zemel, *Solid State Surf. Sci.* **1**, 291 (1969).
16. J. N. Walpole, A. R. Calawa, R. W. Ralston, T. C. Harman, and J. P. McVittie, *Appl. Phys. Lett.* **23**, 620 (1973).
17. H. Holloway and J. N. Walpole, *Prog. Cryst. Growth Charact.* **2**, 49 (1979).
18. A. M. Andrews, J. A. Higgins, J. T. Longo, E. R. Gertner, and J. G. Pasko, *Appl. Phys. Lett.* **21**, 285 (1972).
19. S. H. Groves, K. W. Nill, and A. J. Strauss, *Appl. Phys. Lett.* **25**, 331 (1974).
20. H. M. Manasevit and W. I. Simpson, *J. Electrochem. Soc.* **122**, 444 (1975).
21. R. F. Porter, *J. Chem. Phys.* **34**, 583 (1961).
22. R. F. Brebrick and A. J. Strauss, *J. Chem. Phys.* **40**, 3230 (1964).
23. R. F. Brebrick and A. J. Strauss, *J. Chem. Phys.* **41**, 197 (1964).
24. E. E. Hansen and Z. A. Munir, *J. Electrochem. Soc.* **118**, 983 (1971).
25. D. K. Hohnke and S. W. Kaiser, *J. Appl. Phys.* **45**, 892 (1974).
26. E. G. Bylander, *Mater. Sci. Eng.* **1**, 190 (1966).
27. D. A. Northrop, *J. Electrochem. Soc.* **118**, 1365 (1971).
28. R. F. Bis and J. N. Zemel, *J. Appl. Phys.* **37**, 228 (1966).
29. H. Holloway, D. K. Hohnke, R. L. Crawley, and E. Wilkes, *J. Vac. Sci. Technol.* **7**, 586 (1970).
30. D. L. Smith and V. Y. Pickhardt, *J. Electron. Mater.* **5**, 247 (1976).
31. H. Holloway, unpublished data.
32. D. L. Smith and V. Y. Pickhardt, *J. Electrochem. Soc.* **125**, 2042 (1978).
33. D. L. Smith, *Prog. Cryst. Growth Charact.* **2**, 33 (1979).
34. L. R. Koller and H. D. Coghill, *J. Electrochem. Soc.* **107**, 973 (1960).
35. P. Hudock, *Trans. AIME* **239**, 338 (1967).
36. R. F. Bis, J. R. Dixon, and J. R. Lowney, *J. Vac. Sci. Technol.* **9**, 226 (1972).
37. A. Lopez-Otero and L. D. Haas, *Thin Solid Films* **23**, 1 (1974).
38. K. Duh and H. Preier, *Thin Solid Films* **27**, 247 (1975).
39. I. Kasai, J. Hornung, and J. Baars, *J. Electron. Mater.* **4**, 299 (1975).
40. I. Kasai, D. W. Bassett, and J. Hornung, *J. Appl. Phys.* **47**, 3167 (1976).
41. R. B. Schoolar, J. D. Jensen, and G. M. Black, *Appl. Phys. Lett.* **31**, 620 (1977).
42. R. B. Schoolar and J. D. Jensen, *Appl. Phys. Lett.* **31**, 536 (1977).
43. A. Lopez-Otero, *J. Appl. Phys.* **48**, 446 (1977).
44. A. Lopez-Otero, *Thin Solid Films* **49**, 3 (1979).
45. A. Lopez-Otero, *Appl. Phys. Lett.* **26**, 471 (1975).
46. H. Preier, R. Herkert, and H. Pfeiffer, *J. Cryst. Growth* **22**, 153 (1974).
47. A. Bradford and E. Wentworth, *Infrared Phys.* **5**, 303 (1975).
48. R. K. Pandy, *Solid State Commun.* **15**, 449 (1974).
49. T. F. Tao and C. C. Wang, *J. Appl. Phys.* **43**, 1313 (1972).
50. E. Krikorian, *J. Vac. Sci. Technol.* **12**, 186 (abstr.) (1975).
51. C. Corsi, I. Alfieri, and G. Petrocco, *Appl. Phys. Lett.* **24**, 484 (1974).
52. J. N. Zemel, J. D. Jensen, and R. B. Schoolar, *Phys. Rev. A* **140**, 330 (1965).

53. J. H. Meyers, R. H. Morriss, and R. J. Deck, *J. Appl. Phys.* **42**, 5578 (1971).
54. R. F. Egerton, *Philos. Mag.* [8] **20**, 547 (1969).
55. R. B. Schoolar, *J. Vac. Sci. Technol.* **6**, 918 (abstr.) (1969).
56. R. B. Schoolar, *Appl. Phys. Lett.* **16**, 446 (1970).
57. R. B. Schoolar, *J. Vac. Sci. Technol.* **9**, 225 (abstr.) (1972).
58. H. Holloway, *in* "The Use of Thin Films in Physical Investigations" (J. C. Anderson, ed.), p. 111. Academic Press, New York, 1966.
59. A. Lopez-Otero, *Appl. Phys. Lett.* **29**, 441 (1976).
60. R. S. Allgaier and B. B. Houston, Jr., *Proc. Int. Conf. Semicond., 1962* p. 172 (1962).
61. E. M. Logothetis and H. Holloway, *J. Nonmetals* **2**, 1 (1973).
62. T. O. Fahrinre and J. N. Zemel, *J. Vac. Sci. Technol.* **7**, 121 (1970).
63. H. Holloway and E. M. Logothetis, *J. Appl. Phys.* **42**, 4522 (1971).
64. H. Holloway, *J. Nonmetals* **1**, 347 (1973).
65. D. K. Hohnke and M. D. Hurley, *J. Appl. Phys.* **47**, 4975 (1976).
66. R. F. Bis, E. N. Farabaugh, and E. P. Muth, *J. Appl. Phys.* **47**, 736 (1976).
67. R. E. Callender, *Pap., IRIS Detector Specialty Group Meet., 1973* (unpublished).
68. W. S. Chan, *Infrared Phys.* **14**, 177 (1974).
69. T. J. McMahon, *Pap., IRIS Detector Specialty Group Meet., 1974* (unpublished).
70. A. Lopez-Otero, L. D. Haas, W. Jantsch, and K. Lischka, *Appl. Phys. Lett.* **28**, 546 (1976).
71. R. B. Schoolar, J. D. Jensen, G. M. Black, and D. L. Demske, *Pap., IRIS Detector Specialty Group Meet., 1977* (unpublished).
72. R. S. Allgaier and W. W. Scanlon, *Phys. Rev.* **111**, 1029 (1958).
73. G. A. Antcliffe and J. S. Wrobel, *Mater. Res. Bull.* **5**, 747 (1970).
74. C. M. Wolfe and G. E. Stillman, *Appl. Phys. Lett.* **18**, 205 (1971).
75. V. L. Lambert, *J. Appl. Phys.* **46**, 2304 (1975).
76. D. K. Hohnke, H. Holloway, and M. D. Hurley, *Thin Solid Films* **38**, 49 (1976).
77. e.g. P. W. Kruse, L. D. McGlauchlin, and R. B. McQuistan, "Elements of Infrared Technology." Wiley, New York, 1962.
78. K. W. Nill, A.R. Calawa, T. C. Harman, and J. N. Walpole, *Appl. Phys. Lett.* **16**, 375 (1970).
79. J. N. Walpole and K. W. Nill, *J. Appl. Phys.* **42**, 5609 (1971).
80. K. W. Nill, J. N. Walpole, A. R. Calawa, and T. C. Harman, *in* "The Physics of Semimetals and Narrow-Gap Semiconductors" (D. L. Carter and R. T. Bate, eds.), p. 383. Pergamon, Oxford, 1971.
81. D. A. Gorski, U.S. Patent 3,969,743 (1976).
82. H. Holloway, unpublished data.
83. E. M. Logothetis, H. Holloway, A. J. Varga, and W. J. Johnson, *Appl. Phys. Lett.* **21**, 411 (1972).
84. C. C. Wang, T. F. Tao, and J. W. Sunier, *J. Appl. Phys.* **45**, 3981 (1974).
85. J. P. Donnelly, T. C. Harman, and A. G. Foyt, *Appl. Phys. Lett.* **18**, 259 (1971).
86. J. P. Donnelly and H. Holloway, *Appl. Phys. Lett.* **23**, 682 (1973).
87. D. Haas, H. Heinrich, W. Jantsch, K. Lischka, A. Lopez-Otero, L. Palmetshofer, and M. Wagenhuber, *Crit. Rev. Solid State Sci.* **5**, 547 (1975).
88. D. P. Mathur, *Opt. Eng.* **14**, 351 (1975).
89. H. B. Morris, R. A. Chapman, and R. L. Guildi, *Pap., IRIS Detector Specialty Group Meet., 1974* (unpublished).
90. H. B. Morris, R. A. Chapman, R. L. Guildi, and D. W. Bellevance, *Pap., IRIS Detector Specialty Group Meet.*, (unpublished).

91. P. LoVecchio, M. Jasper, J. T. Cox, and M. Garber, *Infrared Phys.* **15**, 295 (1975).
92. J. Bardeen, F. J. Blatt, and L. H. Hall, in "Photoconductivity Conference" (R. G. Breckenridge, B. R. Russel, and E. E. Hahn, eds.), p. 146. Wiley, New York, 1956.
93. J. O. Dimmock, in "Semimetals and Narrow-gap Semiconductors" (D. L. Carter and R. T. Bate, eds.), p. 319. Academic Press, New York, 1970.
94. N. Piccioli, J. M. Besson, and M. Balkanski, *J. Phys. (Paris)* **33**, 119 (1972).
95. N. Piccioli, J. M. Besson, and M. Balkanski, *J. Phys. Chem. Solids* **35**, 971 (1974).
96. I. H. Malitson, *J. Opt. Soc. Am.* **42**, 684 (1964).
97. R. A. Smith, "Semiconductors," p. 303. Cambridge Univ. Press, London and New York, 1959.
98. J. J. Goedbloed and J. Joosten, *Electron. Lett.* **12**, 363 (1976).
99. H. Holloway, *J. Appl. Phys.* **50**, 1386 (1979).
100. E.g., O. S. Heavens, "Optical Properties of Thin Solid Films." Butterworth, London, 1955.
101. A. I. Golovashkin, *Sov. Phys.—JETP (Engl. Transl.)* **21**, 548 (1965).
102. W. Shockley, *Bell Syst. Tech. J.* **28**, 435 (1949).
103. C. T. Sah, R. N. Noyce, and W. Shockley, *Proc. IRE* **45**, 1228 (1957).
104. S. M. Sze, "Physics of Semiconductor Devices." Wiley, New York, 1969.
105. H. Holloway, *J. Appl. Phys.* **49**, 4264 (1978).
106. I. Melngailis and T. C. Harman, *Appl. Phys. Lett.* **13**, 180 (1968).
107. W. van Roosbroeck and W. Shockley, *Phys. Rev.* **94**, 1558 (1954).
108. W. Shockley and W. T. Read, *Phys. Rev.* **87**, 835 (1952).
109. P. R. Emtage, *J. Appl. Phys.* **47**, 2565 (1976).
110. E. R. Washwell and K. F. Cuff, *Proc. Int. Conf. Semicond., 1964* p. 11 (1965).
111. K. Lischka and W. Huber, *J. Appl. Phys.* **48**, 2632 (1977).
112. F. Stern, *Solid State Phys.* **15**, 299 (1963).
113. T. S. Moss, G. J. Burrell, and B. Ellis, "Semiconductor Opto-Electronics," p. 58. Wiley, New York, 1973.
114. N. Piccioli, Thesis, University of Paris, 1971.
115. H. Heinrich, W. Huber, K. Lischka, and A. Lopez-Otero, *J. Vac. Sci. Technol.* **13**, 919 (abstr.) (1976).
116. L. H. DeVaux, H. Kimura, M. J. Sheets, F. J. Renda, J. R. Balon, P. S. Chia, and A. H. Lockwood, *Infrared Phys.* **15**, 271 (1975).
117. M. R. Johnson, R. A. Chapman, and J. S. Wrobel, *Infrared Phys.* **15**, 317 (1975).
118. H. Preier, *Infrared Phys.* **18**, 43 (1978).
119. T. X. Hoai and K. H. Herrmann, *Phys. Status Solidi B* **83**, 465 (1977).
120. W. H. Rolls and D. V. Eddols, *Infrared Phys.* **13**, 143 (1973).
121. B. R. Pagel and R. L. Petritz, *J. Appl. Phys.* **32**, 1901 (1961).
122. H. Holloway and K. F. Yeung, *Appl. Phys. Lett.* **30**, 210 (1977).
123. R. F. Egerton and C. Juhasz, *Thin Solid Films* **4**, 239 (1969).
124. J. D. Jensen and R. B. Schoolar, *J. Vac. Sci. Technol.* **13**, 920 (1976).
125. T. S. Sun, N. E. Byer, and J. M. Chen, *J. Vac. Sci. Technol.* **15**, 585 (1978).
126. T. S. Sun, S. P. Buchner, N. E. Byer, and J. M. Chen, *J. Vac. Sci. Technol.* **15**, 1292 (1978).
127. S. P. Buchner, T. S. Sun, W. A. Beck, N. E. Byer, and J. M. Chen, *J. Vac. Sci. Technol.* **16**, 1171 (1979).
128. R. Longshore, M. Jasper, B. Sumner, and P. LoVecchio, *Infrared Phys.* **15**, 311 (1975).
129. R. W. Grant, J. G. Pasko, J. T. Longo, and A. M. Andrews, *J. Vac. Sci. Technol.* **13**, 941 (1976).

130. J. F. Butler, *Pap., IRIS Detector Specialty Group Meet., 1972* (unpublished).
131. H. Holloway, A. J. Varga, K. F. Yeung, A. E. Asch, and D. A. Gorski, unpublished data.
132. R. B. Schoolar, private communication.
133. M. Lanir, C. C. Wang, and A. H. B. Vanderwyck, *Appl. Phys. Lett.* **34,** 50 (1979).
134. J. L. Streete, *Appl. Opt.* **7,** 1545 (1968).
135. J. M. Lloyd, "Thermal Imaging Systems," p. 200. Plenum, New York, 1975.
136. D. K. Hohnke and H. Holloway, *Appl. Phys. Lett.* **24,** 633 (1974).
137. H. Holloway, D. K. Hohnke, K. F. Yeung, A. E. Asch, and D. A. Gorski, unpublished data.
138. A. E. Asch, private communication.
139. E. M. Logothetis and H. Holloway, *J. Appl. Phys.* **43,** 256 (1972).
140. D. K. Hohnke, H. Holloway, K. F. Yeung, and M. D. Hurley, *Appl. Phys. Lett.* **29,** 98 (1976).
141. R. E. Callender, R. E. Flannery, and D. R. Kaplan, *Pap., IRIS Detector Specialty Group Meet., 1976* (unpublished).
142. R. T. Bate, D. L. Carter, and J. S. Wrobel, *Phys. Rev. Lett.* **13,** 180 (1968).
143. A. G. Foyt, W. T. Lindley, and J. P. Donnelly, *Appl. Phys. Lett.* **16,** 335 (1970).
144. A. M. Andrews, J. T. Longo, J. E. Clarke, and E. R. Gertner, *Appl. Phys. Lett.* **26,** 438 (1975).
145. A. J. Noreika, M. H. Francombe, W. J. Takei, R. N. Ghoshtagore, and J. L. Wentz, *Pap., IRIS Detector Specialty Group Meet., 1977* (unpublished).
146. H. Holloway, M. D. Hurley, and E. B. Schermer, *Appl. Phys. Lett.* **32,** 65 (1978).
147. H. Holloway and A. D. Brailsford, unpublished work.
148. H. Holloway, *Thin Solid Films* **58,** 73 (1979).
149. H. Holloway and G. Jesion, unpublished work.
150. G. Kramer and M. A. Levine, *Appl. Phys. Lett.* **28,** 101 (1976).
151. J. F. Skalski, *Proc. IEEE* **53,** 1792 (1965).
152. D. Lisle and J. C. Anderson, *Solid-State Electron.* **12,** 735 (1969).
153. W. J. Johnson, H. Holloway, and A. J. Varga, unpublished work.
154. A. E. Asch, D. A. Gorski, P. I. Zappella, and H. Holloway, *Pap., IRIS Detector Specialty Group Meet., 1975* (unpublished).
155. A. E. Asch, Final Report, Contract DAAK02-74-C-0181 (1975).
156. D. A. Gorski, P. I. Zappella, and A. E. Asch, *IEEE Trans. Parts, Hybrids, Packag.* **php-11,** 312 (1975).
157. W. A. Beck, S. P. Buchner, N. E. Byer, and T. S. Sun, *Appl. Phys. Lett.* **35,** 163 (1979).
158. M. Green and M. G. Albrecht, *Thin Solid Films* **37,** L57 (1976).
159. T. F. Deutsch, D. J. Ehrlich, and R. M. Osgood, Jr., *Appl. Phys. Lett.* **35,** 175 (1979).

The Universal Dielectric Response: A Review of Data and Their New Interpretation

A. K. JONSCHER

Chelsea College
University of London
London, England

I. Introduction	206
II. Definitions and Basic Relations	209
III. Methods of Presentation of Dielectric Data	217
IV. Basic Mechanisms of Polarization	223
V. Empirical Classification of Loss Characteristics	226
1. Complex Susceptibility Representations	227
2. The Loss Peaks	229
3. The Universal Dielectric Response	229
4. Time Domain Response	230
5. Response of Charge Carriers	231
6. Extreme Values of the Exponent n	232
7. The Temperature Dependence of Dielectric Response	233
VI. Survey of Experimental Data	235
1. The Debye Response	236
2. Broadened Debye Response with Symmetric Peaks	236
3. Asymmetric α Peaks	238
4. Dipolar β Peaks	246
5. General Analysis of the Loss Peaks	251
6. The "Universal" Response without Loss Peaks in Carrier Systems	256
7. "Lattice" Response	267
8. Strong Low-Frequency Dispersion	270
9. Time-Domain Response	272
10. Charge Injection into Solids	278
11. Time Response to Photoinjection	279
12. Contribution of Charge Carriers to Polarization	281
13. Conclusions Regarding the Experimental Evidence	283
VII. Currently Accepted Interpretations	287
1. Historical Background	287
2. Distributions of Relaxation Times	288
3. Hopping Electronic Systems	290
4. Ionic Conductors	292
5. Diffusive Models	292

      6. Correlation Function Approach 293
      7. Interfacial Phenomena and Maxwell–Wagner Effects 294
      8. Conclusions Regarding the Accepted Interpretations 295
VIII. The Many-Body Interpretation 296
      1. The Screened Hopping Model 296
      2. The Loss Peak in the Screened Hopping Model....................... 301
      3. The Application of the Ising Model.................................... 302
      4. Ngai's Infrared Divergence Model 303
      5. Dissado and Hill's Analysis of Two-Level Systems 305
  IX. Concluding Comments 312
      References 314
      Note Added in Proof 317

I. Introduction

Electrical conduction in disordered nonmetallic films under steady-field, i.e., under direct current (DC), conditions was the subject of an earlier review in the present series (*1*). The principal interest there was the transport of hopping-charge carriers, mainly electrons, which is typically encountered in nonmetallic disordered and glassy materials. In that mode of conduction, carriers have to traverse the entire distance in the medium between two electrodes and the rate-limiting processes are the most difficult hopping transitions between localized states distributed randomly in the material. Thus the theory of DC conductivity involves the calculation of the appropriate average of different jump probabilities and the conduction process is strongly influenced by both temperature and high electric fields, which enhance the most difficult transitions.

By comparison, the flow of electric current under time-varying fields, whether sinusoidally alternating, step function, or otherwise, presents an essentially different problem. Oppositely charged particles closely bound together in the form of *dipoles,* which do not contribute to DC conduction, now give rise to the flow of current by changing their orientation in the external field. This dipolar contribution may be supplemented by movements of hopping charges over limited distances in the material, reversing on alternate half-cycles in the alternating field, or only moving as far as the nearest "difficult" transition in the case of a step function field. Thus the alternating current (AC) conduction is dominated by the easiest hopping transitions and the AC conductivity is therefore much higher than the DC conductivity.

The behavior of materials under transient or alternating fields, i.e., their *dielectric response,* forms the subject of this chapter. The dielectric response may be characterized in one of two basically equivalent ways:

(1) as the *time dependence* of the polarization or of the polarizing current under step-function excitation, or (2) as the *frequency dependence* of polarization under alternating field excitation, with either the radian frequency ω or the circular frequency $f = \omega/2\pi$ measured in hertz.

The available frequency spectrum spans an enormous range, shown schematically in Fig. 1, from the highest frequencies of the electromagnetic spectrum, X rays, through the ultraviolet (UV) and visible to the infrared (IR), and ultimately the microwave region down to, say, 10 GHz. Below this follow the radio, audio, subaudio, and finally very low frequencies, the lower end of which may be set for practical purposes at 10^{-6} Hz, although this is not a rigid figure. As a very gross approximation we may say that the dielectric response in the "optical" range at frequencies in excess of, say, 100 GHz, is determined by phonon and quantum processes—lattice vibrations and excitations of individual atoms and molecules, etc.—but is not influenced to any large extent by the much slower processes arising from molecular dipolar rearrangements and from motions of charge carriers in the lattice.

While the dielectric response in the optical region presents many interesting problems, this chapter is concerned with "low-frequency" dielectric processes, which have received very little coordinated theoreti-

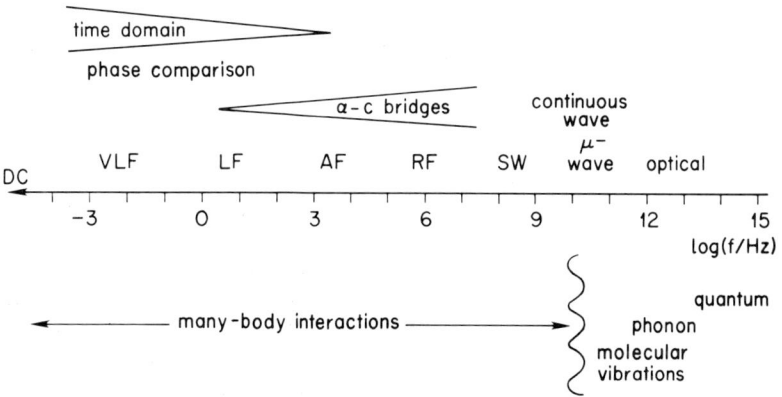

FIG. 1. A schematic representation of the frequency range from optical to very-low-frequency (VLF), through microwave, short-wave (SW), radio frequency (RF), audio frequency (AF), and low-frequency (LF). The tapering symbols indicate decreasing importance of the time domain and AC bridge methods of measurements, respectively, at high and low frequencies. The nature of the principal physical mechanisms is indicated. The present review is concerned with the frequency range below the wavy line at around 10 GHz.

cal attention, even though the technological and scientific importance of the behavior of materials in this range of the spectrum is undeniably high. There exists a vast amount of experimental data for all manner of materials, spanning large ranges of frequency and often extending also over a range of temperatures, but there is no coherent presentation of the totality of the information, nor even of significant parts of it. For the most part, individual materials or groups of materials are treated in isolation and no attempt is made to view their properties in the perspective of a more general approach.

We set out here to provide a unified presentation of the dielectric response of solids in this low-frequency range—a presentation informed by the recently developed concept of the "universality" of dielectric behavior. This should go some way in the direction of ordering a hitherto incoherent body of experimental material and should provide a convenient starting point for a new theoretical approach based on the concept of many-body interactions as the dominant mechanism in the dielectric response of condensed matter. In this sense our present approach departs radically from the accepted interpretations, which are based for the most part on the Debye model of dielectric relaxation.

Section II provides some definitions of dielectric concepts and gives a minimal mathematical framework essential for the proper understanding of the theory and for effective handling and interpretation of the experimental data.

Section III provides a brief introduction into the various methods of presentation of dielectric data and should help to clarify some of the confusions arising from the existence of various "traditions" of presentation in different schools of dielectric work, often separated according to various materials under study.

Section IV gives a brief outline of the principal physical mechanisms of polarization, with particular reference to the classical Debye process.

Our principal objective having been defined above, Section V sets out an empirical classification of the principal types of dielectric responses with a view to facilitating a rapid appraisal of a wide range of results that may, at first glance, appear to be completely unconnected.

The actual experimental data are then discussed in the light of this classification in Section VI, which leads us to the concept of the universality of the dielectric response (2). It is our deliberate policy to begin with experimental evidence, since the intention is not to prejudge the question of the theoretical interpretation before having all the relevant experimental information in front of us. Section VII follows then with a discussion of the principal existing theoretical approaches to the interpretation of dielectric data and with their critical examination in the light of the experimental evidence. This leads to the conclusion that none of these theories is adequate to account plausibly for the totality of the observed phenomena, while at the same time it is suggested that it is very unlikely that several different approaches should somehow give the same empirically established universal behavior.

Finally, Section VIII gives the proposed new approach to the interpretation of dielectric phenomena in solids and, to some extent, in condensed matter in general, based on the recognition of the role of many-body interactions between charges and dipoles as the one common or universal feature in all the materials under consideration. This is where we make the complete break with the Debye philosophy, and within the framework of the "non-Debye" approach we are then discussing the different aspects of the experimental data.

We offer no apology for the inclusion of this chapter in a volume of the *Physics of Thin Films*—most dielectric measurements are taken on samples in the form of films ranging in thickness, according to circumstances, from molecular and atomic dimensions up to truly "bulk" values of many micrometers and even millimeters.

In addition to the limitation of the frequency spectrum under review to low frequencies as defined above, it is proposed to exclude from detailed discussion the very wide topic of time-dependent charge transport in dielectrics under space-charge-limited conditions—this topic is adequately reviewed elsewhere (*3–5*) and involves essentially different concepts. Our discussion is therefore confined predominantly to linear behavior in the amplitude of the applied signal; nonlinear behavior is only mentioned where it constitutes a small perturbation of the predominantly linear response. We likewise exclude the general topic of the field dependence of the dielectric permittivity as in hyperpolarization and nonlinear optical behavior. Similarly, we do not propose to deal with the ferroelectric response—a major field in its own right, well covered in specialist literature—only passing mention will be made of it here. Yet another important but specialized topic that is omitted is the electret (*6*).

II. Definitions and Basic Relations

The response of a dielectric medium to an applied electric field E consists in the appearance of a net dipole moment per unit volume, which is the polarization P. The external manifestation of this polarization is the appearance at the surface of this material—assumed for simplicity to be in the form of a planar slab of thickness w—of a charge $Q_m = P$. Assuming that the field E is produced by the application of a potential difference between two metallic electrodes attached to the opposite faces of the slab, the total charge Q at the electrodes will consist of the component Q_m due to the dielectric medium and a component Q_0, which would be there even in the absence of the material medium and is due to the free space between the electrodes:

$$Q = Q_0 + Q_m = D \tag{1}$$

where the field vector D is defined as the *dielectric induction*. Most dielectric measurements consist essentially in the determination of the charge Q in response to an applied potential difference across a slab of dielectric material.

While the determination of the relation between D and E in the limit of steady-state or static conditions is of some theoretical interest, its practical measurement in the case of most solid dielectrics is so difficult and often ambiguous on account of the DC conductivity, that we do not propose to enter into this aspect of dielectric behavior here. From our point of view, the essence of dielectric behavior consists in the time dependence of $Q(t)$ under excitation by a time-varying electric field $E(t)$. For the purpose of definition, we choose initially a *delta function* excitation at time $t = 0$, $\delta(t)$, and the strength of this delta function is given by the product of an amplitude of the field E and its duration Δt, in the limit when $\Delta t \to 0$.

We now define the *dielectric response function* $f(t)$ as the response of the polarization of the medium to the delta function excitation:

$$P(t) = \epsilon_0 (E\,\Delta t) f(t) \tag{2}$$

where $\epsilon_0 = 10^7/4\pi c^2 = 8.854 \times 10^{12}$ F/m is the *dielectric permittivity of free space* and c the speed of light in vacuum. The physical cause of the time dependence of $P(t)$ is the inevitable "inertia" of any material medium that cannot instantaneously follow arbitrarily rapid variations of the exciting electric field.

The function $f(t)$ has the fundamental property

$$f(t) \to 0, \quad \text{as} \quad t \to \infty \tag{3}$$

since we do not expect any permanent polarization after an infinite time without the external field, and the complementary property

$$f(t) \equiv 0, \quad \text{for} \quad t < 0 \tag{4}$$

which implies that there can be no reaction before the cause.

We now make the further assumption that the system is *linear* and therefore the *superposition principle* holds: The response to consecutive excitations is the sum of the responses to the individual excitations. With these assumptions, it is an elementary matter to show that the response of polarization to an arbitrarily time varying function $E(t)$ may be written in the form of the convolution integral

$$P(t) = \epsilon_0 \int_0^\infty f(\tau) E(t - \tau) \, d\tau \tag{5}$$

It is equally clear that if $E(t)$ becomes a step function of amplitude E, then in virtue of the property (4) we have

$$P(t) = \epsilon_0 E \int_0^t f(\tau) \, d\tau \tag{6}$$

We note that the function $f(t)$ has the dimension of reciprocal time. The charge Q_0 due to the response of free space is, by definition, instantaneous, so that

$$Q_0(t) = \epsilon_0 E \tag{7}$$

regardless of the functional form of $E(t)$. The total time-dependent dielectric induction $D(t)$ is the sum of Q_0 and $P(t)$.

Since $D(t)$ manifests itself as a charge $Q(t)$, the evident experimental technique suggests itself of measuring the time dependence of this charge as the electric current $i(t) = dQ/dt = dD/dt$. Taking the case of step function excitation of amplitude E, differentiation of Eq. (7) gives a delta function current, while the differentiation of (6) gives the second component in the expression:

$$i(t) = \epsilon_0 E [\delta(t) + f(t)] \tag{8}$$

This equation forms the basis of the experimental method of measuring $f(t)$ in the time domain (7–9): the dielectric response function is seen to be a suitably scaled response of the charging current $i_c(t)$ to a rising step function field, or of the discharging current $i_d(t)$ to a falling step function excitation. The delta function in (8) is not observable experimentally and is normally converted to an exponential decay due to the finite resistance of the measuring circuit.

Provided that the principle of superposition holds, the charging and discharging currents are seen to be replicas of one another with a change of sign, and this would be rigorously true if no charges were entering or leaving the system via the electrodes, i.e., if the entire polarization was a matter of an internal rearrangement of charges, be it in the form of dipoles or as finite displacements of mobile charges. In real materials, however, the application of a steady field for an infinitely long time is bound to produce the flow of direct current i_0 and this should be added to the charging current, but not to the discharging current. This argument leads us to expect that

$$|i_d(t)| = i_c(t) - i_0 \qquad (9)$$

We see later that this relation is not normally obeyed, due to the onset of injection of charges into the dielectric leading to nonlinear response.

The total charge Q_m arising from the steady-state polarization of the dielectric material must be recovered in the process of discharge, so that

$$Q_m = \epsilon_0 E \int_0^\infty f(t)\, dt \qquad (10)$$

and this clearly shows a further physical limitation on the functional form of $f(t)$.

Time domain measurements are in every sense fundamental to the understanding of the polarization processes and remain the true and valid expressions of physical reality even in the presence of nonlinearities. The interpretation of nonlinear behavior is more profitable in the time domain than in the frequency domain, since we must always bear in mind the fundamental truth that *Mother Nature works in the time domain*. There are, however, serious practical limitations to the use of time domain techniques and the alternative frequency domain approach has therefore become preferred to the point that highly advanced equipment is available commercially for the latter, while it is necessary to build one's own for time domain work.

The basis of the frequency domain response is the application of a harmonically variable field $E(t) = \hat{E} \sin \omega t$ at a radian frequency ω. Returning to the fundamental equation (5) we may apply the Fourier transform to both sides and the physical quantities $E(t)$, $P(t)$, and $D(t)$ become their complex Fourier transforms $\mathcal{E}(\omega)$, $\mathcal{P}(\omega)$ and $\mathcal{D}(\omega)$, respectively, while the convolution integral transforms to the simple product of the form

$$\mathcal{P}(\omega) = \epsilon_0 \chi(\omega) \mathcal{E}(\omega) \qquad (11)$$

This shows the simplicity of the response of a linear system in the frequency domain: Every Fourier component of the driving field $\mathcal{E}(\omega)$ causes its own response of the polarisation component $\mathcal{P}(\omega)$ through an appropriate transfer function $\chi(\omega)$, which is given by the Fourier transform of the dielectric response function

$$\chi(\omega) = \int_0^\infty f(t) \exp(-i\omega t)\, dt \qquad (12)$$

This shows that the *dielectric susceptibility* $\chi(\omega)$ is in general a complex function of frequency

$$\chi(\omega) = \chi'(\omega) - i\chi''(\omega) \tag{13}$$

with the real and imaginary components being, respectively, the cosine and sine Fourier transforms of $f(t)$ and determining the in-phase and quadrature components of the "rotating vector" $\mathcal{P}(\omega)$ with respect to the driving field $\mathcal{E}(\omega)$.

The imaginary component $\chi''(\omega)$ is the *dielectric loss*, since the electric current due to this component is in phase with the driving field, while the real part $\chi'(\omega)$ determines the energy of dielectric polarization stored in the material at the maximum of $P(t)$—this energy is wholly reversible since the current due to this component is entirely in quadrature with the field.

The susceptibility function $\chi(\omega)$ defines the dielectric response of the *material medium* between the electrodes and the complete response of the system is given by the sum of this and of the free space contribution, given by Eq. (1) as the dielectric induction. This becomes in the frequency domain representation

$$\mathcal{D}(\omega) = \epsilon_0[1 + \chi(\omega)]\mathcal{E}(\omega) \tag{14}$$

The complex ratio $\mathcal{D}(\omega)/\mathcal{E}(\omega)$ is defined as the *dielectric permittivity* $\epsilon(\omega)$ so that

$$\mathcal{D}(\omega) = \epsilon(\omega)\mathcal{E}(\omega) = \epsilon_0\epsilon_r(\omega)\mathcal{E}(\omega) \tag{15}$$

where ϵ_r is known as the relative dielectric permittivity or the dielectric constant.

The unity in the brackets of Eq. (14) represents the contribution of free space that is purely real. The entire loss comes from the material medium and we may write for the real and imaginary components of the complex permittivity

$$\epsilon_r'(\omega) = 1 + \chi'(\omega), \qquad \epsilon_r''(\omega) = \chi''(\omega) \tag{16}$$

Equation (12) defines two functions $\chi'(\omega)$ and $\chi''(\omega)$ in terms of a single physically measurable function $f(t)$. It is a simple matter to invert the transforms to obtain

$$f(t) = (2/\pi) \int_0^\infty \chi'(\omega) \cos \omega t \, dt \tag{17}$$

$$= (2/\pi) \int_0^\infty \chi''(\omega) \sin \omega t \, dt \tag{18}$$

This shows that the knowledge of the *frequency spectrum* of either $\chi'(\omega)$ or $\chi''(\omega)$ in the *infinite frequency range* is sufficient to determine the complete time dependence of $f(t)$. In practice, a finite frequency range is sufficient to determine $f(t)$ in the corresponding range of time, although errors arising from the limited range of integration demand that one should have $N + 2$ decades of frequency to obtain N decades of time and vice versa.

Recalling the properties of Fourier transforms, we note from Eq. (13) that $\chi'(\omega)$ is an even function of frequency. The static value of the susceptibility is obtained by setting $\omega = 0$:

$$\chi(0) = \int_0^\infty f(t)\, dt \tag{19}$$

and this corresponds directly to Eq. (10). Similarly, $\chi''(\omega)$ is an odd function of frequency and has therefore a series expansion of the form

$$\chi''(\omega) = \sigma_0/\omega + a\omega + \text{higher odd powers} \tag{20}$$

where the leading term that gives a singularity at the origin arises from a finite value of the DC conductivity and does not therefore belong properly to the dielectric response, but is necessarily present in experimental measurements.

Physical considerations require that both $\chi'(\omega)$ and $\chi''(\omega)$ should vanish at infinitely high frequencies—no physical system can follow such rapid excitations.

Equations (13), (17), and (18) show that the functions $\chi'(\omega)$ and $\chi''(\omega)$ are intimately interrelated so that it is possible in principle to "eliminate" $f(t)$ and express them directly in terms of one another, as Hilbert transforms *(10)*, which are known as the *Kramers–Kronig relations* and may be represented in the following form of double-sided integrals:

$$\chi'(\omega) = \frac{1}{\pi} \int_{-\infty}^{\infty} \frac{\chi''(x)}{x - \omega}\, dx \tag{21}$$

$$\chi''(\omega) = -\frac{1}{\pi} \int_{-\infty}^{\infty} \frac{\chi'(x)}{x - \omega}\, dx \tag{22}$$

where the integrals are taken as the Cauchy principal values, i.e., ignoring any imaginary contribution from the singularity in the integrand. Making use of the parity relations $\chi'(\omega) = -\chi'(-\omega)$ and $\chi''(\omega) = \chi''(-\omega)$, we may transform these integrals to the single-ended form for positive values of ω only:

$$\chi'(\omega) = \frac{2}{\pi} \int_0^\infty \frac{x\chi''(x)}{x^2 - \omega^2} dx \qquad (23)$$

$$\chi''(\omega) = -\frac{2\omega}{\pi} \int_0^\infty \frac{\chi'(x)}{x^2 - \omega^2} dx \qquad (24)$$

The proper derivation of these relations requires the use of the theory of the functions of the complex variable and lies beyond the scope of our review (11), but their practical significance is considerable and their availability as computer programs is invaluable as a check on the consistency of experimental measurements or, more important still, as a means of obtaining one component in a frequency region in which for some reason this component cannot be determined accurately, while the other is measurable without difficulty (12). Thus, if the low-frequency behavior is dominated by DC conductivity [Eq. (20)], it is possible to determine the true dielectric loss unambiguously from the values of $\epsilon'(\omega)$.

The integral relations (21)–(24) remain valid if $\epsilon(\omega)$ is used instead of $\chi(\omega)$, since the Hilbert transform of a constant is identically zero.

One immediate consequence of Eq. (23) is obtained by setting $\omega = 0$:

$$\chi(0) = \frac{2}{\pi} \int_{-\infty}^\infty \chi''(x) \, d(\ln x) \qquad (25)$$

which relates the *polarization increment* arising from a particular mechanism to the area under the loss curve on a logarithmic frequency scale (Fig. 2). This illustrates the important principle that polarization is linked inseparably with loss and any polarization mechanism must therefore be

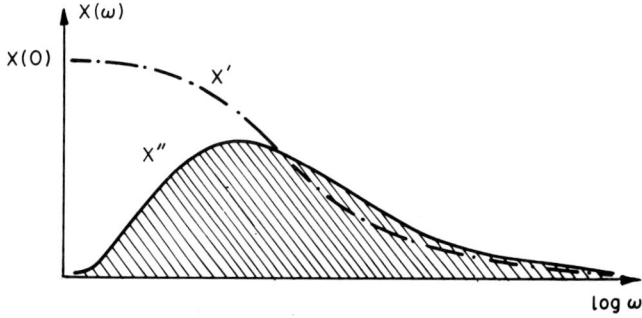

FIG. 2. The significance of the relation between the area of the curve of $\chi''(\omega)$ vs. $\log_e \omega$, represented by the shaded area, and the dielectric increment, Eq. (25).

associated with a loss peak falling at some characteristic frequency ω_p determined by the speed of response of this mechanism.

In any particular material there may exist several polarization mechanisms, each characterized by its own frequency dependence. In these situations the response of the material is obtained by summation of the contributions of the individual mechanisms, labeled here by the index α,

$$\epsilon(\omega) = \epsilon_0 \left\{ 1 + \sum_\alpha [\chi'_\alpha(\omega) - i\chi''_\alpha(\omega)] \right\} \quad (26)$$

and this is shown schematically in Fig. 3. It often happens that the losses of some mechanisms become insignificant in certain frequency ranges below the respective loss peaks, so that the particular $\chi(0)$ values may be added to ϵ_0 to give an effective *high-frequency permittivity* for all the *lower frequency* mechanisms, and Eq. (26) becomes

$$\epsilon(\omega) = \epsilon_\infty + \epsilon_0 \sum_{\alpha \geq \alpha_1} [\chi'_\alpha(\omega) - i\chi''_\alpha(\omega)] \quad (27)$$

where α_1 is now the first mechanism with a nonvanishing loss in the frequency range under consideration. The practical determination of ϵ_∞ may conveniently be performed with the use of the Kramers–Kronig relations on the basis of the loss data.

For all mechanisms of interest in the present investigation the function $\epsilon'(\omega)$ is monotonically decreasing with frequency.

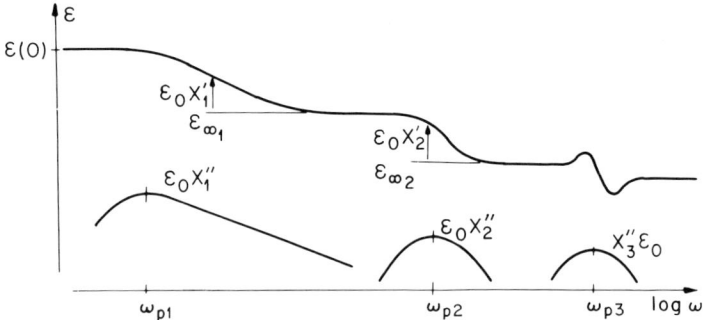

FIG. 3. The relation between the dielectric permittivity $\epsilon'(\omega)$ and the various contributions of the susceptibilities $\chi'_\alpha(\omega)$, with the corresponding dielectric losses $\chi''_\alpha(\omega)$. The respective values of ϵ_∞ are indicated for mechanisms 1 and 2.

Since $\epsilon''(\omega)$ determines the energy lost per radian, the energy loss per unit time, i.e., power loss, is given by the *AC conductivity*

$$\sigma(\omega) = \sigma_0 + \epsilon_0 \omega \chi''(\omega) \tag{28}$$

where we have added the DC conductivity, which gives rise to the singularity in Eq. (20).

III. Methods of Presentation of Dielectric Data (13, 14)

A parallel-plate capacitor of area A and separation w between plates, filled with homogeneous dielectric medium of complex permittivity $\epsilon(\omega)$, has a complex capacitance

$$C(\omega) \equiv C'(\omega) - iC''(\omega) = (A/w)\epsilon(\omega) \tag{29}$$

and the complex *admittance* of this capacitor is

$$Y(\omega) \equiv Y'(\omega) + iY''(\omega) = i\omega C(\omega)$$
$$= \omega C''(\omega) + i\omega C'(\omega) = G(\omega) + i\omega C'(\omega) \tag{30}$$

The equivalent electric circuit of this admittance is therefore a parallel combination of a conductance $G(\omega) = \omega C''(\omega)$, which is a function of frequency and is related to the dielectric loss in the material of the capacitor, and of an ideal, i.e., lossless capacitor $C'(\omega)$, as shown in Fig. 4. If the lossy dispersive capacitor contains in addition a DC conductance

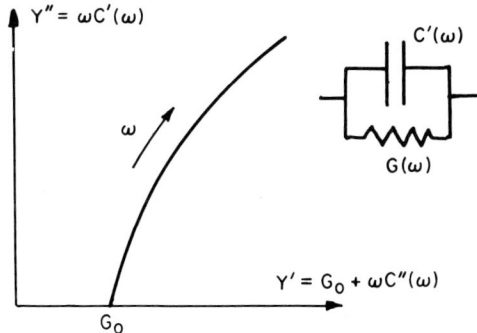

FIG. 4. The use of the complex admittance diagram to separate out the direct current conductance G_0 from the complex AC properties represented by an equivalent parallel circuit.

G_0 due to some internal conduction mechanism in the material, then G_0 may be considered to be in parallel with $G(\omega)$, by analogy with Eq. (28). At $\omega = 0$ the admittance is purely real and is equal to G_0 and it is possible to characterize the response of the dielectric by plotting the complex admittance diagram with the frequency as the implicit variable, as shown in Fig. 4. On the other hand, it may be more instructive to plot $G(\omega)$ as a function of frequency which is entirely sufficient for the characterization of the material, in view of the Kramers–Kronig relations, although it is both helpful and customary to plot $C'(\omega)$ as well.

It often happens that the experimentally unavoidable leads, electrodes, etc., constitute an effective series resistance R_s with the lossy capacitor in question, as in Fig. 5. This may particularly be the case if the capacitor in question is a Schottky barrier in series with the bulk resistance of the semiconducting material on which the barrier was induced by the influence of the electrode or by other interfacial processes (15), or the space charge region of a p–n junction. In either case the presence of the series resistance tends to obscure the true nature of the dielectric response of the material in question. The complex admittance of a series combination may be difficult to visualize readily and the much more appropriate technique is that of plotting the equivalent *complex impedance*

$$Z(\omega) = 1/Y(\omega) = R_s + 1/[i\omega C(\omega)] \tag{31}$$

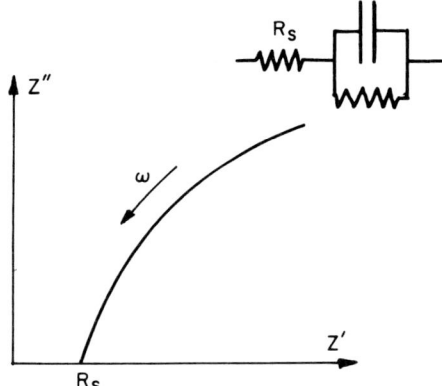

FIG. 5. The use of the complex impedance diagram to determine a series resistance R_s that is independent of frequency and is placed in series with a circuit containing conductance and capacitance, which may themselves be frequency dependent.

which immediately resolves the series resistive element and enables us to isolate the complex capacitance of the other element.

It should be clearly borne in mind that the admittance of the series combination of Eq. (31) may bear little relation to the true admittance of the capacitor itself, the complex capacitance of which has to be obtained by inverting the impedance from which R_s has been subtracted:

$$C(\omega) = \{i\omega[Z(\omega) - R_s]\}^{-1} \tag{32}$$

The proper procedure in evaluating the dielectric data consists, therefore, in the evaluation of the complex impedance diagram in order to ascertain if there is a discernible series element, resistive or capacitive, that has to be subtracted. In the event of the series element being a resistance R_s, the second stage consists in the inversion of $Z(\omega) - R_s$ to obtain the admittance of the capacitor proper and to evaluate any parallel DC conductivity G_0 that may be present, which must be subtracted to obtain the "true dielectric response"

$$C(\omega) = \frac{1}{i\omega}\left(\frac{1}{Z(\omega) - R_s} - G_0\right) = A/w \, \epsilon(\omega) \tag{33}$$

where the complex permittivity $\epsilon(\omega)$ so determined has the singularity σ_0/ω due to the DC conductivity removed from it.

We note, in particular, that a series combination of an ideal, i.e., loss-free capacitor and a frequency-independent resistor has the admittance

$$Y(\omega) = i\omega C_0[1 + i\omega C_0 R_0]^{-1} \tag{34}$$

which corresponds to the ideal Debye response (Section IV). Failure to account correctly for the series resistance leads, therefore, to a spurious high-frequency dependence of the equivalent conductance $G(\omega) \propto \omega^2$ and the corresponding dielectric loss $\chi''(\omega) \propto \omega$, which has often been mistaken for a Debye mechanism.

Yet another experimental artefact that may lead to spurious results is the presence of a series capacitor, e.g., arising from a Schottky barrier, with a parallel combination of a capacitor and conductor representing the "bulk" dielectric whose properties we are investigating. Figure 6 shows the simple case where the two capacitances are both ideal, i.e., dispersion free, and the conductance G_p in parallel with the second is also independent of frequency. On the very reasonable assumption that the "barrier" capacitance is much larger than the volume one, $C_s \gg C_p$, on account of the much smaller thickness involved, and also ignoring any conductance in parallel with C_s, we obtain the characteristic frequency dependence shown in Fig. 6, with the approximate values of the various parts of the

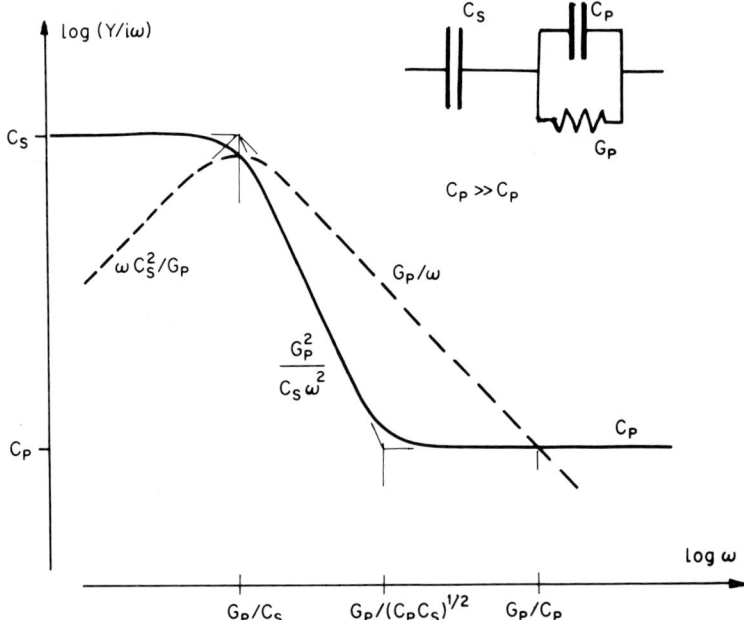

FIG. 6. The frequency dependence of the effective complex permittivity in the presence of a series capacitor C_s representing a barrier with a parallel G_p–C_p circuit representing the bulk material properties. C_s, C_p, and G_p are taken to be independent of frequency. The diagram relates to the case $C_s \gg C_p$.

characteristic. While the detailed formulas are somewhat involved, the important point to note is that the effective permittivity, for the series combination of barrier and volume treated as though it were a uniform dielectric medium, is given by the imaginary component of admittance divided by ω, $\epsilon'_{\text{eff}} = \text{Re}(Y/i\omega) = Y''/\omega$. Similarly, the effective loss is given by $\epsilon''_{\text{eff}} = \text{Im}(Y/i\omega) = Y'/\omega$. Figure 6 shows therefore that the components of the effective permittivity give a region of strong dispersion, in which the loss goes as $1/\omega$ and the permittivity as $1/\omega^2$. The frequency range spanned by the region of strong dispersion and the corresponding variation of the effective parameters may amount to several decades and may, in any given experimental situation, correspond to the total available range of measurements.

This brings us to an important point regarding the presentation and interpretation of dielectric data. By and large, and with one important exception described in Sections V and VI as the strong low-frequency dispersion, the values of $\epsilon'(\omega)$ do not vary by many orders of magnitude. If the experimental data should lead us to such a conclusion, there is a strong presumption of the presence of a series capacitor and it is advisable to check carefully the impedance diagram for any indications to this effect.

A practical problem frequently arising with the interpretation of the complex impedance and admittance diagrams is the rapid variation of the absolute magnitudes with frequency, which makes it difficult to represent on the same diagram the high- and low-frequency components. In such situations it is convenient to use logarithmic coordinate plots (16), which naturally accommodate a wide range of values. An off-set such as R_s in the Z diagram or G_0 in the Y diagram is immediately evident as a steep drop at the relevant value. Figure 7 shows an example of this type of behavior, where a series capacitance due to a strongly voltage-dependent barrier region is present with a voltage-independent, i.e., linear, bulk region. Each of these regions separately would be represented by semicircular Z plots, which in the logarithmic representation have the characteristic shape of a line of slope $\frac{1}{2}$ rapidly bending over downward into a vertical. If the circle is offset from the origin, then there is also a vertical drop at the lower end of the logarithmic representation.

The point should be made that the complex impedance contour is of little use other than as a means of determining the presence of a series resistance, since its shape gives only a very coarse indication of the nature of the dielectric response of the medium. Examples of this are given later. Likewise, the complex admittance diagram has the advantage of separating out the DC conductance, thereby facilitating subsequent analysis of the proper dielectric response.

The only really meaningful manner of presenting dielectric data is the plot of $\epsilon'(\omega)$ and $\epsilon''(\omega)$ vs. frequency, preferably in a logarithmic representation. Alternatively, a plot of $\chi'(\omega)$ may be more informative.

A parameter often quoted, particularly in engineering applications, is the loss angle δ between the $\mathscr{D}$ and $\mathscr{E}$ vectors in the complex plane—this angle gives the measure of the amount of energy loss in the system and it is customary to quote its tangent

$$\tan \delta = \epsilon''(\omega)/\epsilon'(\omega) \tag{35}$$

From the standpoint of the fundamental analysis of the dielectric properties the loss angle is not a useful quantity since it consists of the ratio of two parameters, each of which is of interest in its own right. Only in very low-loss materials, in which $\epsilon'(\omega) \simeq \epsilon_\infty$, does $\tan \delta$ reflect effectively the variation of $\epsilon''(\omega)$.

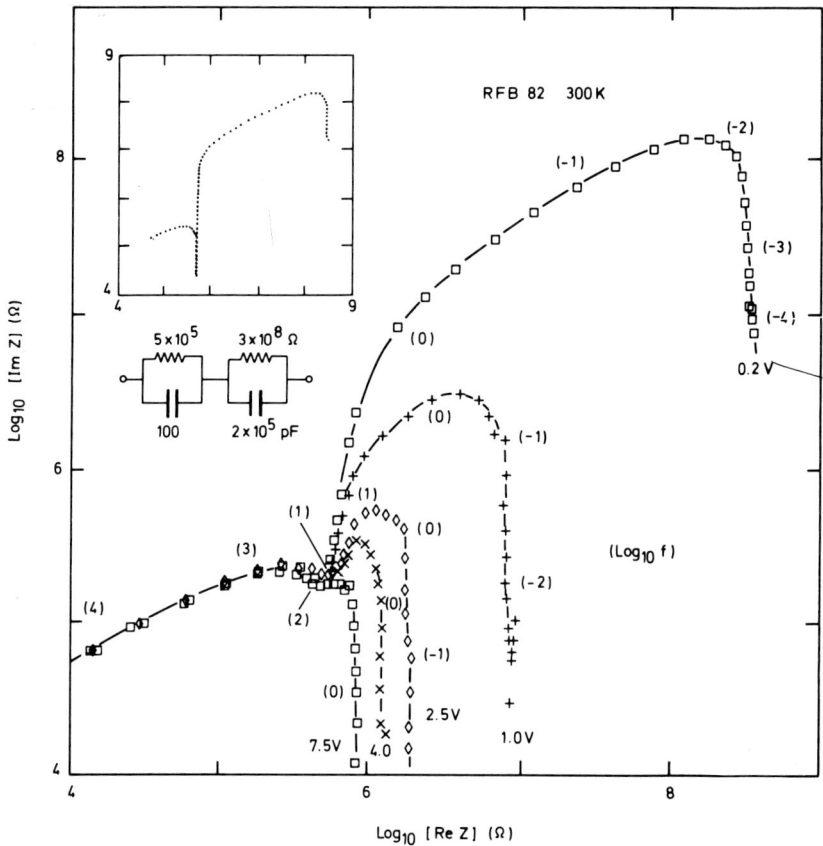

FIG. 7. The logarithmic representation of the complex impedance plot facilitating the display of a large range of values of Z on one diagram. Data refer to a sputtered film of STAG glass with Al electrodes. The series barrier impedance "collapses" as the signal amplitude increases. RMS signal shown in volts, frequencies in hertz. Inset shows the representation of a matching equivalent circuit with the indicated parameters. From Jonscher and Frost (16).

Another manner of representing the dielectric data that is deeply rooted in tradition in some branches of the subject is the complex ϵ plot, known as the *Cole–Cole diagram* (17). This was originally conceived in the context of the Debye response (cf. Section IV) for which it should represent a semicircular arc. Like all complex contours it suffers from the disadvantage that the frequency variable is implicit and this representation is not very useful except as a rapid means of "fingerprinting" dielectric data.

A time domain representation of the dielectric response function $f(t)$ or the discharge current $i_d(t)$ constitutes, in principle, an excellent means of conveying the essential information about the response of a material. The principal limitation lies in the featureless appearance of this type of graph when plotted as log $i_d(t)$ − log t, compared with, for example, the log $\epsilon''(\omega)$ − log ω plots, which tend to show easily recognizable "topographical" features such as the loss peaks. The corresponding feature in the transient current plot is a gentle change of slope that is much less striking to the eye. The time domain graphs do come into their own, however, in situations where nonlinearities arise as a result of, for example, space charge injection (some examples of this are given in Section VI,8).

IV. Basic Mechanisms of Polarization

Since all matter is composed of positive and negative charges, the application of an external electric field gives rise to a slight relative displacement of the centers of gravity of the positive and negative charge distributions and this is the cause of polarization. The so-called *electronic polarization* arising from the relative displacement of the valence shell electrons with respect to the atomic nuclei is so rapid, of the order of 10^{-15} sec, that for our purposes the response is instantaneous and contributes to ϵ_∞. The same is true of *ionic polarization*, which is due to the relative displacements of positive and negative ions in ionic or partially ionic lattices with response times of the order of 10^{-13} sec. As far as we are concerned both these mechanisms simply contribute to ϵ_∞, and this is relatively insensitive to temperature because the mechanisms in question do not involve thermal excitation (*18*). This behavior is contrasted with dipolar behavior below.

The archetypal dielectric response is represented by the *Debye mechanism*, which may correspond to any one of the following three physical situations: (1) a freely floating isolated inertialess molecular dipole in a viscous medium (*19*); (2) an isolated charge oscillating thermally between two preferred localized sites represented by a potential double well (Fig. 8); (3) the analogous case of an inertialess dipole constrained to assume one of two discrete orientations in space. A fourth, essentially spurious situation in which Debye response is obtained is the series $R-C$ combination discussed in Section III.

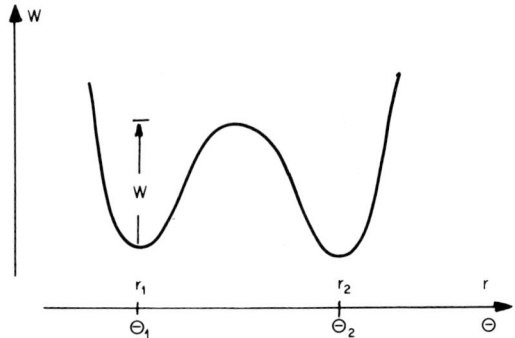

FIG. 8. The potential double well for angular orientation of a dipole with preferred orientations Θ_1 and Θ_2, and the corresponding diagram for transitions of a point charge between preferred positions r_1 and r_2. W is the activation energy of the process.

Quite generally, the Debye response is obtained in all situations in which the decay of polarization from an initial value after sudden removal of the exciting field follows the simple first-order rate equation with a time constant τ:

$$dP/dt = -P/\tau \tag{36}$$

which has the solution $P(t) = P_0 \exp(-t/\tau)$, where P_0 is the initial value. This gives the time domain response function

$$f(t) = (P_0/\tau) \exp(-t/\tau) \tag{37}$$

The Fourier transform of this function gives the frequency domain expression for the complex susceptibility from Eq. (12):

$$\chi(\omega) = \frac{\chi(0)}{1 + i\omega\tau} = \chi(0) \left(\frac{1}{1 + \omega^2\tau^2} - i \frac{\omega\tau}{1 + \omega^2\tau^2} \right) \tag{38}$$

The time constant τ is known as the *Debye relaxation time* and is related to the viscosity of the medium in case (1), is equal to the reciprocal natural frequency of oscillation in cases (2) and (3), and is equal to the product RC in the series resistance–capacitance case. In general, τ is strongly temperature dependent and usually has a clearly defined activation energy W:

$$\tau(T) = 1/\omega_p = \tau_\infty \exp(+W/kT) \tag{39}$$

where ω_p denotes the loss peak frequency and τ_∞ is a suitable preexponential factor. W is the height of the barrier in the case of the double well, is the activation energy of viscosity in case (1), and may be related to the activation energy of the series resistance R.

A more detailed analysis of rotating dipoles and hopping charge carriers reveals the temperature dependence known as the *Curie law:*

$$\chi \propto 1/T \tag{40}$$

contrasting strongly with the nearly temperature-independent lattice polarization.

The functional form of Eq. (38) is shown schematically in Fig. 9 in the log–log representation. The loss peak is symmetric in log ω, its width at half-height being given by

$$\lambda_D = 1.144 \quad \text{decades} \tag{41}$$

We note that peaks corresponding to different temperatures are displaced along the frequency axis by amounts corresponding to the temperature dependence of $\tau(T)$, while the peak amplitude may vary somewhat, but usually much less than exponentially.

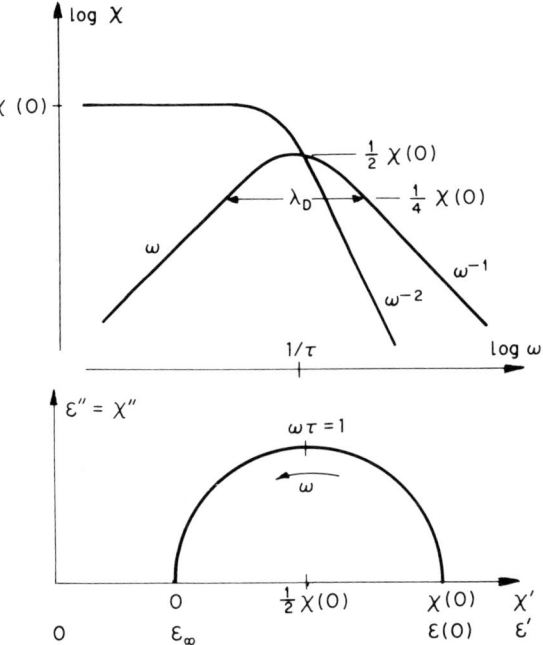

FIG. 9. The frequency dependence of the real and imaginary components of the complex susceptibility corresponding to an ideal Debye system, indicating the characteristic width at half-height, $\lambda_D = 1.144$ decades. The lower diagram shows the corresponding complex χ diagram and the change of coordinates to convert to the complex ϵ diagram.

It is possible to eliminate ω between the real and imaginary components of ω in Eq. (38):

$$\chi'(\omega) = \chi(0)/(1 + \omega^2\tau^2), \qquad \chi''(\omega) = \chi(0)\omega\tau/(1 + \omega^2\tau^2) \qquad (42)$$

and to obtain a direct relation between them:

$$\chi'^2 - \chi(0)\chi' + \chi''^2 = 0 \qquad (43)$$

which corresponds to a semicircle passing through the origin in the complex χ plane (Fig. 9). By adding the value ϵ_∞ we convert this graph to that of complex ϵ, which is the classical Cole–Cole diagram.

Apart from permanent dipoles, any charges present in the dielectric medium must, in principle, contribute to dielectric polarization. Taking first the limiting case of *free carriers in metals or in crystalline semiconductors*, we may define a *mean free time between collisions* τ_c and the related *mobility* $\mu = q\tau_c/m^*$, where q is the charge and m^* their effective mass. The AC electrical conductivity in this model becomes

$$\sigma(\omega) = \sigma_0/(1 + \omega^2\tau_c^2) \qquad (44)$$

where σ_0 is the DC conductivity. This leads to a *negative* dielectric loss using Eq. (28):

$$\chi''(\omega) = -\frac{\sigma_0}{\epsilon_0}\frac{\omega\tau_c^2}{1 + \omega^2\tau_c^2} \qquad (45)$$

Since typically $\tau_c \le 10^{-14}$ and since the space charge relaxation time $\tau_s = \epsilon_0/\sigma_0 \simeq 10^{-11}/\sigma_0$ is generally much larger than τ_c in not highly conducting materials, this expression becomes

$$\chi''(\omega) \simeq -(\tau_c/\tau_s)\omega\tau_c \ll 1 \qquad (46)$$

up to frequencies in the short microwave region. This means that the free carriers effectively contribute only to the DC conductivity and have a negligible influence on the dielectric response proper in nonmetallic solids. In metals, of course, this is no longer the case, in view of the much larger value of σ_0 and the effective permittivity may become negative.

By contrast with free carriers, *localized* or *bound* carriers behave like Debye dipoles, at least to the extent to which they may be considered to be independent of one another, as already mentioned above (*20, 21*).

V. Empirical Classification of Loss Characteristics

The frequency response of real dielectrics appears to be very complicated, especially when viewed across the entire spectrum of materials, but we are now in the position to outline certain general types of behavior into which it is possible to divide most known spectra. The frequency dependence is the most fundamental characteristic and in this respect the imaginary component $\epsilon''(\omega)$, i.e, the dielectric loss, is the more sensitive indicator, being an odd function of frequency and shows more features than the even function $\epsilon'(\omega)$. We shall be referring exclusively to the log $\epsilon''(\omega)$ − log ω representation. In outlining the various types of response we also give the corresponding complex χ Cole–Cole diagrams, which are very familiar to many people.

1. Complex Susceptibility Representations

We have already described the pure Debye response resulting from identical but noninteracting dipoles or charges [Eq. (38); Figs. 8 and 9]. Although this mechanism has dominated the philosophy of dielectric interpretations for over half a century, it is hardly ever seen in practice either in solids or in liquids, which is not surprising in view of the fact that they must involve strong many-body interactions, i.e., the very mechanism not taken into account in the Debye model.

The closest approximation to the archetypal Debye response found in some dielectric systems is a slightly broadened loss peak with the width parameter in the range $1.1 \leq \lambda/\lambda_\mathrm{D} \leq 2.0$, which may be described purely formally by the Cole–Cole expression (*17*)

$$\chi(\omega) = \chi(0)/[1 + (i\omega\tau)^{1-k}] \qquad (47)$$

This expression signifies the *mathematical operation* of rotating the semicircle through the angle $k\pi/2$, and the resulting λ parameter may be calculated as follows for a typical range of values of k:

k	$k\pi/2$	λ
0	0	1.144
0.05	4.5	1.244
0.1	9.0	1.354
0.2	18.0	1.606
0.3	27.0	1.919
0.4	36.0	2.323
0.5	45.0	2.872

This situation is represented in Fig. 10.

Experimentally one seldom finds peaks with λ values in excess of 1.4–1.6, i.e., with $k > 0.1$–0.2, that are symmetric in log ω. A much more common departure from the simple Debye form of loss is a peak that is

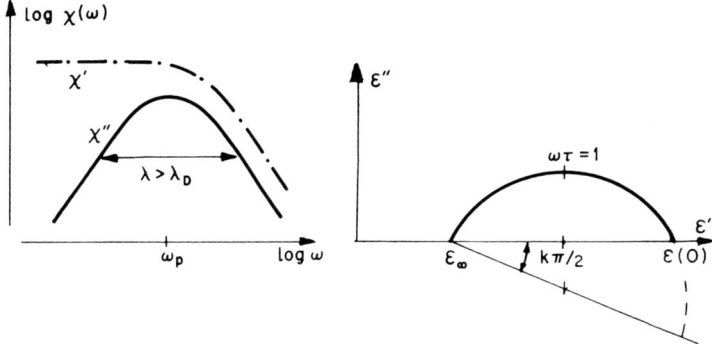

FIG. 10. The frequency dependence of the real part (—·—) and the imaginary part (—) of the complex susceptibility $\chi(\omega)$ corresponding to the Cole–Cole expression (47). The diagram on the right gives the corresponding complex ϵ representation, which is given by a circular arc tilted by an angle $k\pi/2$.

asymmetric in log ω and may be characterized by a wider range of λ parameters. This corresponds in the complex χ plane to a pear-shaped arc, as shown in Fig. 11, with the following formal expression suggested by Davidson and Cole (22):

$$\chi(\omega) = \chi(0)/(1 + i\omega\tau)^{1-n}, \qquad \text{with } n < 1 \qquad (48)$$

With these two purely formal functional representations approximating the empirical response of dielectric materials, it was inevitable that they

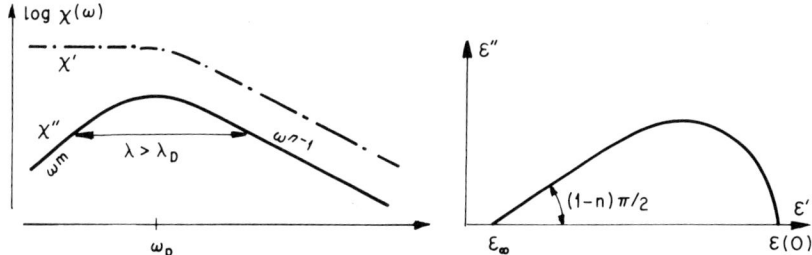

FIG. 11. The frequency dependence of the real part (—·—) and the imaginary part (—) of the complex susceptibility $\chi(\omega)$ corresponding to the Cole–Davidson formula (48). The diagram on the right shows the complex ϵ representation with the exponent n determining the slope of the high-frequency part in both diagrams. This form actually corresponds to the value of the exponent $m = 1$, which is only seldom found experimentally, so that a more general expression is required, which is described in Section VIII,6.

should become combined to give an expression with two parameters n and k, giving a correspondingly wider scope for fitting of experimental results (23). Similarly, considerable mathematical ingenuity went into developing purely formal expressions for the dielectric response functions $f(t)$ (24, 25), but all these represent essentially formal approaches, bearing no relation to the actual physical situation in hand and we do not pursue this path any further.

2. The Loss Peaks

The asymmetric loss peaks of the type discussed above, with $\lambda \leq 2$, are frequently found in liquids and also in materials showing the presence of a glass transition at a temperature T_g, at temperatures in excess of T_g. These peaks are known as the α peaks, by contrast with the so-called β peaks, which are generally much broader and occur in "solidlike" situations, in particular, in glassy materials below the glass transition temperature. Where α and β occur in the same material, there is a basic difference between them in the temperature dependence of $\omega_p(T)$, as discussed in Section VI.

3. The Universal Dielectric Response

We now note that the high-frequency limit of both expressions (47) and (48) is of the form

$$\chi(\omega) \propto (i\omega)^{n-1} = [\sin(n\pi/2) - i\cos(n\pi/2)]\omega^{n-1} \qquad (49)$$

which is shown schematically in Fig. 11 and which we shall refer to as the *universal law of dielectric response*. The characteristic feature of this law of frequency dependence is the fact that the real and the imaginary components of the complex susceptibility are the same functions of frequency, so that their ratio is frequency independent:

$$\chi''(\omega)/\chi'(\omega) = \cot(n\pi/2) \qquad (50)$$

This is consistent with the fact that the high-frequency branch of the complex χ plot is a straight line and is also a direct consequence of the Kramers–Kronig relations—the "universal" function ω^{n-1} has the unique property that it remains invariant, except for a constant, under the Hilbert transformation. The constancy of this ratio is in complete contrast with the Debye response, for which it is equal to $\omega\tau$.

Now Eq. (50) has the simple physical significance that the ratio of the energy lost per radian to the energy stored in the system at the peak of the polarization is independent of frequency:

$$\frac{\chi''(\omega)}{\chi'(\omega)} = \frac{\text{energy lost per radian}}{\text{energy stored}} = \cot\frac{n\pi}{2} \tag{51}$$

4. Time Domain Response

It is possible to approximate the experimentally observed loss peaks in many dipolar systems by the expression (26)

$$\chi''(\omega) \propto [(\omega/\omega_p)^{-m} + (\omega/\omega_p)^{1-n}]^{-1} \tag{52}$$

where the exponents m and n are both smaller than unity. This expression suggests that the pre- and postpeak portions of the loss characteristic are power law relations in frequency and this appears to be reasonably well borne out by experiment. It is now instructive to carry out the Fourier transformation of expression (52) to obtain the time domain response function in the approximate form

$$f(t) \propto [(\omega_p t)^n + (\omega_p t)^{m+1}]^{-1} \tag{53}$$

This expression has the advantage of bringing out the physically significant feature that the time domain response is made up of two sequential decay processes (27). The first one obeys the relation

$$f(t) \propto t^{-n}, \quad \text{with} \quad n < 1 \tag{54}$$

which applies at short times and corresponds to the high-frequency part of the response (52), and is known as the Curie–von Schweidler law (28–30), while the second is of similar functional form but has logarithmic slope steeper than unity. This is shown schematically in Fig. 12; its physical significance is discussed in Section VIII, 2.

It should be noted that the various analytical expressions for $\chi(\omega)$ and $f(t)$ quoted above are not necessarily valid in the entire range of frequency or time from zero to infinity. This is sometimes raised as an objection to their use altogether in the interpretation of dielectric response data, on the grounds that they are "unphysical" (31, 32). We suggest that this criticism is not valid—it is not necessary for a given physical law to apply in the entire range of the physical variable involved, but it is perfectly legitimate to use a certain approximation within one range and to approximate the solution by some other function in a different range. A more important

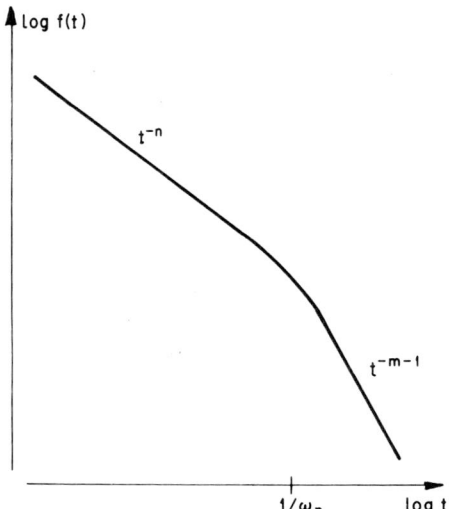

FIG. 12. The time domain response of the discharging current on a sudden removal of a constant field, which defines the response function $f(t)$ and corresponds to the universal relation t^{-n}, followed by a steeper region at longer times. This form of $f(t)$ is given by Eq. (53), and its Fourier transform gives a peak of dielectric loss at a frequency ω_p defined by the time at which the logarithmic slope of $f(t)$ goes through the value of -1.

limitation is set by the fact that even mathematically "correct" functions cannot account for the onset of completely new physical phenomena at the extremities of the range, in particular for optical and quantum processes.

5. Response of Charge Carriers

The dielectric responses described thus far are essentially limited to materials in which the dominant polarizing species are dipoles, especially the permanent ones. Dipoles as such cannot give rise to DC conductivity and this enables the low-frequency prepeak loss response to be measured. If, in addition to the dipoles, there are also present some mobile charge carriers giving rise to DC conductivity, the corresponding loss term σ_o/ω eventually becomes dominant and has to be suitably subtracted from the measured loss, as described in Section III, but the remaining response is then essentially dipolar in nature.

An important new form of dielectric response is obtained when hopping-charge carriers, electrons, ions, or polarons become the dominant polarizing species. It is very common in this situation not to find any loss peak at all down to the lowest frequencies, or at least to the onset of DC conduction, although quite often a second "universal" power law may appear, giving a response of the form

$$\chi''(\omega) = (\omega/\omega_c)^{n_1-1} + (\omega/\omega_c)^{n_2-1} \qquad (55)$$

with n_1 and n_2 representing the two slopes shown in Fig. 13 and ω_c being a characteristic frequency.

The widespread occurrence of the power law (49) and (55) and the correspondingly common Curie–von Schweidler law in the time domain (54) in all types of solids (8, 33–35) has led to the name "universal" dielectric response being proposed, with a typical range of values of the exponent n, or of the higher frequency value n_2, between 0.6 and 0.95.

6. Extreme Values of the Exponent n

We now come to two limiting examples of the universal law, corresponding to values of $n \simeq 1$ at high frequencies and values $0 < n < 0.3$, say, at the lower frequencies, respectively. The former gives an essentially frequency-independent or "flat" loss and is commonly observed in low-loss materials, in accordance with the Kramers–Kronig relations (50). This behavior is later identified with the response of the dielectric

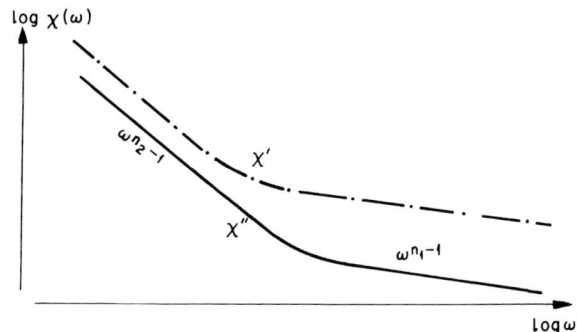

Fig. 13. Schematic representation of the dielectric response as found in many systems in which the dominant contribution to polarization comes from hopping-charge carriers of electronic or ionic nature. The response represents the summation of two universal laws with different values of the exponents n, the lower frequency one being smaller. The real part $\chi'(\omega)$ is parallel to the imaginary part $\chi''(\omega)$ and their ratio is related to the slope n through the universal relation (51).

lattice or *matrix* as distinct from *randomly distributed* dipoles or localized charge carriers giving rise to the values of n in the range 0.6–0.95.

The latter form of low-frequency response corresponds to *both ϵ' and ϵ'' rising steeply* toward low frequencies, as shown in Fig. 14, a behavior not properly documented in the past and loosely labeled as the Maxwell–Wagner response. We have established that this is a proper limiting case of the universal behavior and we discuss it in more detail in Section VII within the general framework of the many-body formalism. It should not be confused with the effect of a series capacitor, discussed in Section III, nor with the presence of a DC conductivity that gives rise to a steeply rising loss with a *constant ϵ'*.

This completes the general characterization of the various types of dielectric response that we have given here in order to facilitate the appraisal by the reader of the experimental results given in Section VI. It only remains now to briefly mention the effects of temperature.

7. The Temperature Dependence of Dielectric Response

It is a very general property of all dielectric response that a change of temperature causes predominantly a *lateral shift in frequency*, with

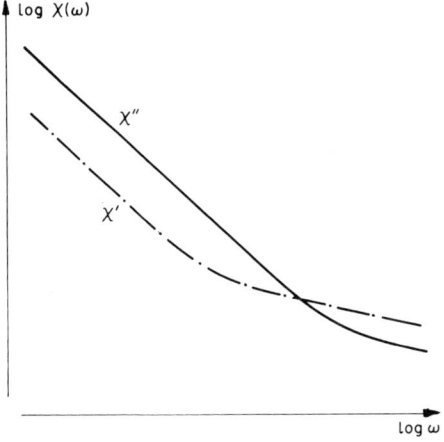

Fig. 14. The anomalous low-frequency dispersion of a dielectric system with a large carrier density. Both the real and the imaginary components of the complex susceptibility show a strong dispersion according to the universal law with a small value of the exponent n. At higher frequencies a second universal region sets in with a larger value of n. Note the crossing over of the real and imaginary components.

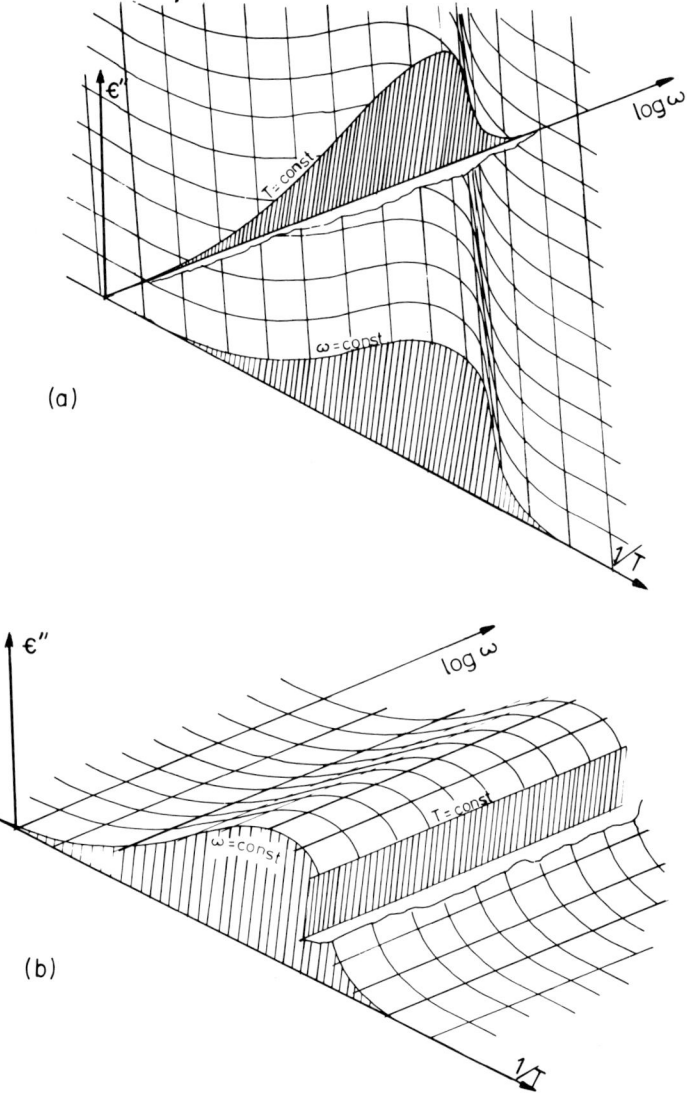

FIG. 15. A schematic diagram of the three-dimensional relation $\epsilon''(\omega, 1/T)$ for the limiting case of an ideal Debye response (a) and for the temperature-dependent but frequency-independent "non-Debye" response (b). The measurement of the temperature dependence at a constant frequency would give very similar results in both cases, from which it could be inferred mistakenly that the material is Debye-like in each case. From Jonscher (*14*).

relatively much less significant change of the amplitude of the response. Likewise, the slopes of the various parts of the loss characteristic, whether involving a loss peak or not, are relatively little affected. The general conclusion is that dielectric processes are determined by *rates of adjustment* to the prevailing field, which are slowed down or accelerated by temperature or, in appropriate circumstances, by pressure or humidity. The technique of bringing into coincidence the $\chi'(\omega)$ and $\chi''(\omega)$ data corresponding to different temperatures by shifting them laterally in the log χ–log ω representation is known as *normalization* and constitutes a powerful means of analyzing experimental results (*36*) giving the values of the activation energies involved in the rate processes and also providing a significant increase in the credibility of the data, since more points become available for the determination of empirical laws.

In the case of the Debye response, taking Eqs. (38) and (39) it is evident that the complex χ and its components χ' and χ'' are functions of a single variable

$$\xi = \omega \exp(W/kT) \tag{56}$$

such that the plot of $\chi(\omega,T)$ vs. ln ω is identical to that vs. W/kT (Fig. 15a). Thus, for the simple Debye case, a complete characterization of the dielectric response is obtained by taking *either* the temperature dependence at constant frequency *or* the frequency dependence at constant temperature. The former procedure is very commonly employed for rapid assessment of materials since it is generally much simpler experimentally. It is shown, however, that in cases where the behavior departs seriously from the Debye response, which coresonds to the majority of materials, temperature dependence ceases to be a reliable guide to the frequency response and it becomes essential to measure the complete dependence $\chi(\omega, T)$ in function of two independent variables. It often happens that temperature dependence shows well-defined loss peaks while frequency response remains substantially "flat," i.e., "non-Debye," as shown schematically in Fig. 15b).

VI. Survey of Experimental Data

We now proceed to review a representative selection of the available experimental evidence, arranged according to the principal categories described in Section V. This procedure tends to "mix" very different materials and may therefore appear to be confusing, but we suggest that it is, in fact, preferable to the more common procedure of grouping together related materials. The classical procedure therefore tends to lose whatever common features may exist between the responses of different materials. The evident existence of such common features, which is demonstrated in the present review, leads to a much better insight into the physical nature of dielectric response. The principal point of this chapter is that the similarity of the dielectric response of most solids, or even of condensed matter as a whole, renders very improbable any particular mechanisms proposed for the interpretation of this or that particular result and calls for a radical reappraisal of our theoretical approach.

1. The Debye Response

We have already stated that there are very few examples of pure Debye behavior in solids—ice (37) and tricyclohexyl carbinol (38) may be mentioned as coming very close to it and it is interesting to note that in both these cases polarization is thought to arise from the motion of hydroxyl groups. As neither of these falls into the category of solids of principal interest here we do not dwell on them further. Further discussion of slight departures from the Debye relations will be found in Jonscher (104).

While discussing the Debye response, it is worth returning to the question of the spurious effect of a series resistance mentioned in Sections III and IV. This frequently leads to a mistaken interpretation of hopping-charge polarization as two-center hopping and it is worth noting that if a Debye-like response is suspected, its loss peak frequency should be examined critically to check its plausibility in the given experimental context (13). Very often the values obtained are much too low to be compatible with any simple one-particle excitation, a point discussed further in Section VIII,2.

2. Broadened Debye Response with Symmetric Peaks

The most notable examples of this type of response may be found in the high-frequency (gigahertz region) dispersion of ferroelectrics. Here the complex ϵ plot offers a sensitive means of determining the departures from the Debye response by the extent of tilting of the circular arc. Figure 16 gives some data for triglycine sulfate (39). A similar response is found also in liquid crystals parallel to the orientation of the molecules (40) (cf. Fig. 23).

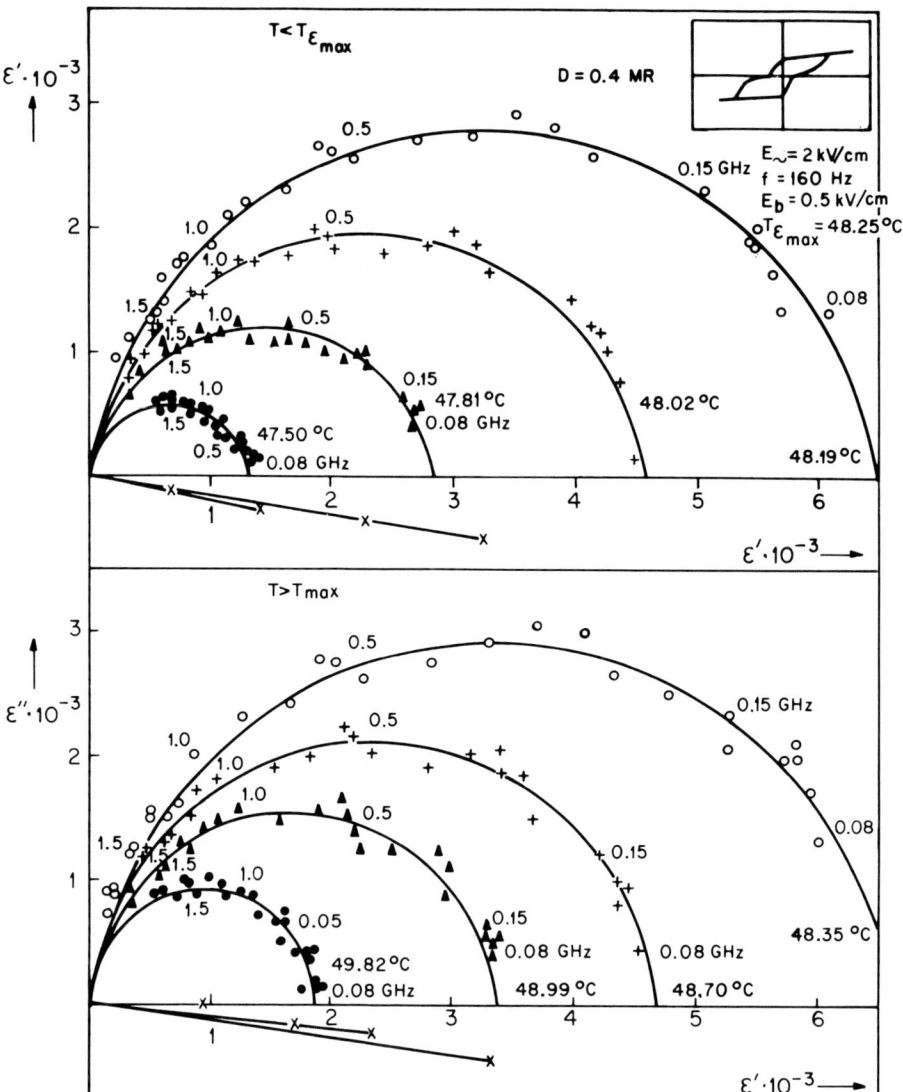

FIG. 16. Complex permittivity diagrams for triglycine sulfate (TGS) below and above the curie temperature. The frequencies indicated are in GHz. The inclination of the lines of centers of the circular arcs gives a measure of departure from the ideal Debye response. From Pawlaczyk (*39*).

3. Asymmetric α Peaks

This form of response differs from the preceding one in the asymmetry of the loss peak—its width may be comparable in both cases. The most notable examples of this type of behavior are to be found in the polar polymers above T_g (*26, 36*) and also in many glasses (*41, 42*). Figure 17 shows the α peak in polydian carbonate with the corresponding normalization, giving $\lambda = 3.2$ (*43*). The asymmetry is clearly visible and the normalization appears to give a single generating curve for all temperatures, indicating the dominance of a single well-defined mechanism. The locus of a reference point is shifted purely horizontally—the amplitude of the loss peak remains strictly constant at all temperatures, again showing that only the *rate process* is affected by temperature. There exists some uncertainty at the lower frequencies in view of the need to subtract the

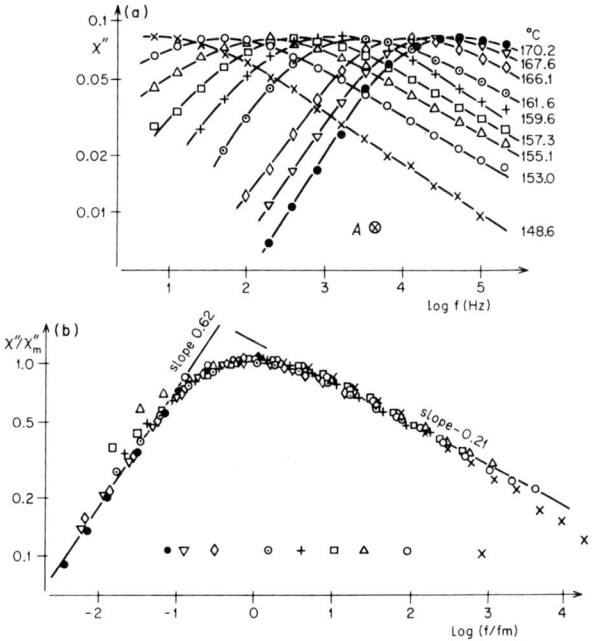

FIG. 17. Dielectric loss peaks for the α relaxation in polydian carbonate for a range of temperatures and the corresponding normalization. All experimental points with the exception of the lowest temperature fall on one generating curve and there is practically no variation of the amplitude with temperature. Original data from Ishida and Matsuoka (*44*), values quoted from Jonscher (*26*).

increasingly important DC contribution. Figure 18 shows the same type of result for polyvinyl acetate (44), which has a sharper peak, λ = 2.0, but there is evidence of a second mechanism setting in at high frequencies. The peak amplitude again remains invariant with temperature. A more complex behavior is seen in Fig. 19 relating to poly-n-butyl methacrylate

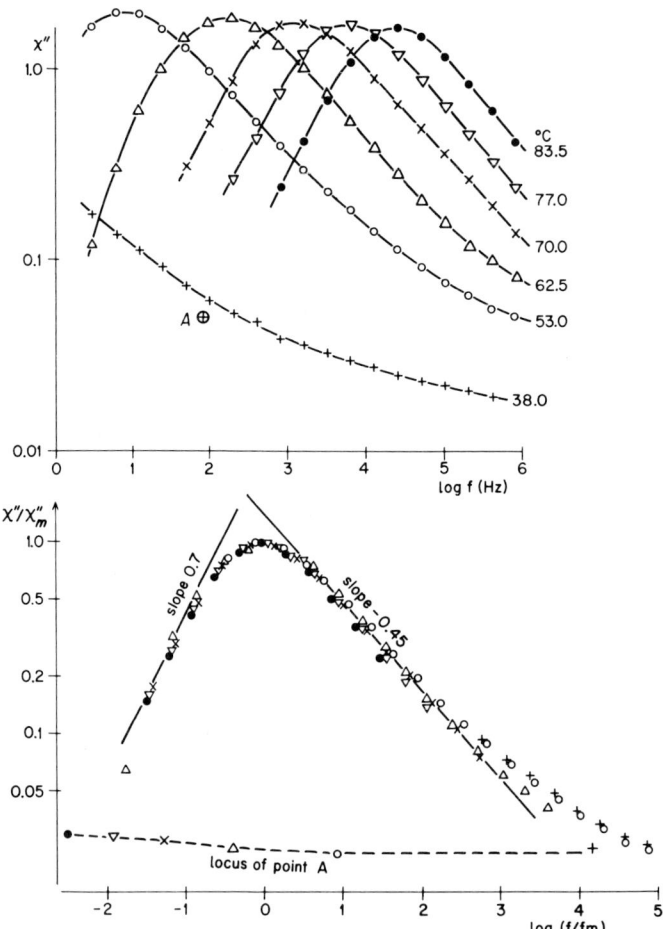

FIG. 18. Dielectric loss peaks for polyvinyl acetate [after Ishida *et al.* (44)] with the corresponding normalization. The normalized plot suggests the existence of a second mechanism at the high-frequency and low-temperature end of the spectrum. Figure taken from Jonscher (26).

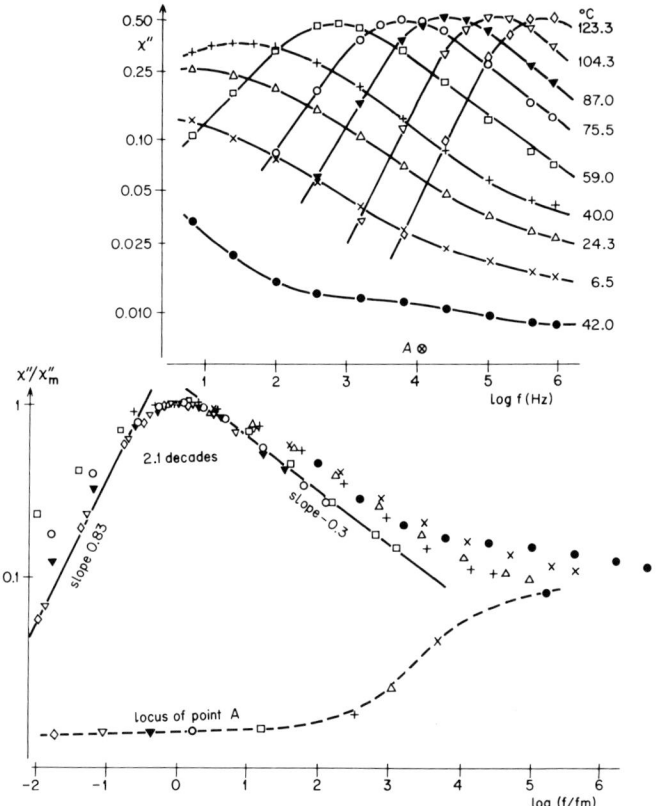

FIG. 19. Dielectric loss peaks (a) and the normalized data (b) for the α relaxation in poly-*n*-butyl methacrylate (after Ishida, *45*). Apart from the main peak of width 2.1 decades, there is clear evidence of a second mechanism at the higher frequency and at lower temperatures, having a different temperature dependence. From Jonscher (*26*).

(*45*), where the results corresponding to lower temperatures again indicate the onset of a different mechanism. The activation plots for the α peaks do not, in general, give straight lines but are convex upward as shown in Fig. 20 and cannot be interpreted as the summation of two activated mechanisms. This has been interpreted as the result of many-body interactions giving rise to these peaks (*46*).

The behavior of glasses shows very similar features to that of polymers—the α peaks may be seen above T_g (an example is shown in Fig.

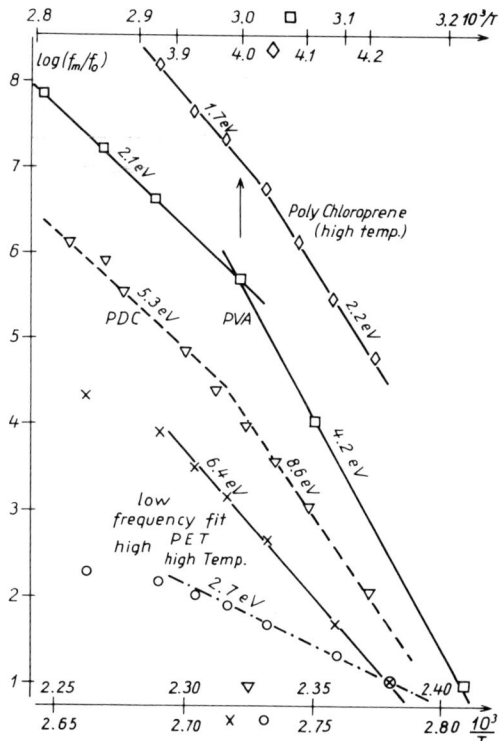

FIG. 20. Activation energy diagrams for the peak frequency $f_m = \omega_p/2\pi$ (with an arbitrary scaling factor f_o) for loss peaks in the α category. The various reciprocal temperature scales refer to the symbols indicated. The straight lines drawn should be taken as tangents to continuous curves, rather as in Fig. 25. The relevant energies are indicated in electron volts. From Jonscher (26).

21). An interesting and unexpected example of the α-like peaks is found at cryogenic temperatures in polar polymers or in nonpolar ones with polar impurities (47–50). Figure 22 shows the data for slightly oxidized samples of polyethylene, both as families of curves and as normalized graphs—we stress that we use the term α peak only as characterizing the frequency response and there is no suggestion of any similarity of the basic mechanism with the behavior of glasses and polymers above T_g. The surprising fact here is the very large value of the dielectric loss at temperatures at which any form of activated motion of dipoles or charges

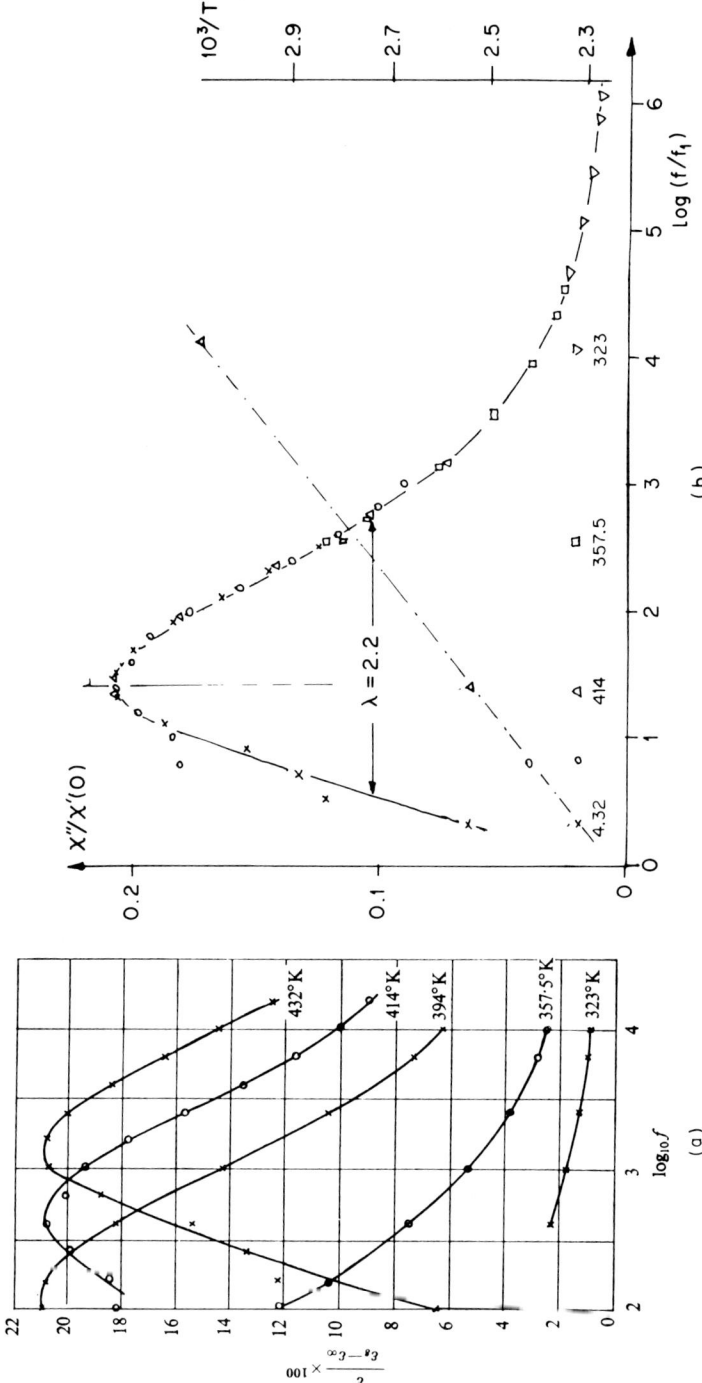

Fig. 21. The α loss peak in a Na_2O–SiO_2 glass at a range of temperatures (a) and the normalization of these data showing the noticeable asymmetry of the peak, which has a width $\lambda = 2.2$ decades (b). The locus of a representative point is shown and the dot–dash line gives the dependence of the characteristic frequency shift on reciprocal temperature showing a single activation energy. The long "tail" in the linear representation should be noted, which most likely corresponds to the universal law. From Taylor (41).

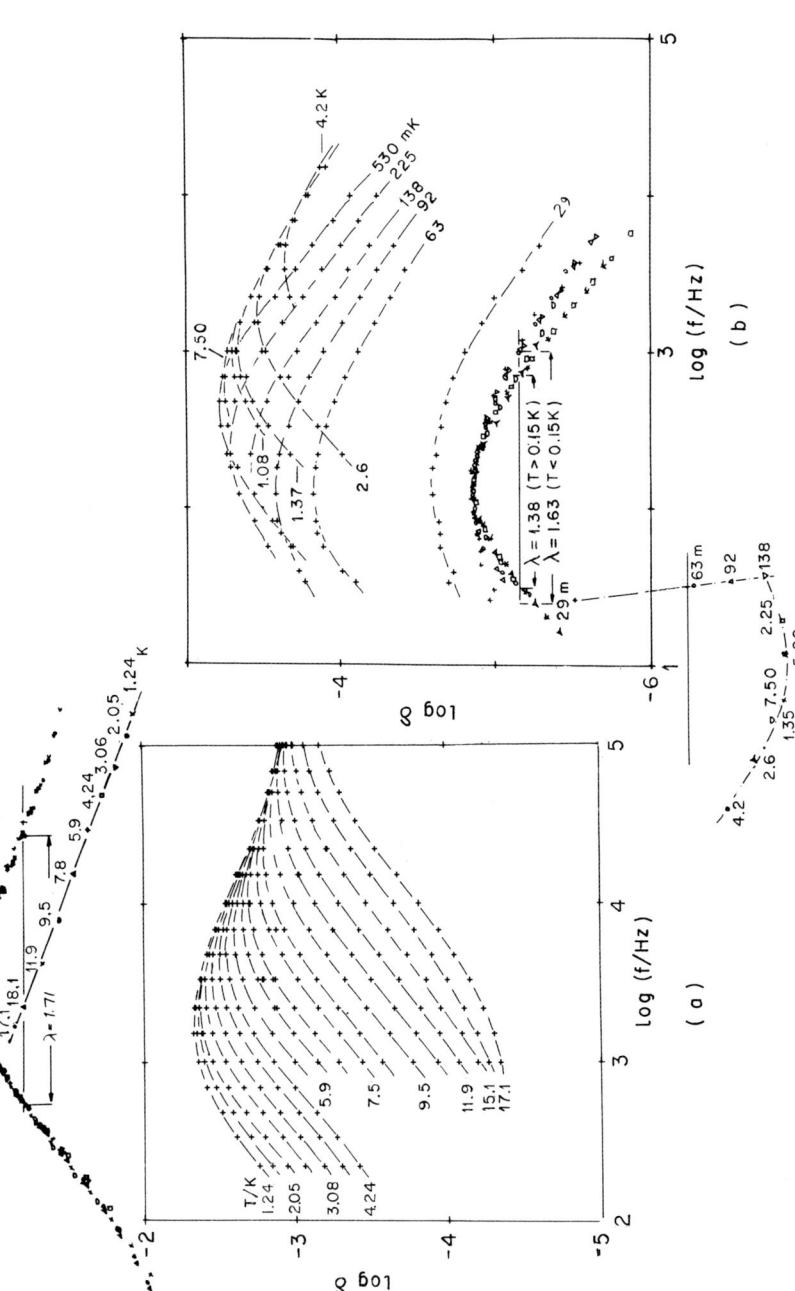

FIG. 22. The dielectric loss peaks for two slightly oxidized polyethylene samples at cryogenic temperatures, based on numerical data from measurements by Gilchrist [(47) for (a), (48) for (b)]. The normalization of the data is also given, together with the corresponding loci of a representative point. (a) A steady trend, indicating a single mechanism in the entire frequency range 1.24–17.1 K, with the amplitude of the loss peaks *increasing* and the frequency ω_p decreasing with decreasing temperature. (b) A similar trend at higher temperatures, which becomes reversed below 150 mK. The loss peak frequencies do not follow the classical Arrhenius activation relation, Eq. (39), which is found at higher temperatures.

243

may be expected to be frozen out. The generally accepted interpretation is in terms of proton tunneling between hydrogen bonds (*50*), but it is clear that many-body phenomena have to enter into play here, or else the frequency response would be purely Debye-like. The characteristic feature of these cryogenic responses is the complete absence of any DC conduction, which makes it much easier to study the low-frequency part of the response.

Distinctly asymmetric peaks are seen in liquid crystals in the direction normal to the orientation of the molecules (*40*) (Fig. 23). These should be contrasted with the behavior parallel to the orientation of the molecules also shown.

Dielectric loss in p–n junctions in single-crystal silicon shows the presence of peaks strongly resembling the dipolar peaks in polymeric materials (*51*) (Fig. 24). We are looking here at the response of a thin depletion region, typically 0.1 μm between bulk extrinsic n- and p-type regions. More recent results on a range of p–n junctions will be found in Ref. (*105*).

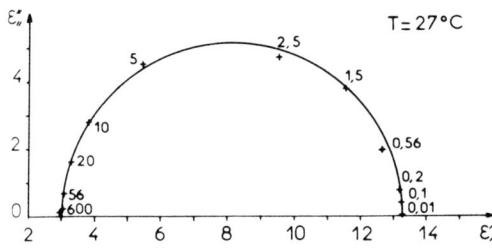

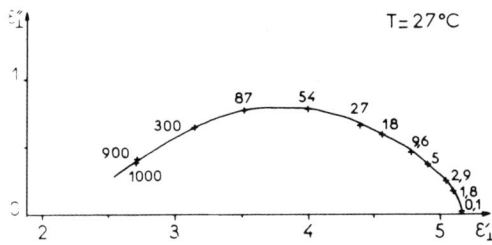

FIG. 23. The complex impedance diagrams of the smectic phase of a liquid crystal, with frequencies given in megahertz. The permittivity parallel to the orientation of the molecules gives a classical Debye plot, characteristic of ferroelectrics at high frequencies. The orientation normal to the molecules gives a typical broadened asymmetric plot. From Druon and Wacrenier (*40*).

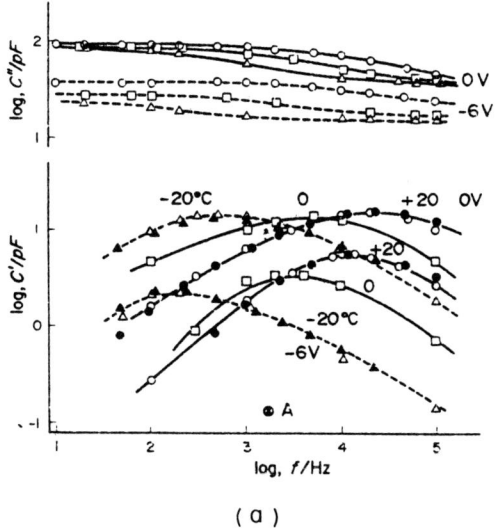

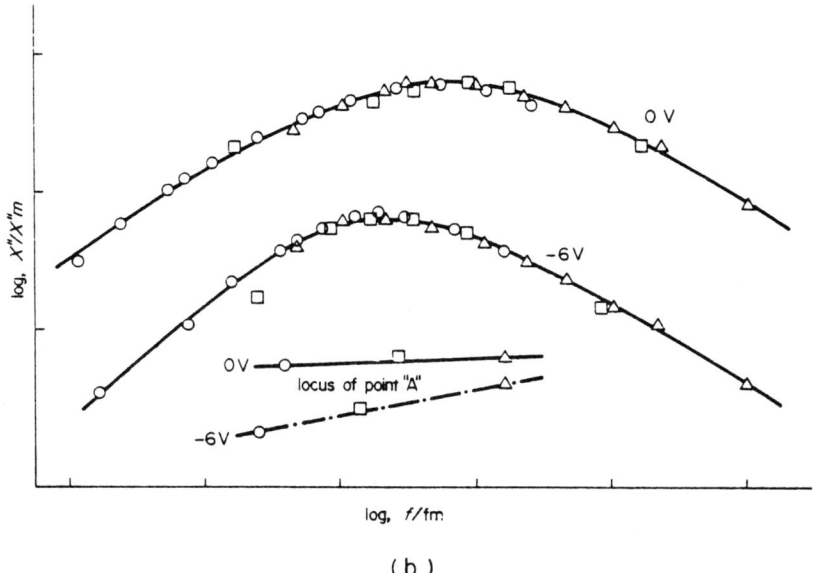

FIG. 24. (a) The dielectric response in the form of the frequency dependence of the complex capacitance of a gold-doped silicon power rectifier p–n junction. The data were taken at temperatures −20, 0, and 20°C and at biases 0 and −6 V. The open points in the loss data are measured values after subtraction of a suitable DC conductance contribution; the solid points were obtained as a check by Kramers–Kronig transformation of the C' data. (b) Normalization of the above data, separately for each of the two values of bias. The width parameter is $\lambda = 2.0$. From Barsony and Jonscher (51).

4. Dipolar β Peaks

These represent a type of response that may to some extent merge with the higher-temperature α-type peaks—in the context of polymeric and glassy materials these occur below T_g, but for our purposes we include under this heading all very broad, $\lambda > 3$ or so, and strongly asymmetric loss peaks. These peaks usually show a single value of activation energy, as shown in Fig. 25, differing in this respect from the α peaks.

A typical example is shown in Fig. 26, where normalization gives a single generating curve with only very slight variation of peak amplitude with temperature. The width in this instance is $\lambda = 6.5$, i.e., almost six

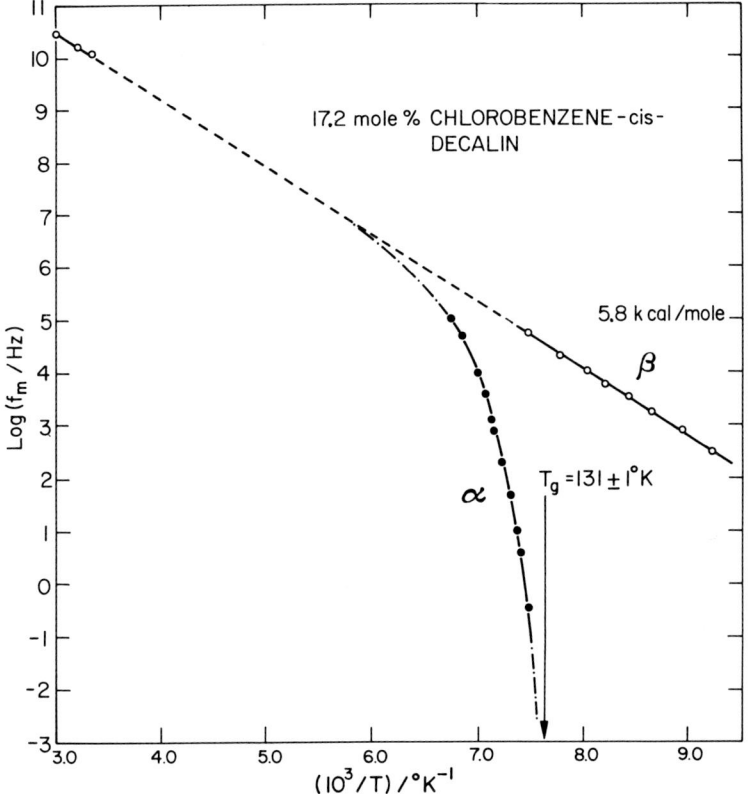

Fig. 25. The frequency of loss peak maximum for chlorobenzene–cis-Decalin system showing the glass transition and demonstrating the different types of activation plots for the α and β peaks. From Johari (46).

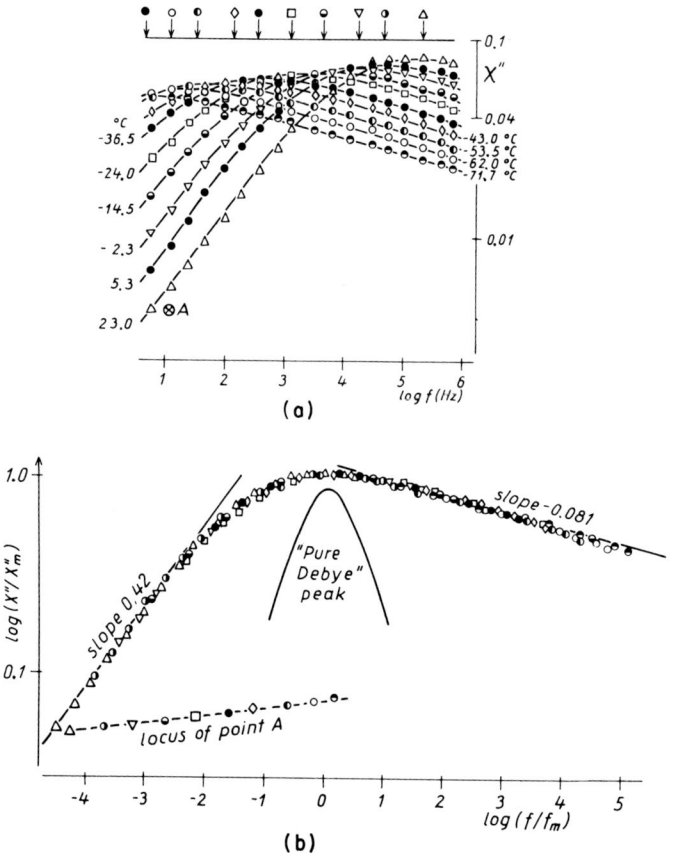

FIG. 26. (a) Dielectric loss peaks below the glass transition for 5% crystalline polyethylene terephthalate. After Ishida *et al.* (*44*). (b) The normalized loss curve from (a), showing the locus of the representative point A. All the experimental points fall on a single generating curve. The shape of a pure Debye peak is indicated. From Jonscher (*26*).

times the pure Debye width. Similar peaks are shown in Fig. 27 for a different polymer, which gives evidence of some complicating features indicating the presence of a second mechanism of polarization.

Figure 28 shows the activation energy plots for the β peaks of a number of polymers, including the ones shown in Figs. 26 and 27. The narrow range of activation energies is remarkable for such different materials. Figure 29 shows a compilation of the normalized loss data for a range of polar polymers, including both α and β peaks, showing the smooth

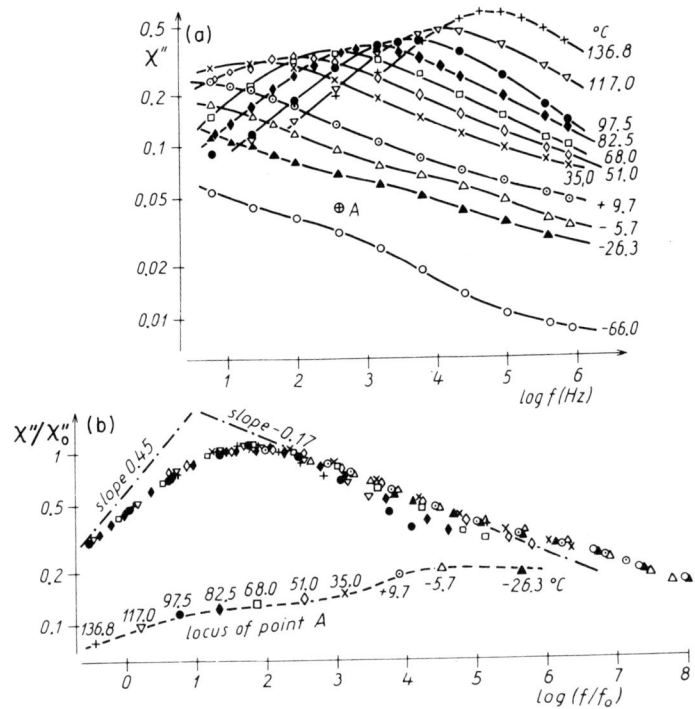

FIG. 27 (a) Dielectric loss peaks in the β relaxation region of polymethyl methacrylate at different temperatures. After Ishida (45). (b) The normalization curve showing the presence of more than one loss mechanism. The break in the activation energy plot of Fig. 28 occurs between 35.0 and 51.0°C. From Jonscher (26).

transition from the narrowest, $\lambda = 2.0$, to the broadest, $\lambda > 6$, as well as the lack of any clear-cut difference from the point of view of the frequency dependence between the two categories. The exponents m and n in the empirical formula (52) may or may not be dependent upon temperature. In all these responses, however, the high-frequency part of the loss characteristic shows the "universal" relation ω^{n-1}, with $n < 1$.

A very interesting class of materials is represented by the clathrates, in which any one of a wide range of "guest" molecules, which may be polar or nonpolar, may be accommodated in regular dodecahedral, i.e., nearly spherical, "cages" formed by H_2O molecules. The dielectric properties of clathrates have received a significant amount of attention and the results, mostly plotted as complex ϵ diagrams, show very flat and asymmetric arcs which in the traditional interpretation would require a very broad

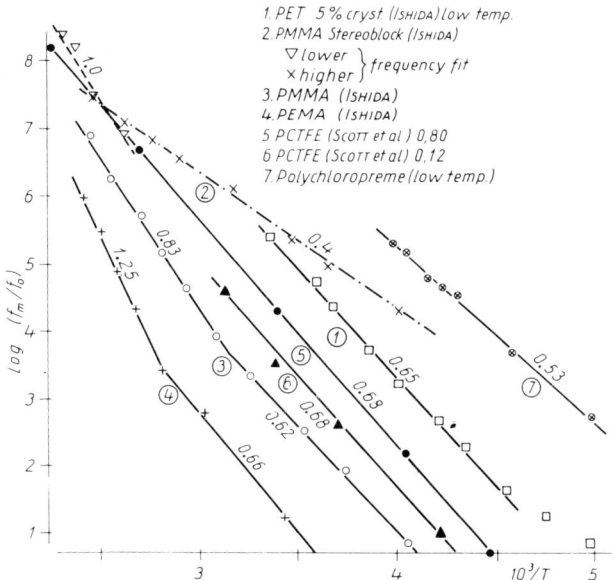

FIG. 28. Activation energy diagrams of the loss peak frequency $f_m = \omega_p/2\pi$ with an arbitrary scaling factor f_0 for the β-type loss peaks in a range of polar polymers. The energies are given in electron volts. From Jonscher (26).

distribution of relaxation times. Typical results are shown in Fig. 30, relating to guest molecules of acetone, trimethylene oxide, and tetrahydrofuran (53).

An example of broad loss peaks in an entirely different type of material is shown in Fig. 31, which relates to an electrolytic capacitor (54). These structures consist of a metallic substrate, usually Al or Ta, which may be planar, deeply etched, or in the case of Ta may consist of a sintered porous body, with a view to increasing the effective area. The metal is anodically oxidized and the structure is completed by a suitable liquid or solid electrolyte, which acts as the counterelectrode. The dielectric properties of planar capacitors are very similar to those made with deeply etched or porous bodies, suggesting clearly that the observed characteristics are not the result of the geometrical shape of the material, as sometimes suggested.

A dramatic example of the breadth of a loss peak and of the low-frequency behavior of loss at $\omega < \omega_p$ is shown in Fig. 32, relating to the low-temperature response of a silica glass (55). This is made possible by

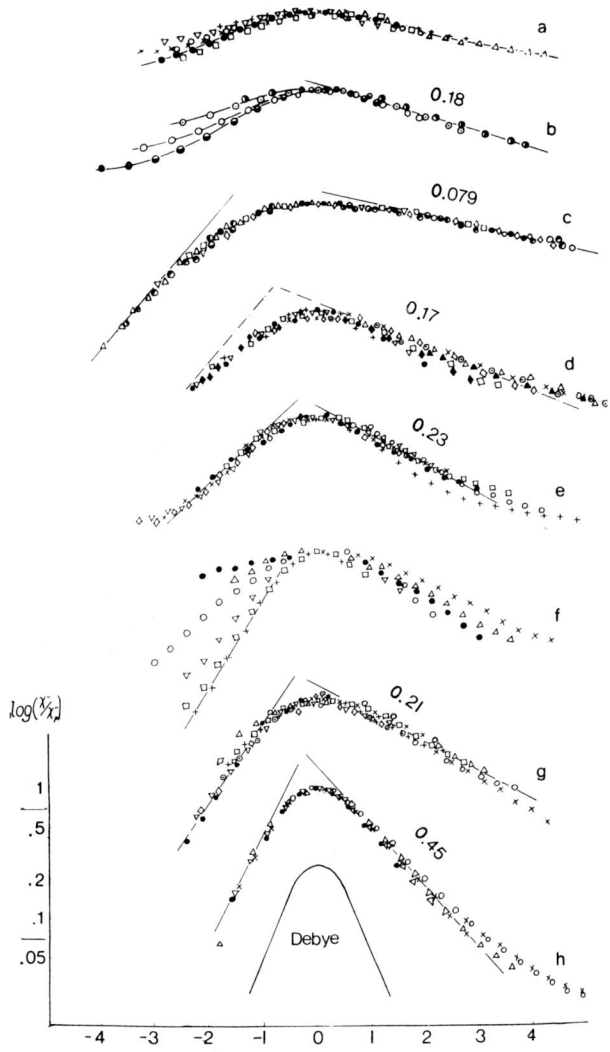

FIG. 29. A compilation of dielectric loss data for a range of polymeric solids, plotted as log χ'' vs. log ω. Different symbols correspond to various temperatures, the data having been normalized by lateral displacement. The data for different materials have been displaced vertically for clarity. The numbers give the values of the slopes $1 - n$. The shape of a pure Debye peak is shown for comparison. (a) Polychloroprene, low-temperature data; (b) PCTFE; (c) polyethylene terephthalate; (d) polymethyl methacrylate (Fig. 27); (e) polychloroprene, high-temperature data; (f) polyethyl methacrylate; (g) polydian carbonate (Fig. 17); (h) polyvinyl acetate (Fig. 18). Further particulars and source references in (26).

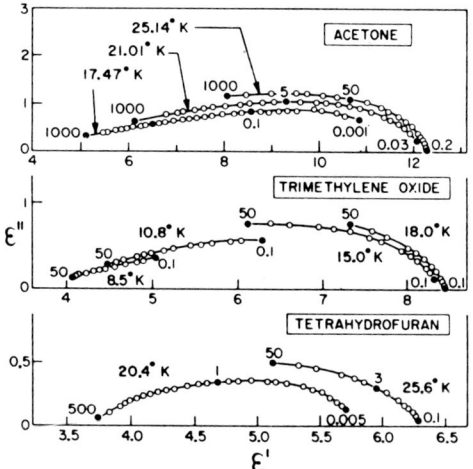

FIG. 30. Low-temperature complex permittivity plots for clathrates containing different guest molecules as indicated. Temperatures in K, frequencies in kHz. The strongly flattened and asymmetric shape of these plots should be noted. From Gough et al. (53).

the virtual freezing out of all DC conductivity in the material, which provides an experimental confirmation of the validity of the postulated form of the first term in the empirical expression (52).

Similar broad loss peaks may be seen in some glasses below T_g, but there are many more examples of the response without a loss peak, as disucssed in Section VI, 6.

5. General Analysis of the Loss Peaks

Hill (56) has shown recently that the exponents m and n in the empirical relationship (52) are independent of one another. He analyzed the data for 50 different dipolar materials, mostly solids, although some were liquids, and plotted the exponents m against $1 - n$ as a point for each material (Fig. 33). The same diagram shows the single point corresponding to the ideal Debye response, with coordinates (1, 1), the diagonal line corresponding to the symmetric Cole–Cole and Fuoss–Kirkwood empirical relations, and the upper side corresponding to the asymmetric Cole–Davidson formula for which the leading slope gives m = 1, while the exponent $1 - n$ can take a range of values. Also plotted as the dot–dash

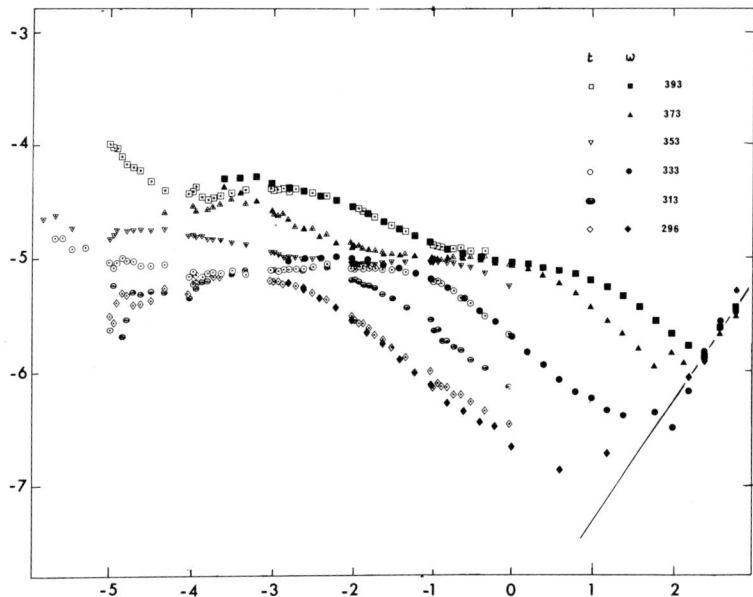

Fig. 31. The dielectric loss data for an electrolytic capacitor of 50 μF in sintered Ta. The solid symbols have been measured directly in the frequency domain, the open ones in the time domain and have been transformed using the Hamon approximation. The unit slope at high frequencies arises from the series resistance in the liquid electrolyte. Two loss peaks are clearly visible. The good agreement between the time and frequency domain data is due to the small amplitude of the measuring signals. No such agreement is obtained at larger amplitudes because of the nonlinearity of the system. From Meca and Jonscher (54).

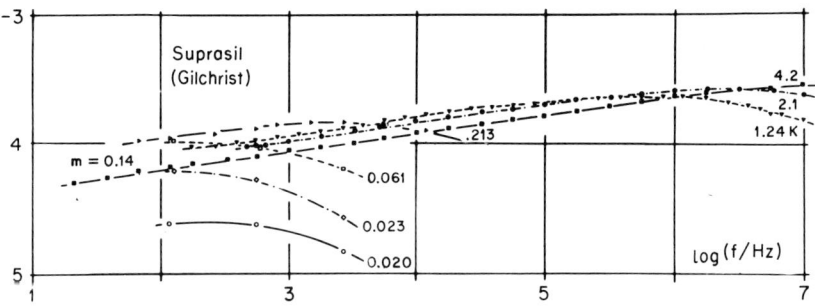

Fig. 32. The dielectric loss spectrum of silica glass at cryogenic temperatures, showing the loss behavior below the loss peak frequency. From Frossati et al. (55).

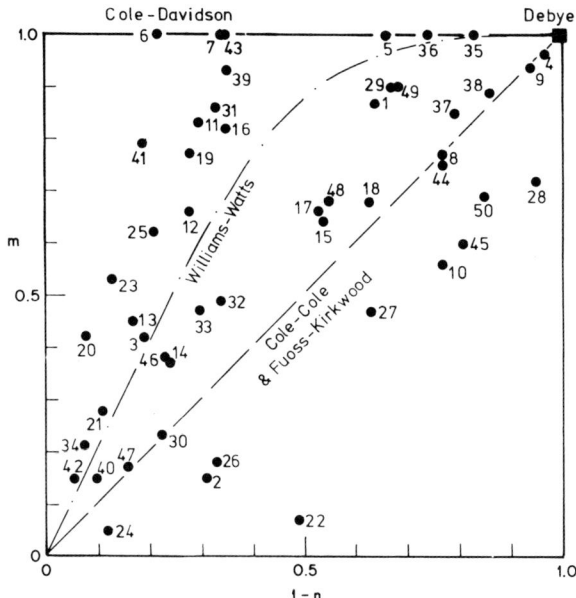

FIG. 33. A plot of the exponents m and $1 - n$ of the empirical relation (52) for a number of dipolar materials. The key to the numbers is given in Table I along with the references. After Hill (56).

TABLE I

DIELECTRIC PROPERTIES (SEE FIG. 33)

Material	m	$1 - n$	Plot no.	Refs
Pentachlorotoluene	0.87	0.64	1	1
Suprasil glass, $T \leq 4.2$ K	0.15	0.31	2	1a
Poly-γ-benzyl-L-glutamate	0.42	0.19	3	2
Water at 293 K	0.96(3)	0.96(8)	4	3
2,4,6-Tri-*tert*-butylphenol	1.0	0.66	5	6
Menthol (liquid)	1.0	0.22	6	7
Butyl stearate, $f > 10^9$ Hz	1.0	0.34	7	8
H$_2$O at 181 K			8	4
D$_2$O, $T \leq 262.3$ K			9	5
Butyl stearate, $f < 10^9$ Hz			10	8
Poly-*n*-butyl methacrylate			11	9
Polymethyl methacrylate			12	9a
Polymethyl methacrylate			13	9
Polypropylene			14	10

TABLE I (continued)
Dielectric Properties (see Fig. 33)

Material	m	$1 - n$	Plot no.	Refs.
Amorphous polyacetaldehyde			15	11
Polyethylene, $4.2 \leq T < 18$ K			16	1a
Oxidised high density polyethylene, $T < 4.2$ K			17	1a
Deuterated high density polyethylene, $T < 4.2$ K			18	1a
Polyethylene terephthalate				
α			19	12
β			20	12
Polyvinyl fluoride			21	13
Polyvinylidene fluoride				
α			22	14
β			23	14
γ			24	14
Polydian carbonate			25	15
Polyacronitrile			26	16
n-Docosyl bromide			27	17
Neo-hexanol			28	18
5-Methyl-3-heptanol				
α			29	19
β			30	19
3-Methyl-3 heptanol				
α			31	19
β			32	19
Cyclohexyl chloride in polystyrene				
α			33	20
β			34	20
Tricyclohexyl carbinol				
$f > 10^9$ Hz			35	21
$f < 10^9$ Hz			36	21
Picric acid			37	22
Pinacol hydrate			38	23
Chlorobenzene-pyridine, 43.4%				
α			39	24
β			40	24
Chlorobenzene-cis-decalin,				
α			41	19
β			42	19
Benzene, $f > 10^9$ Hz			43	25
p-methoxyphenylazoxy-p'-butylbenzene,				
nematic phase			44	26
isotropic phase			45	26

TABLE I (continued)

DIELECTRIC PROPERTIES (SEE FIG. 33)

Material	m	$1 - n$	Plot no.	Refs.
Polychloroprene				
α			46	27
β			47	27
$BaTi_6MgO_{16}$			48	28
$AgNa(NO_2)_2$			49	29
Natural quartz containing impurity ions			50	30

1. A. Turney, *Proc. Inst. Electr. Eng.* **46**, 100 (1953).
1a. J. le G. Gilchrist, personal communication.
2. K. Hikichi, K. Saito, M. Kanedo, and J. Furuichi, *J. Phys. Soc. Jpn.* **19**, 577 (1964).
3. P. R. Mason, J. B. Hasted, and L. Moore, *Adv. Mol. Relaxation Processes* **6**, 217 (1974).
4. S. R. Gough and D. W. Davidson, *J. Chem. Phys.* **52**, 5442 (1970).
5. G. P. Johari and S. J. Jones, *Proc. R. Soc. London, Ser. A* **349**, 467 (1976).
6. R. J. Meakins, *Trans. Faraday Soc.* **52**, 320 (1956).
7. H. Cachet and J.-C. Lestrade, *C.R. Hebd. Seances Acad. Sci.* **259**, 541 (1964).
8. J. S. Dryden, *J. Chem. Phys.* **26**, 604 (1957).
9. Y. Ishida, *J. Polym. Sci., Part A-2* **7**, 1835 (1969).
9a. G. Williams, *Trans Faraday Soc,* **60**, 1556 (1964).
10. R. N. Work, R. D. McCammon, and R. G. Saba, *J. Chem. Phys.* **41**, 2950 (1964).
11. G. Williams, *Trans. Faraday Soc.* **59**, 1397 (1963).
12. Y. Ishida, M. Matsuo, and K. Yamafuji, *Kolloid-Z.* **180**, 108 (1962).
13. Y. Ishida and K. Yamafuji, *Kolloid-Z.* **200**, 50 (1964).
14. Y. Ishida, M. Watanabe, and K. Yamafuji, *Kolloid-Z.* **200**, 48 (1964).
15. Y. Ishida and S. Matsuoka, *Polym. Prepr., Am. Chem. Soc., Div. Polym. Chem.* **6**, 795 (1965).
16. Y. Ishida, O. Amano, and M. Takayanagi, *Kolloid-Z.* **172**, 129 (1960).
17. J. S. Dryden and S. Das Gupta, *Trans. Faraday Soc.* **51**,1661 (1955).
18. G. P. Johari, *Ann. N.Y. Acad. Sci.* **279**, 117 (1976).
19. G. P. Johari and M. Goldstein, *J. Chem. Phys.* **55**, 4245 (1971).
20. M. Davies and J. Swain, *Trans. Faraday Soc.* **67**, 1637 (1971).
21. R. J. Meakins, *Trans. Faraday Soc.* **52**, 320 (1956).
22. R. J. Meakins, *Trans. Faraday Soc.* **51**, 371 (1955).
23. R. J. Meakins, *Prog. Dielectr.* **3**, 170 (1961).
24. G. P. Johari and M. Goldstein, *J. Chem. Phys.* **53**, 2372 (1970).
25. E. Constant and A. Lebrun, *J. Chem. Phys.* **61**, 166 (1964).
26. J. P. Parnieux, A. Chaporon, and E. Constant, *J. Phys. (Paris)* **36**, 1143 (1975).
27. M. Matsuo, Y. Ishida, K. Yamafuji, M. Takayanagi, and F. Irie, *Kolloid-Z.* **201**, 89 (1965).
28. J. S. Dryden and A. D. Wadsley, *Trans. Faraday Soc.* **54**, 1574 (1958).
29. K. Gesi, *Ferroelectrics* **4**, 245 (1972).
30. E. H. Snow and P. Gibbs, *J. Appl. Phys.* **35**, 2368 (1964).

line is the locus corresponding to the empirical Williams–Watts time domain expression (57)

$$f(t) = (d/dt)\{\exp[(-t/\tau)]^{n+1}\}$$

The immediate conclusions that may be drawn from this representation are (1) the exponents m and $1 - n$ are evidently independent of one another and must therefore correspond to *different* physical processes; (2) none of the widely quoted empirical formulas is capable of representing the experimental data with any degree of accuracy—indeed, the experimental points seem to *avoid* these empirical relations; (3) the Debye relation is not found in *any* of these materials, which include water and heavy water.*

It is worth noting that all points except one falling on the top line corresponding to $m = 1$ relate to measurements in which the loss peak falls in the high range of frequencies, 10^8-10^9 Hz. The significance of this feature becomes apparent in the theoretical discussion of Section VIII,5.

6. The "Universal" Response without Loss Peaks in Carrier Systems

We now come to examine the dielectric response of the very wide range of materials in most of which there is good reason to expect that this is dominated by low-mobility charge carriers hopping between localized sites. These charges may be electrons or polarons (20) and they may also be ions (2) and it is customary to represent the data in the form of AC conductivity plots $\sigma(\omega)$, as defined by Eq. (28).

Figure 34 (2, 58–70) shows the representation of $\sigma(\omega)$ for a range of 13 widely different materials on a common frequency axis, but with a vertical translation between the individual sets of data, to avoid confusion (6). It is evident that the exponents n in the universal relation $\sigma(\omega) \propto \omega^n$ vary between 0.6 and virtually unity and may be slightly temperature dependent. The same data are then shown on a common scale of conductivity in Fig. 35, which brings out the strikingly narrow range of the absolute values of AC conductivity, i.e., of the dielectric loss, bearing in mind the wide range of materials to which these data refer. In particular, the organic insulator anthracene, the classic ionic conductor β-alumina, amorphous silicon, and stearic acid all have values of $\sigma(\omega)$ within two decades of one another.

It is noteworthy that none of the materials shown in Fig. 34 reveals the presence of significant loss peaks, which in this representation manifest themselves by a slope that is initially greater and then smaller than unity. Only the anthracene and the β-carotene show a weak loss peak around

* More extensive data may be found in R. M. Hill, *J. Mater. Sci.* (in press).

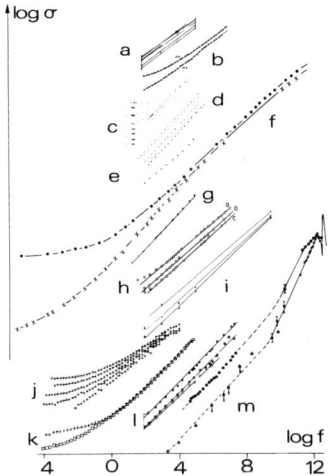

FIG. 34. The compilation of AC conductivity data for a range of materials arranged on a common log f basis (Hz) but displaced vertically along the log σ axis for clarity. Data sets denoted by one letter are on a common log σ scale. (a) Single-crystal Si in the impurity hopping range by electrons, at 3.0 (bottom), 4.2, 8.0, 12.0 K (58); (b) single-crystal β alumina at 77 and 87 K—a classical fast ion conductor by Na^+ ions (59); (c) glow discharge deposited amorphous silicon 84–295 K (60); (d) a range of chalcogenide glasses at 293 K (61); (e) single-crystal anthracene with 1 M saline solution as contact, 294 K (62); (f) single-crystal anthracene (crosses) and carotene (dots) at 294 K (63); (g) TNF–PVK (64); (h) P_2O_5–FeO–CaO glasses at 300 K (65); (i) V_2O_5–P_2O_5 glass at three temperatures (66); (j) evaporated amorphous silicon monoxide at 211–297 K (67); (k) 9-layer stearic acid film between Al and Au electrodes in the dark ($\square$) and with UV light ($+$) (68); (l) three amorphous samples (top to bottom) As_2Se_3, Se, and As_2S_3. The measurements taken at 300 K are believed to be "bulk" data not affected by electrode effects (69); (m) two samples of As_2Se_3 at 300 K extending to far infrared and showing the steeply rising ω^2 region above 10 GHz characteristic of lattice vibration processes (70). From Jonscher (2).

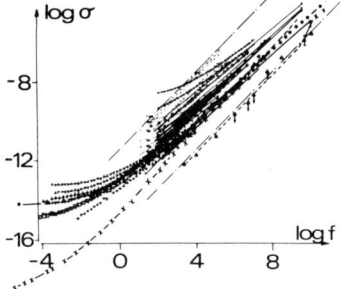

FIG. 35. The compilation of AC conductivity data from Fig. 33 placed on a common log σ (Ω cm)$^{-1}$ scale and showing the remarkably narrow range of absolute values of AC conductivity for a very wide range of different materials, all obeying the universal law (57). The upper and lower dot–dash lines correspond, respectively, to frequency-independent loss $\chi'' = 10$ and 10^{-3}. This is a corrected version of a diagram originally published in Jonscher (2).

1 MHz, superimposed on the general "flat" loss trend. Some of the data show a clear tendency to DC conductivity, although some of the evidence may have to be reexamined in the light of the evidence quoted in Section VI,8.

A very interesting question poses itself in this context regarding the relative values of the DC and AC conductivity in different materials. In many cases it is evident that there is very little relation between these two parameters, σ_0 being determined by extrinsic factors such as doping with foreign impurities, while $\sigma'(\omega)$ is much less sensitive to this. In any case, the two show completely different temperature dependence. Figure 36 is taken as an example of the apparent correlation between the DC conductivity σ_0 and the AC conductivity $\sigma(\omega)$ for a range of chalcogenide glasses (20). While there is some evidence that such a correlation is apparent in glasses evaporated in the form of thin films (71), the bulk materials shown in Fig. 36 may be interpreted in a different manner, as

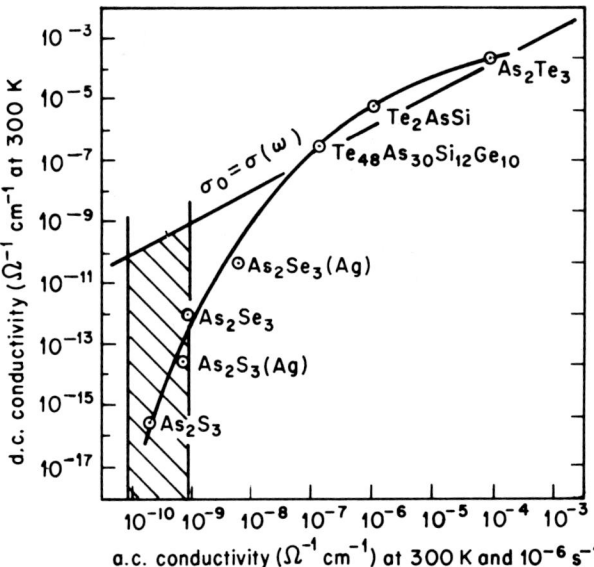

FIG. 36. The relationship between the AC conductivity $\sigma(\omega)$ and the DC conductivity σ_0 for a range of chalcogenide glasses quoted by Mott and Davis (20) with the line $\sigma_0 = \sigma(\omega)$ above which one would not expect any points, and with the band of values corresponding to a lattice contribution to the AC conductivity with frequency-independent values of relative loss $\epsilon''(\omega)/\epsilon_0 = 10^{-3}$ and 10^{-2}. From Hill and Jonscher (72).

has been done by Hill and Jonscher (72). They point out that the expression for the AC conductivity

$$\sigma(\omega) = \sigma_0(T) + A(T)\omega^{n(T\subset)} \tag{57}$$

implies that the total AC conductivity $\sigma(\omega)$ cannot be smaller than σ_0. Drawing the line $\sigma(\omega) = \sigma_0$ we find that this goes comfortably through the three top points. Hill and Jonscher then point out that the AC component represented by the second term in (57) must be taken to include the contribution of the *dielectric lattice*, in addition to that of the *hopping-charge carriers*, while most conventional interpretations seem to overlook this basic fact. The shaded region in Fig. 36 corresponds to a frequency-independent relative dielectric loss $\epsilon''(\omega)/\epsilon_0$ equal to 10^{-3} and 10^{-2}, corresponding to a typical range of values. Since this loss due to the lattice is, by definition, independent of the DC conductivity, the region in question corresponds to a vertical strip in the diagram, showing that the three lowest points related to glasses whose electronic conductivity was too low to show against the lattice contribution. By contrast, the highest three points may be taken as corresponding to glasses that are sufficiently DC conducting for there to be no significant AC contribution at the frequencies of measurement.

The fundamental point is that DC conductivity is determined by the most difficult transitions in complete percolation paths between the electrodes, while the AC conductivity or dielectric loss is determined by the easiest local movements of charges. Thus, it is natural to expect that the absolute magnitudes of these two parameters may not be closely related, and it is evident that their temperature dependence is completely different.

In view of the as yet incomplete understanding of both the DC and the AC conductivity (72), it is premature to give a definite opinion on this subject, apart from stating the fairly obvious point that more correlation may be expected in "dense" carrier systems with higher conductivity than in comparatively more "dilute" systems with low conductivity, since more very difficult transitions may be expected in the latter than in the former.

There is clear evidence, in some cases, of the temperature dependence of the exponent n, but it is difficult to obtain experimental data of sufficiently high quality to enable one to derive reliably the law of temperature dependence. One reason for this difficulty is the need to have at least four to six powers of ten of frequency without interference from any other mechanism. One of the better sets of data in this respect is the result for silicon monoxide shown in Fig. 34j, where n varies from 0.63 at 241 K to 0.56 at 297 K, but even in this case it is difficult to determine $n(T)$ with any certainty.

A significant result is shown in Fig. 34l and refers to measurements on Se, As_2Se_3, and As_2S_3 (69). These three amorphous materials are seen to have very similar values of $\sigma(\omega)$, both regarding the absolute magnitude and the frequency dependence, and yet there are reasons to believe that the densities of localized levels in the forbidden gap are very different, selenium having a relatively "clean" gap, while the other two have large densities. This result suggests therefore that the density of localized levels is not the determining factor in the AC conductivity.

To these examples of the universal response of materials that do not show any loss peaks, we may add the remarkable behavior of supercooled aqueous solutions of salts such as LiCl, which show very similar responses (73).

The behavior of ceramic and single-crystal Al_2O_3 is shown in Figs. 37 and 38, respectively (74, 75). Both materials show essentially similar trends—the first appears to show two regions of slope as in Eq. (55) and in the latter the same trend is much more pronounced, giving the strong dispersion of both $\epsilon'(\omega)$ and $\epsilon''(\omega)$, which are discussed in more detail in Section VI,7. An important point to be noted in relation to Fig. 37 is the good extrapolation of the low-frequency trend at microwave frequencies, where there are certainly no contact effects at play—this proves that the observed universal behavior is not an artefact of some contact phenomena.

The high-frequency response of ferroelectrics, with its characteristic nearly Debye behavior has already been described in Section VI,1. That response was obtained under conditions where the applied field was changing too rapidly for the ordered dipoles to be able to follow. It is therefore interesting to note that in the low-frequency region the response

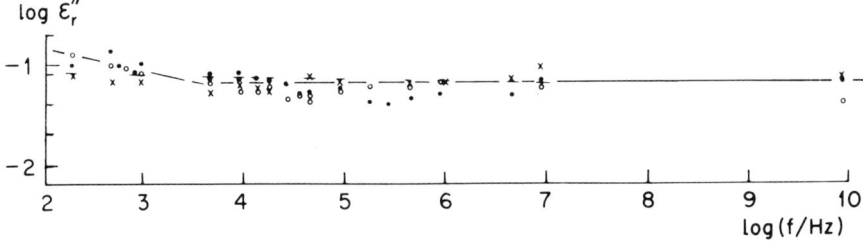

FIG. 37. The dielectric loss of ceramics of compositions Si_3N_4, 5 wt.% MgO/Si_3N_4 and $Si_2Al_4O_4N_4$, measured at 300 K. From Thorp and Sharif (74).

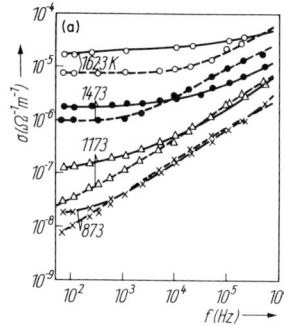
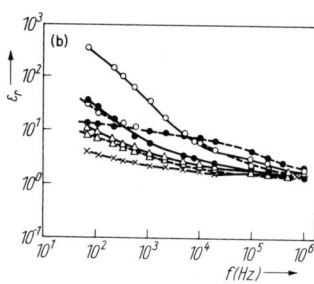

FIG. 38. The frequency dependence of the electrical conductivity and of the relative permittivity over a range of temperatures for single-crystal alumina. The strong dispersion of permittivity and the slowly variable AC conductivity at low frequencies should be noted. From Kizilyalli and Mason (75).

follows the universal law, at least in the case of TGS, both below and above the Curie temperature (76), as shown in Fig. 39. There is no evidence in these data of any loss peak at the lower frequency end of the range extending over four decades. The fact that the same loss relation is obeyed in the ferro- and paraelectric states is highly significant even though it is too early yet to give a detailed interpretation of this behavior.

Mention was made earlier of the universal response in β alumina—an ionic conductor with Na^+ ions as charge carriers in a two-dimensional planar array. Many other ionic conductors, both of the conventional and of the "fast ion" type, obey very similar laws, even though this is not commonly recognized under the guise of the prevailing complex impedance representation. We have pointed out that the usually observed tilted circular arc Z plots are direct evidence for the universal response of these materials (77, 78). Figure 40 gives an example of this type of behavior for a sample of hollandite $K_{1.8}Mg_{0.9}Ti_{7.1}O_{16}$ for a range of temperatures (79). At 77 K the response is clearly of the universal type, while at higher temperatures a loss peak is superimposed on the universal trend and at even higher temperatures a strong low-frequency dispersion sets in, as described in Section VI,8 (79–82).

An extension of the range of temperature down to 5.2 K reveals the striking fact that the dielectric loss of hollandites becomes virtually independent of temperature below 77 K (81). This is shown in Fig. 41, which gives the frequency dependence of both the real and the imaginary parts of the complex permittivity for a sample of Hollandite different than that in Fig. 40. While showing some departures from the response at

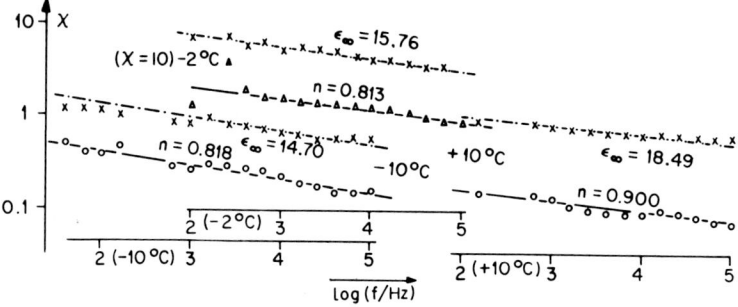
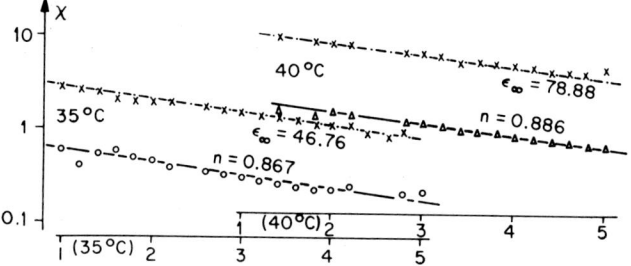
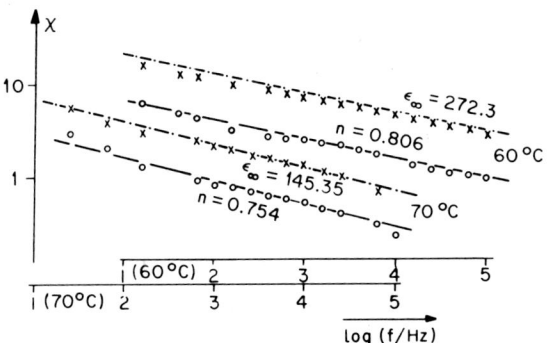

Fig. 39. Dielectric loss and the real part of the susceptibility for triglycine sulfate in the "low-frequency" region in a range of temperatures below and above the Curie temperature, $T_c = 50°C$. The continuous lines define the slopes of the $\chi''(\omega)$ points as measured directly; the dot–dash lines have been drawn in Kramers–Kronig compatible positions; and the values of ϵ_∞ have been calculated to place the $\chi'(\omega)$ data at the highest frequency on the calculated line. The remaining $\chi'(\omega)$ points were then calculated from the measured $\epsilon'(\omega) = \chi'(\omega) + \epsilon_\infty$. The various sets of data are displaced in frequency and amplitude for clarity. Note that the universal relation applies across the T_c. From Jonscher and Dube (76).

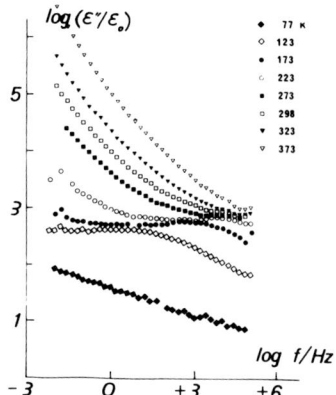

FIG. 40. The frequency dependence of the dielectric loss of the ionic conductor hollandite of the composition $K_{1.8}Mg_{0.9}Ti_{7.1}O_{16}$ in a range of temperatures. The data for 77 K give a very good agreement with the universal law over seven decades of frequency. At 123 K a clear loss peak is superimposed over the universal trend and this peak moves with temperature to higher frequencies. At the lowest temperatures there is a strong low-frequency dispersion, which is shown in more detail in Fig. 41. From Jonscher et al. (79).

higher temperatures, the general trend in Fig. 41 is the same as in Fig. 40, except that the points below 77 K show very little change down to 5.2 K. The other very remarkable feature is the presence and persistence down to the lowest temperatures of the strong low-frequency dispersion, already mentioned in the context of higher temperature response. The fact that this is seen to persist down to cryogenic temperatures strongly supports the view that this dispersive mechanism is inherent to the "universal" dielectric response and is not simply the outcome of interfacial barrier polarizations, e.g., the classical Maxwell–Wagner effect (81).

The fact that fast ionic conductors, and ionic conductors in general, obey the same universal law extends the generality of this law of dielectric response, as well as providing a much more plausible explanation of the observed behavior than the currently accepted interpretations in this field.

An extension of the experimental data down to molecular thicknesses is illustrated in Fig. 42, which gives loss data for multilayers of stearic acid from one monolayer thickness—2.5 nm—up to 13 monolayers, which may be effectively a "bulk" sample (83). The frequency dependence of loss shows two clearly differentiated regions, as in Fig. 37, the low-frequency response following the universal law with $n \simeq 0.7$ while the high-frequency data correspond to $n = 1$ within experimental error. The

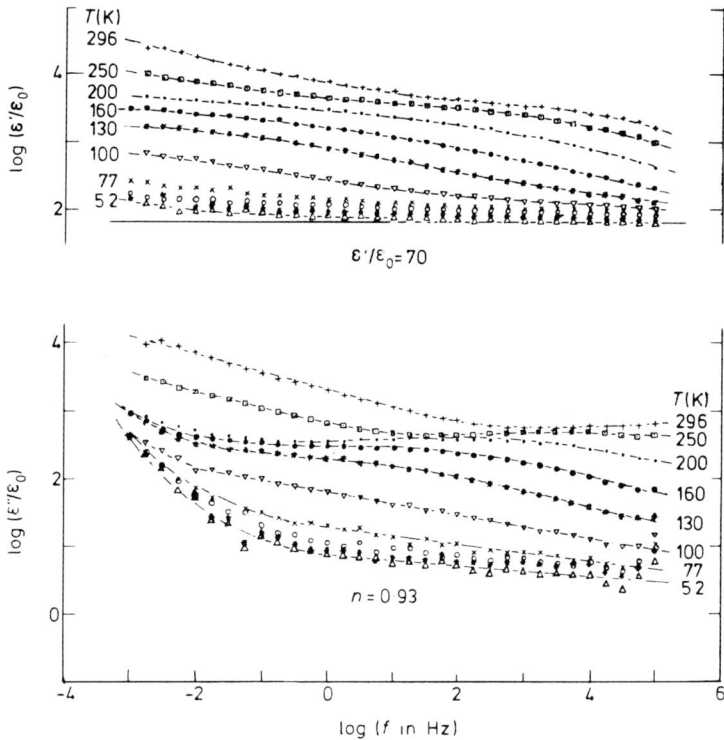

FIG. 41. The frequency dependence of the complex dielectric permittivity of the ionic conductor of the hollandite family $K_{1.6}Mg_{0.8}Ti_{7.1}O_{16}$ in the log–log representation, with temperature as parameter. The temperature range overlaps that in Fig. 40 but extends down to 5.2 K. The symbols for the lowest temperatures are ×, 77 K; ○, 60 K; *, 30 K; △, 5.2 K. The upper diagram represents the real part, with the horizontal line corresponding to the value 70. The lower diagram represents the imaginary part. From Ref. (81).

important point at the present stage is that the nature of the response is unaffected by the thickness down to monolayer structure. Figure 43 shows the temperature dependence of loss for a nine-layer film and it is clear that the two frequency regions show different temperature dependences. A "fine structure" on the high-frequency part is also apparent and is shown more clearly in Fig. 43b, which gives the temperature dependence at constant frequency in the "flat" region of the frequency spectrum. Three loss peaks are clearly visible in the temperature graph and these correspond approximately to the known temperatures of phase transformations in the stearic acid lattice. This is the reason for regarding

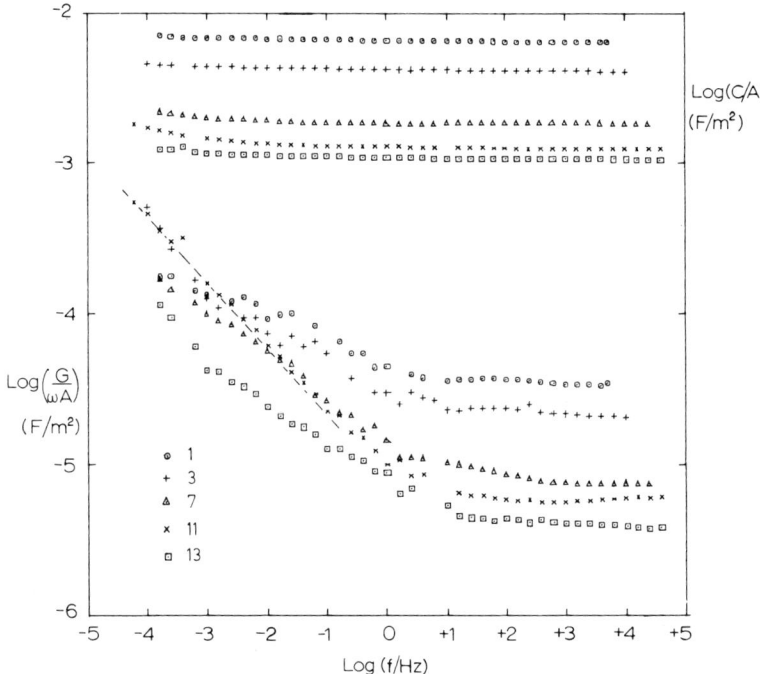

FIG. 42. The frequency dependence of the capacitance C and the effective loss G/ω for stearic acid films of 1, 3, 7, 11, and 13 monolayers thickness, between aluminium electrodes. Except for a change of scale, these correspond to ϵ' and ϵ'' for this material. The direct current has been subtracted where appropriate. From Careem and Jonscher (83).

this part of the frequency response as arising from the lattice of the material, rather than from some carrier or dipolar species present accidentally in the lattice.

An even more dramatic visualization of the frequency and temperature dependence of loss is found in the detailed representation of Fig. 44. Here the fine structure is clearly visible in the frequency plot and the apparently aimless variation with temperature becomes much clearer in the contour map of constant loss vs. frequency and $1/T$. There is a clearly discernible loss peak, which shifts with frequency and temperature in a manner similar to that of a Debye peak (Fig. 15a), and a second one that corresponds to a frequency-independent peak in temperature (Fig. 15b).

This survey of the universal response without clearly defined loss peaks in frequency dependence—as distinct from temperature dependence—is far from complete and it would be possible to cite many more data. It is

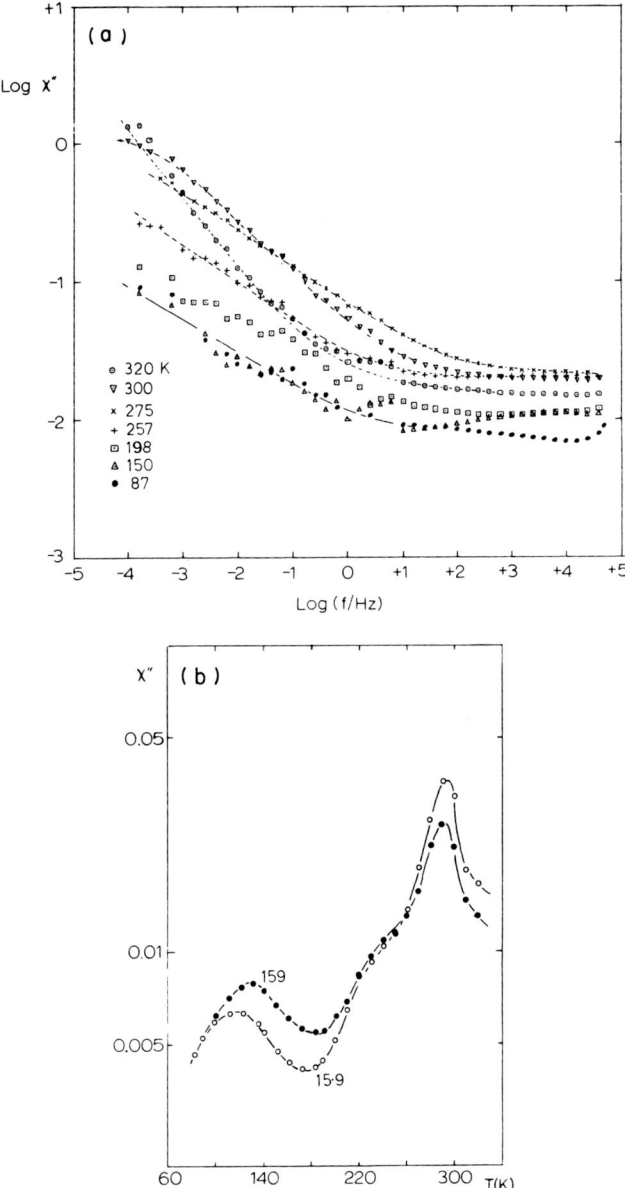

FIG. 43. (a) The temperature dependence of the dielectric loss of a 9 layer stearic acid film between aluminum electrodes. Two different regimes are clearly visible, having different dependences on both frequency and temperature. (b) The temperature dependence at constant frequencies in the "flat" higher frequency region of (a), showing clearly the presence of three peaks, one of which at least is independent of frequency. From Careem and Jonscher (*83*).

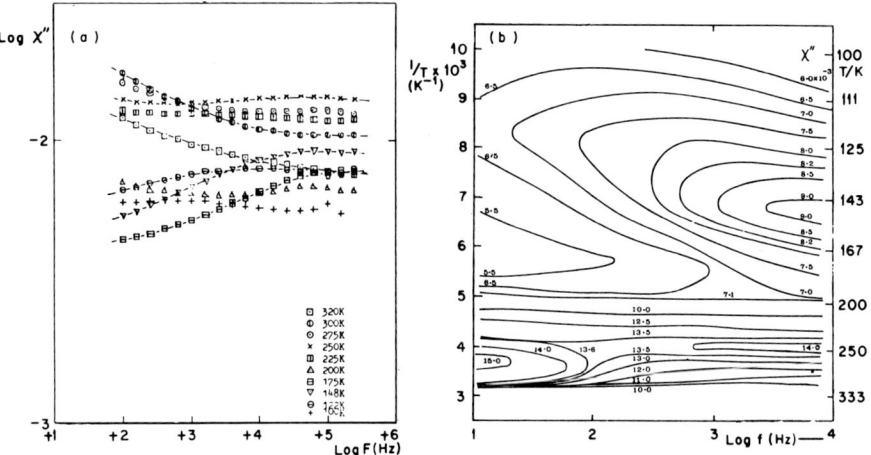

FIG. 44. (a) Detailed loss measurements in a 13 layer stearic acid film with Al electrodes and at ten temperatures in the range 100–320 K. The loss is almost "flat" over 3½ decades of frequency. (b) Contour map of constant loss vs. log f and $1/T$ showing one frequency-dependent and one frequency-independent loss peak. From Careem and Jonscher (83).

hoped, however, that even this brief review will convince the reader that there exists this well-defined category of dielectric response, which appears to be definitely connected with the presence of some species of mobile charge carriers in the dielectric lattice.

7. "Lattice" Response

This brings us to the consideration of a limiting form of dielectric response in which the exponent n becomes effectively equal to unity—the extreme value allowed also from the point of view of the Hilbert transformation. In the case of stearic acid we were able to conclude positively that the relevant branch of the response was due to the lattice, which in this case we envisage as a regular array of permanent induced dipoles. There are many materials in which the loss is flat in frequency over several decades and the Kramers–Kronig relations demand that the level of loss should be very low. The materials in question are therefore generally highly pure with nonpolar lattices. An example of this type of response is shown in Fig. 45, relating to the temperature dependence of loss in high-purity polyethylene with some antioxidant, covering 4–300 K and over two decades in frequency (84). The loss variation with tempera-

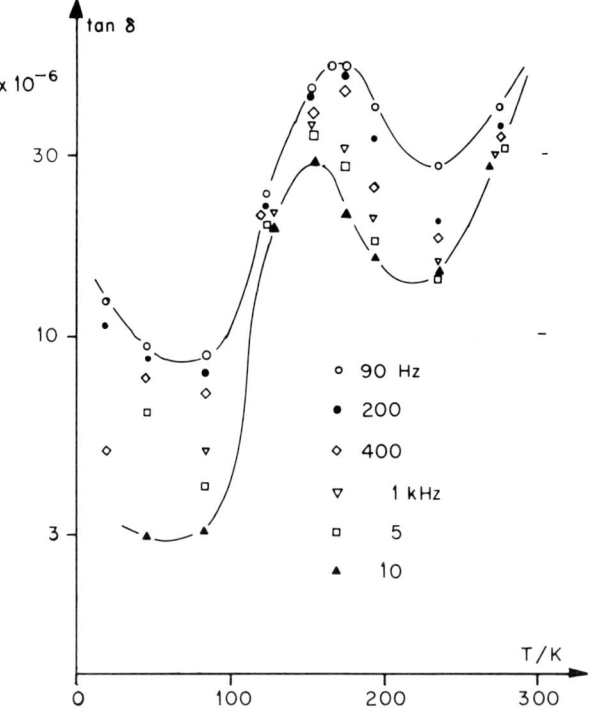

FIG. 45. The temperature dependence of loss at various frequencies for polyethylene with antioxidant. The variation with temperature is stronger than with frequency, suggesting the importance of "lattice" phenomena. From unpublished data (84).

ture covers over one order of magnitude, while that with frequency at any one temperature is much smaller—a situation somewhat similar to that shown in Fig. 43. We are unable, on the basis of the scant evidence at our disposal, to determine with certainty that this is the lattice response, but there are many other examples where the situation is much clearer.

There are some good examples of "flat" losses in frequency at cryogenic temperatures—as well as the α-like peaks quoted in Fig. 22. Figure 46 shows the loss data at 4.2 K for pure polyethylene, which is completely flat, as well as polyethylene with antioxidant, which gives a markedly peaked response (85). The same authors also report that some other samples containing a different type of antioxidant did not show any peaks but gave the same flat response, and so it is not apparently simply the question of the presence of additives. Figure 47 (86) shows the data for a range of polymeric solutions, some of which are remarkably flat in

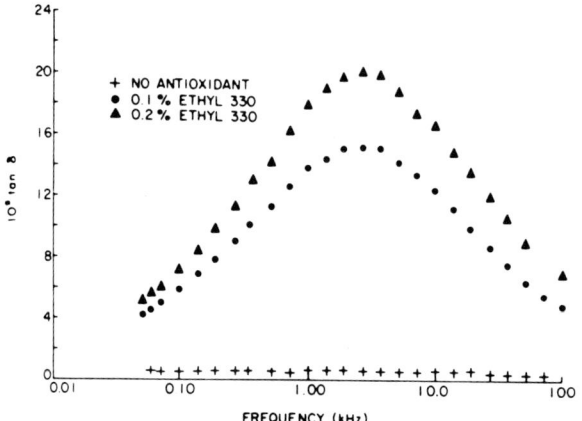

FIG. 46. The dielectric loss of polyethylene with and without antioxidant at 4.2 K. The pure sample has a "flat" loss with frequency; the two with antioxidant give approximately symmetric peaks with width $\lambda = 2.1$. From Thomas and King (85).

frequency at low temperatures. The lattices in this instance are polar, by contrast with the polyethylene in Fig. 46, and the losses appear to be significantly higher, so that one concludes that these lattice losses are directly related in magnitude to the polarizability of the lattice.

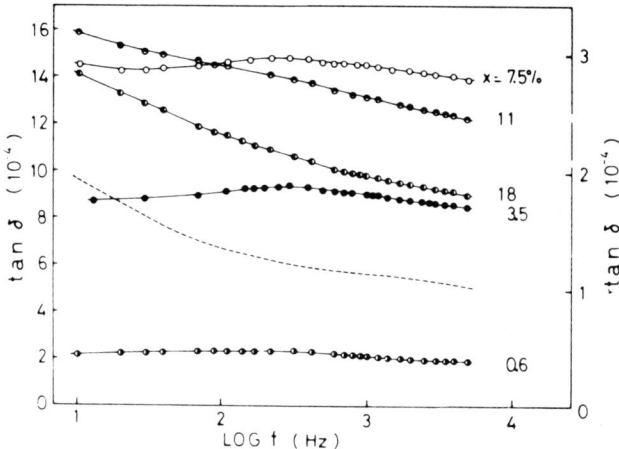

FIG. 47. The frequency dependence of dielectric loss, measured at 4.2 K, of various ethylene–vinyl alcohol copolymers with various comonomer fractions x of vinyl alcohol (left scale), together with the loss of low-density polyethylene (dashed curve, right scale). From Yano et al. (86).

The "lattice" response poses an interesting question regarding the extent to which any of the observed tendency of the exponent n to approach unity with the lowering of temperature may be due simply to progressive "freezing out" of the thermally activated dipolar or charge carrier movement, leaving behind the lattice contribution, for which the exponent is equal to unity. Unfortunately, it is difficult to answer this question because of the poor quality of the experimental data mentioned earlier.

8. Strong Low-Frequency Dispersion

We conclude our survey of the frequency response of the dielectric loss by turning to the strongly dispersive low-frequency behavior sketched in Fig. 14. We believe this to be typical of many materials, the common feature of which is the presence of strong carrier polarization arising from high densities of low-mobility charge carriers (82). It appears that the first specific account of this type of behavior was given by Jonscher and Frost (16), who found it in chalcogenide glasses of the composition Si–Te–As–Ge.

Figure 48 shows the relevant frequency response data in the form of $\sigma(\omega)$ with temperature as parameter. The exponent is of the order 0.2 and a correspondingly strong dispersion of the permittivity $\epsilon'(\omega)$ was also observed.

A good example of this type of response was found in an unusual dielectric "medium"-loose sand with variable degree of humidity of the air circulating in the sample (87). Figure 49 (87, 88) shows the original data points from measurements by Shahidi together with the normalization of both $\epsilon'(\omega)$ and $\epsilon''(\omega)$ giving values of $n = 0.22$ and reasonably good agreement with Kramers–Kronig relations. The remarkable point here is that it is possible to normalize data for variable humidity in the same manner as for variable temperature.

A good example of strong dispersion was found in the ionic conductor (79) hollandite at high temperatures, but it is remarkable that even down to cryogenic temperature there is a strongly pronounced low-frequency dispersion. Figure 50 shows the results of normalization of $\epsilon'(\omega)$ and $\epsilon''(\omega)$ in the higher temperature range, giving a value $n = 0.09$ with reasonably good agreement with Kramers–Kronig relations. The figure also shows the complex impedance diagram resulting from the presence of this strong dispersion, as predicted by Jonscher (78).

The materials in Figs. 48 and 50 are "bulk" rather than thin films and we conclude this account of low-frequency dispersion with an example of

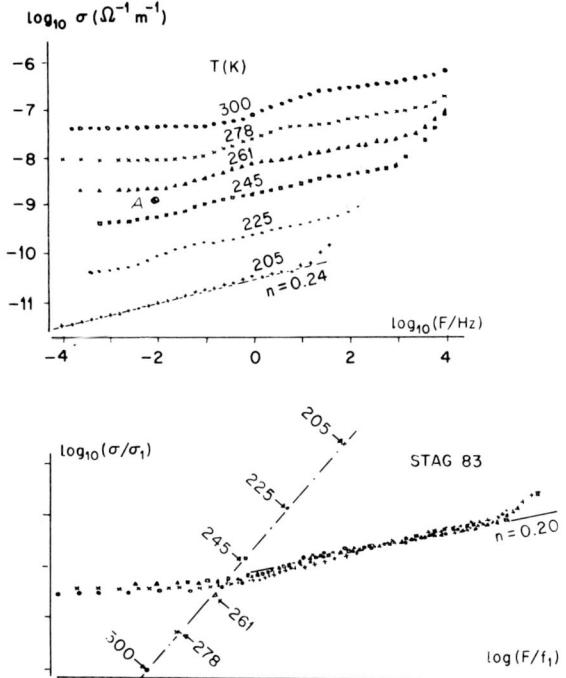

FIG. 48. The upper diagram shows the frequency dependence of the conductivity of a STAG chalcogenide glass for a range of temperatures. The slight kinks at the higher temperatures are the result of the overlap of the response of a barrier, as shown in Fig. 7, but the main body of the data relate to bulk response of the glass. The lower diagram shows the result of normalization with the locus of the representative point A now having unit slope. The arrows indicate the positions of the points on the assumption of a simple activated process. From Jonscher and Frost (16).

a biological membrane—egg lecithine with cholesterol in a saturated solution of $MgCl_2$ at two temperatures (89), Fig. 51. At the lower temperature the response is very similar to the examples given above while the higher temperature data show a slope corresponding to $n = \frac{1}{2}$ believed to be due to diffusive transport of ions in a neutral electrolyte— the so-called Warburg mechanism (90–92).

Several other examples of strong dispersion may be quoted. We have already commented on the alumina results in Fig. 38 and the response of amorphous silicon in Fig. 34c may be considered to be similar. A range of tree leaves and branches were found to give the same type of behavior in the radio-frequency range (93). In general, it appears that the behavior

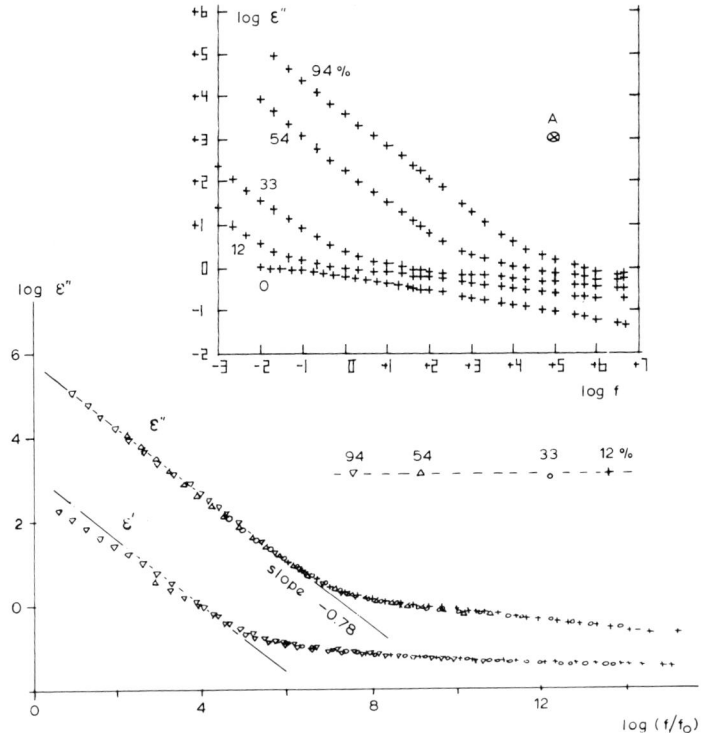

FIG. 49. The dielectric permittivity of dry and humid sand, based on measurements by Shahidi (*87, 88*). The inset shows $\epsilon''(\omega)$ with the indicated levels of relative humidity of the circulating air in the sample as parameter. The lower two diagrams give the normalized data for $\epsilon'(\omega)$ and $\epsilon''(\omega)$ for the four humidities shown. The two diagrams are displaced vertically for clarity, but the two slopes are drawn in Kramers–Kronig compatible positions. The high-frequency part has $n = 0.8$, the low-frequency part $n = 0.22$. The dot–dash line gives the locus of the representative point A. From Jonscher (*82*).

under consideration is found quite commonly in many different materials and that it has failed to attract detailed analysis, apart from vague references to the Maxwell–Wagner mechanism discussed in Section VII,7.

9. Time Domain Response

The time domain response of a dielectric in the linear regime is, strictly speaking, a Fourier transform of the corresponding frequency domain

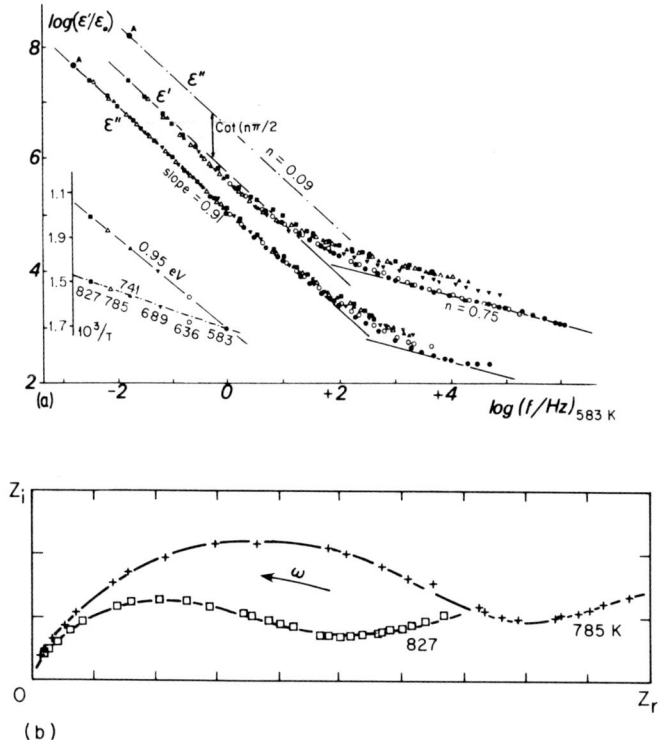

FIG. 50. The normalized data for a Hollandite sample over a range of temperatures, giving ϵ' and ϵ'' separately. The two sets are displaced for clarity and the position of the reference point A indicates the extent of the shift. The dot–dash line gives the correct position of ϵ'' with respect to ϵ'. The inset shows the temperature dependence of the characteristic frequency shift. (b) The corresponding complex impedance plots at two temperatures. From Jonscher et al. (79).

response, so that to that extent it is unlikely to introduce anything essentially new to the characterization of the phenomena in question. There are, however, important considerations that make the knowledge of the time domain response, with or without the frequency domain, an important means of furthering our understanding of the dielectric behavior of materials.

The first point to note is that, as stated in Section II, *Mother Nature works in the time domain,* i.e., time-resolved response to a step function

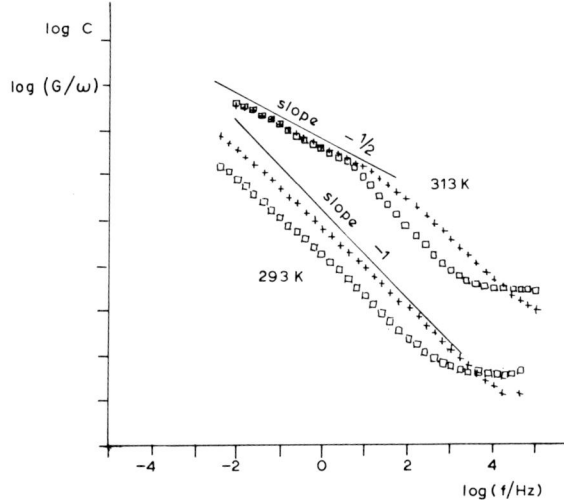

FIG. 51. The real part (□) and the loss (×) of the complex permittivity of an organic membrane egg lecithine + cholesterol (1:1) in a saturated solution of $MgCl_2$ at two temperatures. The two sets of data are displaced vertically by two decades to avoid overlapping. Limiting slopes of -1 and $-\frac{1}{2}$ are indicated. From measurements by Szundi (89).

excitation is a more immediate record of the behavior of the material than the essentially artificial harmonic response. The latter is "natural" in connection with resonance phenomena, but of these we do not encounter any in the frequency region under consideration in the present review, while in the context of essentially viscous responses at lower frequencies this advantage disappears. As against this, we argue that a proper understanding of the loss peaks requires a look at the time domain response. Similarly, in the presence of any nonlinearity in the system, the frequency domain response becomes difficult if not impossible to interpret, since it automatically rejects all higher harmonics. Even if these were taken into account there is a considerable amount of uncertainty with regard to the treatment of nonlinear harmonic responses (94, 95). On the other hand, the time domain presents to us directly the response of the system—we may or may not ultimately be able to understand what is going on, but at least the information given to us is directly related to the natural processes.

An ideal Debye system should have an exponential time domain response, Eq. (37), but just as in the frequency domain we hardly ever see the classical response given by Eq. (38), so also in the time domain it became recognized early on that most systems responded with the relation (54), which is known as the Curie–von Schweidler law (*28*). The principal interest in the present review will not be in the response of linear systems, except insofar as this will be necessary to the understanding of loss peaks, but we concentrate attention on the nonlinear time domain response.

As described in Section II, one of the criteria of linearity is the constancy of the difference between the charging and the discharging currents, which should be equal to the constant DC current i_0. Any result in which the discharge current is actually higher than the charging current over part of the time response is clearly in violation of the linearity postulate. An example of this is shown in Fig. 52, giving the charging and discharging currents in stearic acid multilayers under various conditions of average field and temperature. While under "moderate" conditions the two currents are at least approximately consistent with linearity—although closer examination of the data reveals even then significant departures from it—we observe evident crossing over under the more "extreme" conditions of higher temperatures and fields (*96*). This is interpreted as the result of the injection of space charge into the system during the charging period, with the subsequent withdrawal of this space charge contributing an initial excess discharging current. It should be noted that this contribution is relatively small in the present instance—the system remains to a first approximation linear in the applied field.

A different form of nonlinearity that is readily observable in the time domain response is the rising ratio $i_c(t)/V_0$, where V_0 is the amplitude of the applied signal. For a dipolar response this ratio should remain practically constant up to the highest fields that may be sustained in normal dielectrics before the occurrence of breakdown. Any departure from constancy is therefore interpreted as being due to the onset of nonlinear carrier responses and these may or may not affect also the ratio $i_d(t)/V_0$.

Even more extreme examples of nonlinear behavior may be seen in such systems as electrolytic capacitors, which are very nonlinear on account of the various electrochemical interactions in them (Fig. 53 gives an example of this) (*54*).

Before proceeding further with the presentation of recent evidence relating to the photoinjection of charges into dielectrics, we discuss briefly the principles of charge injection, in general.

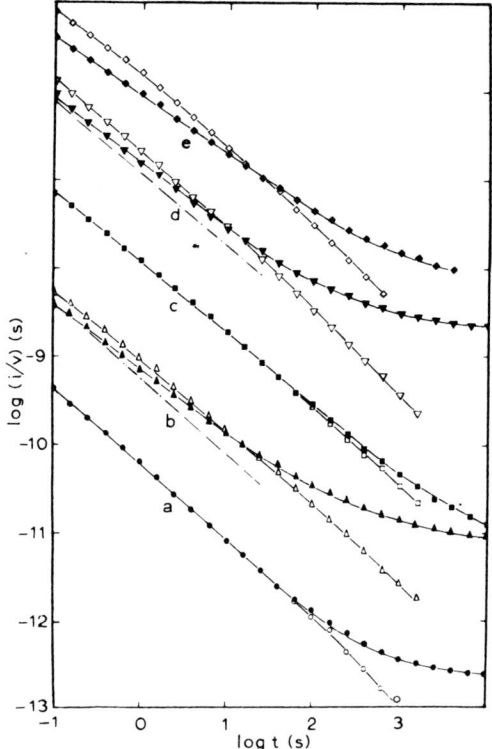

FIG. 52. The logarithmic representation of the charging currents $i_c(t)$ (full symbols) and discharging currents $i_d(t)$ (open symbols) for Al–stearic acid–Al sandwich structures. Curves (a)–(d) relate to a sample of 11 monolayers thickness, curves (e) to 13 monolayers. The temperature and average field were as follows:

Curves	a	b	c	d	e
Average field ($\times 10^7$ V/m)	6.8	6.8	1.1	9.1	4.8
T(K)	250	300	300	300	320

The currents are scaled by the applied voltage V and successive sets of curves are displaced vertically by one decade with respect to one another for clarity. The dot–dash lines on curves (b) and (d) represent the correct relative positions of the lower-stress characteristics given by lines (a) and (c), respectively. The conductance scale refers to curve (a), Jonscher and Careem (96).

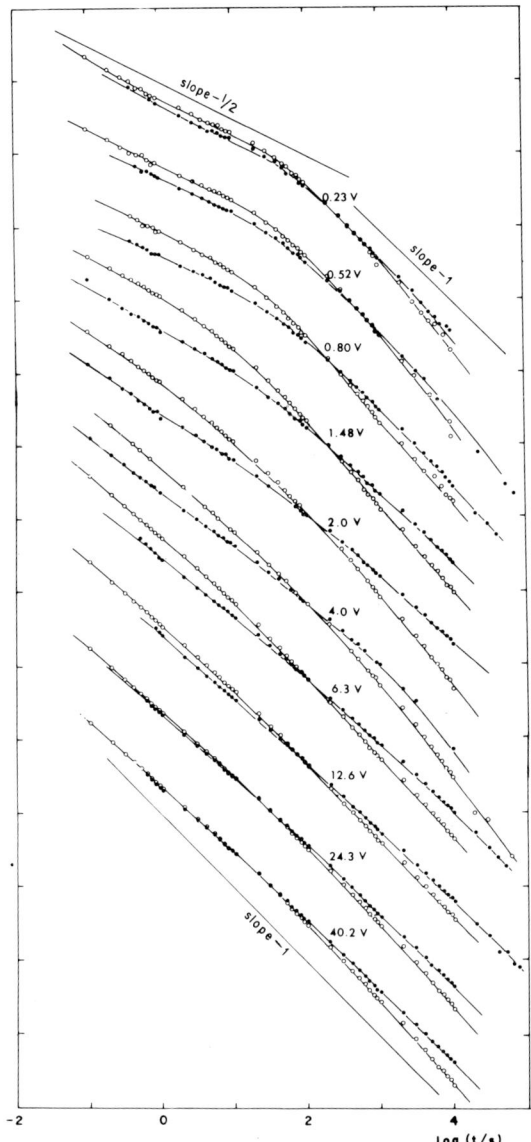

FIG. 53. The charging and discharging currents (solid and open points, respectively), for sintered powder tantalum electrolytic capacitor with various amplitudes of the polarizing voltage. The slopes of $-\frac{1}{2}$ and -1 are indicated. At lower voltages the exponent is $n = \frac{1}{2}$ at short times and $n > 1$ at longer times, and the currents i_c and i_d cross over progressively more as the voltage increases. At still higher voltages the slopes become more uniform and the crossing over disappears. This is a clear example of nonlinear behavior. The individual sets of curves are displaced vertically for clarity by one decade. From Meca and Jonscher (54).

10. CHARGE INJECTION INTO SOLIDS

This topic presents many ramifications and it is desirable to set out some general principles. We may distinguish, in particular, between (97): *external* injection from electrodes, p–n junctions, etc., and *internal* injection through photogeneration or thermal generation within the material. In addition, it is possible to distinguish between three principal regimes of injection:

(1) Electrical neutrality is preserved during injection, either because injection of minority carriers is neutralized by the majority carriers—p–n junction injection—or because charges of both species are being uniformly generated—photoinjection.

(2) Electrical neutrality is not preserved but the dielectric response of the host lattice is instantaneous on the time scale of the transport processes of the injected charges carriers.

(3) Electrical neutrality is not preserved and the host lattice has a dielectric response time comparable to or even much longer than the response of charge transport.

In case (1) the prevailing electric field in the sample is not disturbed by injection and the problem is one of ambipolar transport (98)—this only arises in semiconductors and does not concern us in the present context. Case (2) is typical of majority charge carrier injection into semiconductors and semi-insulators and has been discussed extensively in the literature (3, 5, 99, 100). The principal point here is that the carriers are assumed to have a certain mobility μ, so that their velocity at any point and time may be written as

$$v(r, t) = \pm \mu E(r, t) \tag{58}$$

where $E(r, t)$ is the instantaneous value of the electric field at any point and the sign of the mobility is always taken to be positive, the $\pm$ sign relating to positive and negative charges, respectively. The local charge densities determine the space charge,

$$\rho(r, t) = \sum_i q_i n_i(r, t) \tag{59}$$

where the summation is taken over all charge species taking into account the sign of their charges q_i. The densities $n_i(r, t)$ are determined by the appropriate continuity conditions with due regard to the boundary conditions. The dielectric displacement is determined by the space charge

$$\text{div } D(r, t) = \rho(r, t) \tag{60}$$

The only missing link in the system of equations is now a relation between D and E. In all theoretical treatments of the transport problem the assumption is invariably made

$$D(r, t) = \epsilon_0 E(r, t) + P(r, t) = \epsilon E(r, t) \tag{61}$$

with a constant value of the permittivity ϵ. This expression is perfectly acceptable in "semiconductor" situations, case (2) above, since the response of the lattice may be taken to be instantaneous on the scale of charge transport. However, in the case of dielectric matrices with manifestly slow responses, i.e., case (3), the formalism of Eq. (61) is inadmissible since it is valid essentially only in the frequency domain under spatially uniform conditions and therefore cannot be used in the solution of a spatially nonuniform time-dependent problem.

The proper solution must consist in the inversion of the convolution integral (5), which properly expresses the polarization as the dependent variable in terms of the field E as the independent variable—what we require in the present instance is the inverse relation, with E as the dependent variable. This problem has not yet been properly resolved mathematically, but for a recent treatment of charge motion in lossy dielectrics see Jonscher (*106*).

11. Time Response to Photoinjection

An example of the type of response that leads us to question the validity of the accepted methods of solving transport problems in dielectrics by the conventional semiconductor analysis, including Eq. (61), is the response to photoinjection of charge carriers. Figure 54 shows the comparison of ordinary dark discharge current $i_d(t)$ resulting from the sudden removal of a steady polarizing field E_0, and of the photodecay current resulting from the sudden removal of a UV excitation while keeping the constant applied potential across the sample. The surprising feature of this experiment is that the response of $i_d(t)$ and of the photocurrent $i_\phi(t)$ is identical independently of temperature and of the magnitude of the applied bias. The exponents n in the Curie–von Schweidler law, which is observed in both cases, are practically identical, regardless of the completely different boundary and initial conditions (*101, 102*) as set out in Table II. Under these conditions it is very difficult to see how the solution of the classical semiconductor transport equations could possibly give the same result as

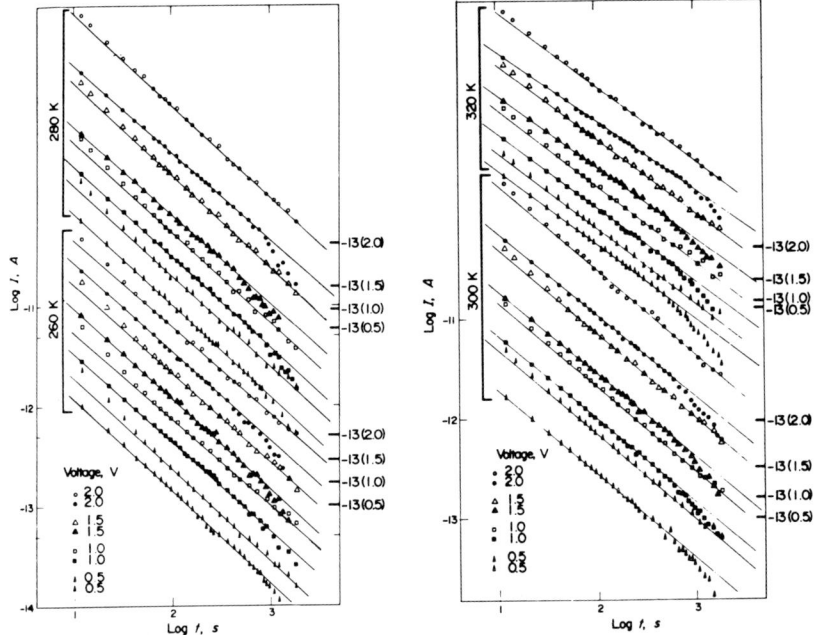

FIG. 54. The dark discharge currents (solid symbols) and the photodecay currents at constant voltage (open symbols) after the removal of an ultraviolet illumination. The readings were taken at 0.5, 1.0, 1.5, and 2.0 V and at temperatures of 260 and 280 K. The left-hand current scale refers to the lowest set of currents; all other sets are displaced vertically for clarity and the markers on the right indicate the position of 10^{-13} A. The strict parallelism between the dark and photodischarge currents should be noted. From Jonscher and Buddhabadana (102).

the dark discharge characteristic function $f(t)$. The answer to this must lie in the strong coupling that exists between the charge carriers and the polarization of the dielectric matrix. The procedure outlined in Section VI,10 would take account of just that coupling.

We propose that the experimental results shown in Fig. 54 lead to the following conclusion: The decay of polarization after the sudden removal of external excitation follows the universal Curie–von Schweidler law, Eq. (54), regardless of the nature of the charge species responsible for the polarization and regardless of the initial and boundary conditions imposed on the system.

The full significance of this conclusion may not be evident at present, but the very simplicity of this extension of the universality of the dielectric response makes it highly plausible.

TABLE II

Boundary and Initial Conditions Applicable to Dark Discharge and to Photorelaxation

Parameter	Dark discharge	Photodecay
Initial field distribution	Substantially uniform	Significantly influenced by injected space charge
Polarization in steady state	Mainly due to dipoles, with some contribution from any dark injected charge carriers at higher voltages and temperatures	In addition to the dipolar contribution, there is a significant polarization due to the photogenerated charge distribution
Change of potential	From finite to zero	No change
Final field distribution	Zero field	Substantially uniform
The change of polarization arises from	Mainly dipolar relaxation	Mainly mobile charge redistribution

12. Contribution of Charge Carriers to Polarization

We have made repeated references to the effects of charge carriers on dielectric polarization, pointing out the remarkable fact that they make a contribution following the same general universal law as the dipoles, except that they tend to give rise to strong low-frequency dispersion, which is normally absent in dipolar systems. We noted in Section VI,11 that charge carriers make an evidently nonlinear contribution to polarization that is clearly visible in the time domain response. It is not always easy, however, to distinguish the charge carrier effects unambiguously in the frequency domain response and the purpose of this section is to outline two experiments in which these processes appear rather strikingly (103).

The first of these relates to the electrolytic capacitor (54) already described in Fig. 31, but this time the measurements were taken with a small signal superimposed on a steady bias (Fig. 55). We note that the strongly dispersive loss characteristic present in the absence of bias, with a slope $-\frac{1}{2}$ characteristic of the Warburg diffusive mechanism, disappears rapidly with the onset of bias, to reach a pure universal law with $n = 0.87$ as the bias exceeds 5 V. Thus, an increase of bias implies a decrease of the dielectric loss in the low-frequency regime.

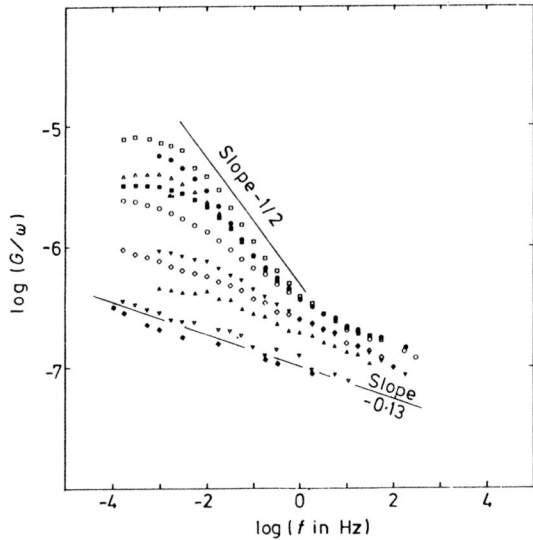

FIG. 55. The frequency dependence of the dielectric loss, expressed as $G/2\pi f$, of a sintered tantalum electrolytic capacitor measured under an alternating voltage signal of 0.2 V and with a steady bias of 0 (□), 0.10 (●), 0.50 (△), 0.75 (■), 1.0 (○), 1.5 (▼), 2.0 (◇), 3.0 (▲), 5.0 (▽), and 9.5 (◆) V. As the steady bias increases, so the dielectric loss at low frequencies decreases rapidly to reach a steady value following the universal law, equation (1), with the exponent $n = 0.87$. From Jonscher et al. (103).

The second experiment was performed on stearic acid, but with a variable amplitude of the applied signal (we note that a similar result was also obtained with a bias superimposed on a small signal, but it was simpler to carry out the measurements with a variable amplitude—this would not have been permissible in the case of the electrolytic capacitor). The results are shown in Fig. 56, giving both the dielectric loss and the real part of the permittivity. It is seen that the tendency to stronger dispersion, visible even at low amplitudes (Figs. 42 and 43), now becomes dramatically enhanced as the amplitude increases, tending to slope -1, which could be interpreted as the onset of direct current conduction. That this is not the case is evident from the data for $\epsilon'(\omega)$, which clearly show a very significant rise, indicating the presence of the strongly dispersive response discussed in Section VI,8.

The interpretation given for this is that in the case of the electrolytic capacitor the dominant charge species are ions that are difficult to inject into solids, so that increasing bias or amplitude tends to sweep them out,

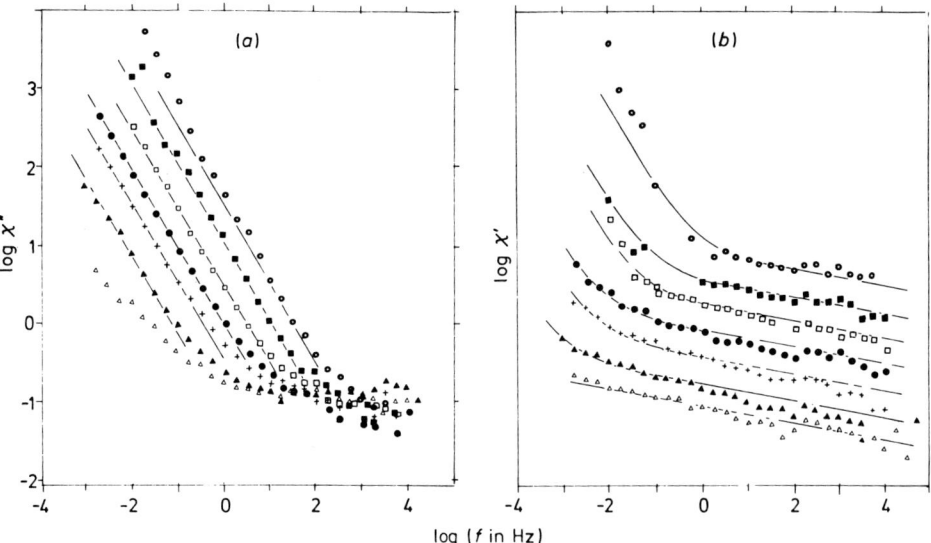

FIG. 56. The frequency dependence of the imaginary part (a) and the real part (b) of the dielectric loss of a 15 layer film of stearic acid sandwiched between two aluminum electrodes, measured at 343 K, for applied signals of amplitude 2.0×10^8 (○), 1.3×10^8 (■), 1.2×10^8 (□), 8×10^7 (●), 6×10^7 (+), 1.6×10^7 (▲), and 2.6×10^6 (△) Vm^{-1}. The straight lines in (a) and the ultimate low-frequency curve in (b) have the slope -1. The loss data were corrected for series resistance effects (Jonscher, 1978), while the $\chi'(\omega)$ data were obtained by subtracting the same value of ϵ_∞ from all the data measured experimentally. The $\chi(\omega)$ curves in (b) are displaced with respect to one another by $\frac{1}{3}$ of a decade to avoid overcrowding and the straight lines are drawn in corresponding positions, to show that there is no significant effect of electron injection on the high-frequency response. From Jonscher et al. (103).

thereby leaving only the lattice contribution, while in the case of stearic acid the dominant species are electrons, which are easily injected.

13. Conclusions Regarding the Experimental Evidence

We complete this review of the dielectric response of a very wide range of materials by summing up the principal features that were already foreshadowed in Section V and that we are now able to appreciate properly. Despite the apparently very wide diversity of dielectric responses of different materials, we maintain that there exists a sufficient measure of uniformity to warrant a closer analysis that will eventually lead us to a completely new approach to the interpretation of dielectric response.

The most general feature of the experimental data is their broad division into two categories: those with more or less clearly pronounced loss peaks and those without such peaks down to the lowest accessible frequencies. The latter category may involve ordinary DC conductivity, σ_0 overshadowing the loss $\epsilon''(\omega)$, but with $\epsilon'(\omega)$ independent of frequency at low frequencies, or it may involve a strong dispersion of both $\epsilon'(\omega)$ and $\epsilon''(\omega)$.

We have reached a tentative conclusion that the materials showing loss peaks are normally dominated by dipolar processes; those without loss peaks are dominated by charge carrier processes. Charge carriers are also evidently responsible for DC conductivity but the precise correlation between $\epsilon''(\omega)$ and σ_0 is not clear at present. One point appears certain, however; no clear division into dipolar and charge processes is permissible, since physically these two species may overlap to a considerable extent, for instance, in the case of hopping-charge carriers, which may act as either.

The various types of dielectric response are summarized in Fig. 57, which follows the broad classification established in Sections V and VI. We note the virtual absence of the pure Debye response, but slightly broadened symmetric peaks are found in ferroelectrics at sufficiently high frequencies. Still broader and slightly asymmetric peaks, referred to here as the α peaks after the polymer literature, are found in many "liquid" systems such as polymers and glasses above the glass transition temperature, in some impure polymers at cryogenic temperatures, as well as in p–n junctions and in liquid crystals. These peaks are strongly temperature dependent and are not normally simply activated.

The transition from the α peaks to the still much broader β peaks is not necessarily clear-cut in terms of their frequency response. β peaks are very asymmetric and are usually characterized by well-defined activation energies that tend to be smaller than for the α peaks. β peaks are found in many "solidlike" systems, particularly in polymers and glasses below their glass transition temperature, and they are also seen in some p–n junctions and in liquid crystals at right angles to the orientation of the molecules.

Both α and β peaks are characterized on the high-frequency side of the loss peak by a power law relation of their frequency dependence,

$$\chi''(\omega) \propto \omega^{n-1}, \quad \text{with } n < 1 \tag{62}$$

with $0.2 \leq n \leq 0.6$ for α peaks, and $0.6 \leq n \leq 0.9$ for β peaks.

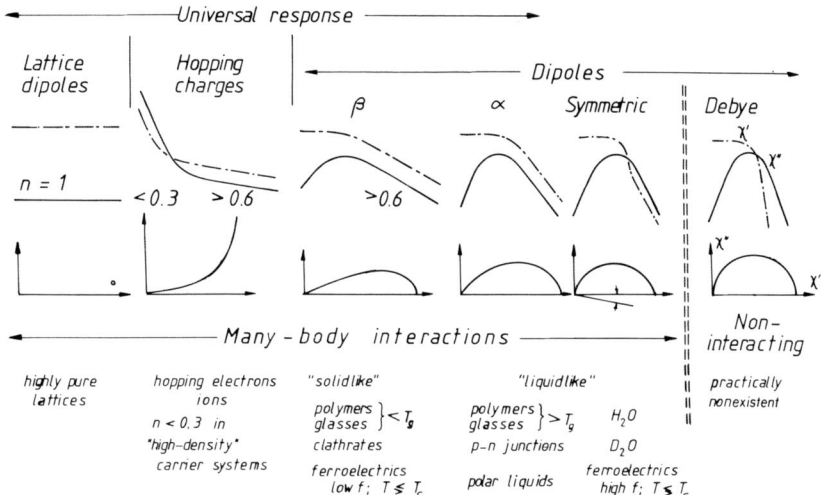

FIG. 57. A schematic representation of the various observed types of dielectric response in the entire range of solids. The upper set of diagrams represent the shapes of the logarithmic plots of $\chi'(\omega)$ (-·-) and $\chi''(\omega)$ (—) ranging from the ideal Debye through the α and β peaks and on to the universal dependence for charge carrier systems. The limiting forms of behavior are represented by the strong low-frequency dispersion with small values of n and by the frequency-independent "lattice" response with $n = 1$. The lower set of diagrams represent the corresponding complex χ plots. The various types of materials obeying the respective types of response are shown and the presumed mechanisms are indicated. Reference (140).

The next type of dielectric response that we associate with the presence of charge carriers of electronic or ionic nature is characterized by the same type of frequency response as the β peaks on the high-frequency side, but no peak is discernible at low frequencies. Sometimes a second region of the same form as Eq. (62), but with a lower value of n can be seen, as in Eq. (55). This type of response is seen in a very wide range of materials of all physical and chemical characteristics and extends normally over very wide range of frequencies. It is quite often found as a background to loss peaks, suggesting the presence of at least two loss mechanisms.

We have distinguished a characteristic type of loss response that is "flat" in frequency, with $n \simeq 1$, and we have associated this with the "lattice" response, i.e., with the response of a regular array of permanent or induced dipoles. This is a new suggestion, which does not appear to have been made explicitly before now, and it is very difficult to explain it in the accepted theories.

Finally, we have formulated a type of low-frequency behavior that is of the same form as Eq. (62) but corresponds to very small values of the exponent n, typically between 0.1 and 0.3. This we associate with the presence of "high" densities of low-mobility charge carriers.

We thus arrive at the conclusion that the empirical power law (62), which we define as the *universal law of dielectric response,* is applicable in a remarkably wide range of physical and chemical situations, corresponding to values of n spanning the entire range $0 < n \leq 1$. This law is applicable independently of:

physical structure—single crystal, polycrystalline, amorphous and glassy;
type of bonding—covalent, ionic, molecular;
chemical type—inorganic, organic, biological;
polarizing species—dipoles, hopping electrons, polarons, ions;
geometrical configuration—from "bulk" dimensions to monomolecular layers, planar and intricate geometries, continuous media, and granular media.

This generality or universality has forced us to look beyond the particular interpretations developed for individual situations and to seek a suitably universal framework within which all these materials would find a common interpretation. In this search we have been encouraged by the remarkably simple property of the universal law, i.e., the constancy of the energy lost divided by the energy stored in the system [Eq. (51)].

We have proposed (2) that the one obvious principle of sufficient generality to be applicable throughout the entire field of condensed matter is the many-body interaction between charge carriers, dipoles, and the "lattice." Although various attempts have been made in the past to take many-body interactions into account in the interpretation of dielectric phenomena, none of them appear to have succeeded in explaining the observed behavior. Part of the reason for this failure may lie in the insufficiently clear recognition of the true nature of the empirical laws of response, and hence our emphasis on a thorough review of experimental facts.

Before proceeding with the exposition of the new many-body approach, we wish to discuss very briefly the principal accepted interpretations of dielectric responses in different materials, since this will make it clear how inadequate these theoretical approaches are in the face of the experimental evidence.

VII. Currently Accepted Interpretations

1. Historical Background

The virtual nonoccurrence of the Debye response in solids and the prevailing strong departures from it, often amounting to behavior that is totally unrecognizable, have been known for over half a century in some cases. A vast literature has arisen dealing with attempts to understand this behavior in terms of the atomic structural properties of the materials in question.

Historically, the first serious attempt to understand the dielectric response of matter has been Clausius' and Mossotti's treatment of the static permittivity $\epsilon(0)$ almost exactly 100 years ago. Having correctly recognized the many-body nature of the phenomenon of polarization, they made a brave attempt to treat it in the static limit by introducing the concept of an effective *local field* resulting from the interaction between the dipoles in the system and the external field. It cannot be said that the resulting "proof" is either intuitively clear or logically convincing. The principal line of the argument is that if the polarizability of an individual atom is α, then the total polarization is obtained by summing the polarizations of the individual atoms in the volume and if, as might have been expected, the result does not give the correct result, the external field has to be "adjusted" by devising an internal field that would give the correct polarization. Thus, instead of acknowledging the fact that the individual α had no relevance to the case of condensed matter with its high density of interacting species, the introduction of an artificial internal field was found as the salvation, which was eventually to lead to frustration and failure.

This criticism is not to detract from the value of the original attempt, which can only be described as visionary, but it does underline the failure of subsequent generations to break away from an apparent blind alley. What is even more serious, however, is that after the initial attempt to deal with the static response on the basis of many-body interactions, the great majority of subsequent treatments of the dynamic response were confined to variations on the theme of the one-particle Debye responses. This fixation on the "Debye philosophy" has weighed very heavily on the development of dielectric thinking in the last 40 years or so and has undoubtedly contributed to the poor reputation of this branch of science among the "respectable" scientists.

An excellent and up-to-date account of the various theoretical interpretations of the classical schools may be found in Böttcher and Bordewijk (*107*).

2. Distributions of Relaxation Times

The first such attempt to reconcile the one-particle Debye response, Eq. (38), with the observed behavior of materials introduced the concept of the distribution function of relaxation times $g(\tau)$, based on the very plausible argument that any material is likely to have sufficient local variation of microscopic conditions to lead to a *distribution of relaxation times* (DRT) (*108–112*). This gives us the formal expression for the complex dielectric susceptibility

$$\chi(\omega) = \int_0^\infty \frac{g(\tau)}{1 + i\omega\tau} d\tau \tag{63}$$

Viewed as an integral transform from the Debye function (38) to the observed response function $\chi(\omega)$ with the kernel $g(\tau)$, this procedure is mathematically without reproach, at least for the type of functions that one is likely to find in practical situations. Physically, it corresponds to the filling of the empirical response with Debye functions, as shown in Fig. 58. The question of the physical plausibility, let alone of any rigorous justification, remains wide open, however and serious interpretational difficulties arise.

First of all, it must be acknowledged that the required range of τ is comparable to the available range of frequency in which the behavior

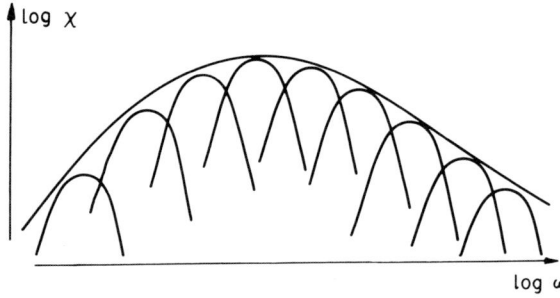

Fig. 58. A schematic representation of the significance of a distribution of relaxation times in application to the interpretation of a "non-Debye" loss peak in terms of a distribution of Debye peaks.

departs from the Debye pattern. There are very few cases indeed in which a range of non-Debye response is bounded at the lower frequency end by a region of $\chi'' \propto \omega$ and at the higher end by $\chi' \propto 1/\omega$. In practically all known cases the range of proven non-Debye behavior is limited only by our inability to make more extended measurements and there are therefore no indications of any limits to the required range of τ, except that this may have to exceed significantly six to ten decades, which is bound to pose serious problems.

In typical "interpretations" of dielectric data, the physical reasoning is set aside and a purely formal procedure of inverting the integral relation (63) is applied to derive the required DRT from the experimentally observed loss. Attempts are sometimes made to justify qualitatively ex post facto the DRT so obtained (113), but this is seldom a satisfactory procedure. Particularly disconcerting, however, is that the majority of researchers are content to state that their results imply a DRT and to leave the matter at that.

One point that is hardly mentioned in all this is the sheer physical implausibility of one-particle responses involving relaxation times of the order of, say 10^{-4}–10^4 sec, which are often required for the interpretation of experimental date. Such slow one-particle movements are inconceivable and the very existence of such slow responses should have led to the recognition of many-body effects. It is interesting to note in this context that there exists one specific case in which the DRT was calculated from first principles and was compared moderately successfully with the observed spectrum (114), but this covers the range 10^9–10^{10} Hz, i.e., frequencies that are perfectly compatible with one-particle oscillations.

The second fundamental difficulty with the DRTs becomes evident when we approach the problem of the interpretation of the dielectric response not from the standpoint of any one narrow class of materials, such as polymers, ceramics, glasses, or organic molecular solids, each of which might conceivably be consistent with a particular DRT, but from the synoptic viewpoint adopted in the present treatment, where a vast range of materials is found to require rather similar types of DRT. Even if the DRT hypothesis were physically admissible, a proof of its general applicability would be required. This to our knowledge has never been attempted, and it constitutes a serious methodological fault.

This state of affairs could only come about because of the far-reaching fragmentation of dielectrics research into narrow compartments dealing with individual materials. An excellent example of this is found in the case of the response of amorphous semiconductors.

3. Hopping Electronic Systems

The hopping of localized electrons or polarons in amorphous semiconductors has been studied extensively in the last 20 years (20) and traditionally the data are being presented in the form of $\sigma(\omega)$ as in Fig. 34. This tends to mask the similarity with the dielectric response of most other materials, and the problem is hardly ever seen in the dielectric context to which it firmly belongs.

In fact, such is the preoccupation of the amorphous semiconductor school with the form of the frequency dependence of the AC conductivity given by Eq. (57), with the exponent $n \simeq 0.8$, that this specific law was taken as the hallmark of hopping conduction, regardless of the fact that this universal law applies to most other systems as well.

The mathematical sophistication involved in the treatment of the AC response of hopping electrons far exceeds anything that has been done in the context of DRTs described above (115–119). However, on closer examination we note the absence of agreement on what should be the appropriate mathematical procedures involved in the averaging over many hopping transitions, including hopping over extended sequences of localized states ultimately leading to DC conduction.

One often-quoted formula gives the dielectric loss of hopping electrons in the form (120)

$$\chi''(\omega) = (\pi/3)e^2kT[N(W_F)]^2\alpha^{-5}[\ln(\nu/\omega)]^4 \tag{64}$$

where $N(W_F)$ denotes the density of states at the Fermi energy, α the decrement of the localized wave function at a typical state, and ν a "suitable" frequency, which may be presumed to fall between typical "lattice" frequencies 10^{13} sec^{-1} and "electronic" frequencies 10^{15} sec^{-1}. The numerical evaluation of data is not helped by the fact that the not very well-defined quantity α occurs in the fifth power and the even less well-determined $N(W_F)$ in the second. However, the more serious objection arises from the fact that formula (64) gives a continuous curvature in the log χ''-log ω plot so that the exponent n defined by the relation (72)

$$n - 1 = d(\log \chi'')/d(\log \omega) \tag{65}$$

is a continuously varying function of frequency, contrary to general experience, which suggests that n is a constant or, if it varies, that it does so in the opposite sense to that predicted by Eq. (64). It can be shown that, setting $\log(\nu/2\pi) = a$, $\log(\omega/2\pi) = x$, Eq. (64) gives

$$1/1 - n(x) = \tfrac{1}{4}(\ln 10)\,(a - x) = 0.576(a - x) \tag{66}$$

This relation is sketched in Fig. 59 with a range of plausible values of the frequency $\nu/2\pi$. The same diagram also indicates the observable range of values of n in typical electronic materials. It is clear that most experimental data fall outside the plausible range of theoretical values. We conclude that expression (64) is not applicable to the majority of experimental data, so that the question may be posed whether it is applicable to any. The same doubt arises with regard to the reliability of the values of $N(W_F)$ derived from this expression.

The fundamental point remains, however, that all these treatments of the frequency dependence of charge hopping amount to little more than another form of DRT—the hopping movements are uncorrelated and therefore cannot properly reflect many-body interactions. Taking now the evidence of Fig. 35, which shows that a very wide range of materials not only follows the universal law, which is well known, but also reveals a remarkably narrow range of absolute values of $\sigma(\omega)$ at any particular frequency, which is not so well known, and taking further into account the similarity of ionic responses, we begin to realize that serious doubts arise with regard to the entire procedure.

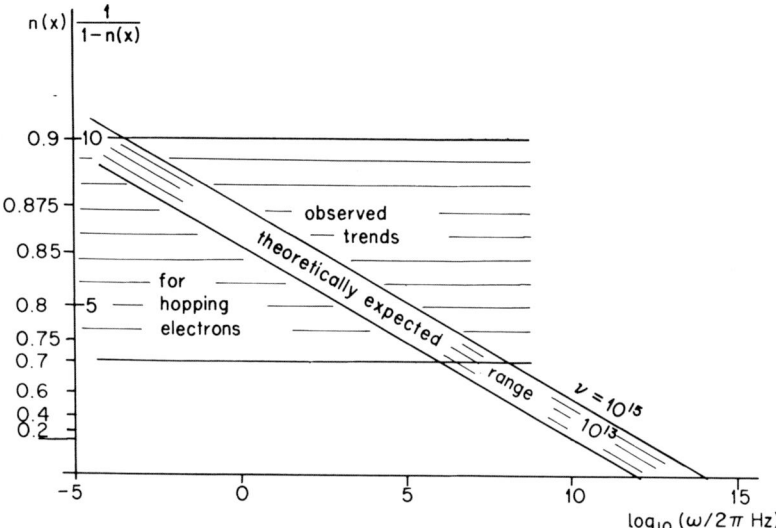

FIG. 59. The theoretically expected frequency dependence of the exponent $n(x)$, defined by Eq. (65) according to formula (64). The experimentally observed trends corresponding to a frequency-independent exponent n are also indicated, including the range of the values observed in electronic hopping materials. The extreme limits for the frequency ν are also indicated. Hill and Jonscher (72).

4. Ionic Conductors

Much less clearly defined notions of DRTs are being invoked in the "interpretation" of the dielectric response of many ionically conducting materials, where the mere existence of inclined circular arc *impedance* diagrams is taken to imply a DRT. One has the impression that the notion has been taken over from the corresponding complex *permittivity* Cole–Cole graph, Eq. (47) and Fig. 10. However, it is clear that inclined circular impedance arcs cannot correspond to inclined circular permittivity arcs, and so it is difficult to envisage the same interpretation. The fact that these impedance arcs are eminently compatible with the universal dielectric response (77, 78) points strongly to the possibility that their interpretation should be sought on the same lines as that of all other materials.

An entirely different form of approach developed to a high level of formal sophistication is widely accepted in the field of ionic conductors, particularly in the so-called fast-ion conductors (121, 122). This is Macdonald's treatment of the incomplete transparency of electrodes to different ionic species, which ascribes the entire frequency dependence of the response to the interfacial processes that may be considered as a variant of the $R-C$ equivalent network approach. It appears, however, that apart from the Warburg impedance with its angle of tilt of 45° (90, 91) which forms a limiting case of this approach, it is difficult to simulate the impedance diagrams with tilt angles in excess of a few degrees. The principal difficulty, however, arises from the uncertainty as to the values of the relevant barrier parameters, which cannot easily be determined independently. There is little doubt that in the case of certain "dense" electrolytes whose DC conductivity is sufficiently high to dominate the bulk dielectric response, Macdonald's approach can give good results, but the situations such as shown in Fig. 40 do not fit easily into this type of analysis.

5. Diffusive Models

We have already mentioned the Warburg mechanism, which represents a brilliant early attempt to interpret the frequency response of liquid electrolytes with blocking electrodes. Warburg's analysis predicts the following form of the complex impedance:

$$Z(\omega) = \frac{kT}{Ase^2 p(2\omega D)^{1/2}} (1 - i) \qquad (67)$$

where A is the area of the electrode, s the valence of the ions taking part in the reaction, D their diffusivity, and p their concentration. Expressed in terms of the equivalent complex permittivity this gives for the diffusive model

$$\epsilon_d(\omega) \propto (1 - i)\omega^{-1/2} \qquad (68)$$

which corresponds to the universal relation with $n = \frac{1}{2}$, which is the special case for which Eq. (50) gives $\epsilon'(\omega) = \epsilon''(\omega)$. While there are evident examples of this type of response, e.g., Fig. 51, it is doubtful if this mechanism can explain the inclined Z arcs in ionic materials, most of which correspond to angles between 15 and 30°.

One significant point regarding the application of this mechanism in solids is the requirement of overall neutrality in the system in order to enable the diffusive processes to dominate over space charge effects. This means that the system must contain comparable densities of oppositely charged ionic species—a condition not always evidently met in solids.

One other attempt may be mentioned to use diffusive mechanisms in the interpretation of dielectric responses—the suggestion that the relaxation of dipoles should not be spontaneous, giving the Debye response, but should be "triggered off" by the arrival at the dipole site of a "defect" moving by a diffusive process (*123, 124*). While the physical nature of the proposed mechanism is far from clear, it represents one of the few attempts to break away from the basic Debye formalism, although again it is difficult to see how it can explain values of n significantly different from $\frac{1}{2}$.

It may be noted that attempts have also been made to extend the internal-field argument into the frequency dependence of dielectric response (*125*) but the result is not very encouraging. Likewise, inertial effects cannot give a significant departure from the simple Debye model.

6. Correlation Function Approach

A large volume of detailed and mathematically advanced analysis has been devoted to the calculation of the correlation functions of dipoles or particles under the action of step function excitation (*126–131*). These amount essentially to the time domain calculation of $f(t)$, under various assumed conditions of interaction. While these techniques have proved to be successful in the treatment of the response of liquids at very high frequencies, the response of solids at lower frequencies could not be treated analytically in a satisfactory manner without the introduction of arbitrary assumptions in order to obtain agreement with the results of measurements. A fair amount of the relevant theory has been developed on purely formal lines, with little reference to the underlying physical reality.

7. Interfacial Phenomena and Maxwell–Wagner Effects

Many dielectric materials are inhomogeneous either on a microscopic or a macroscopic scale—phase separation in glassy materials, microcrystallites in amorphous solids, lamellar and spherulitic formations in polymers, grains in ceramics, or even loose aggregates of powders—and this gives rise to an understandable concern as to the effect of the interfacial phenomena on the dielectric response. Contact barriers, transparency to the various mobile species, accumulation and depletion near interfaces have all received ample treatment in the literature, as they should. However, as time went on, a habit seems to have developed of putting down almost any phenomenon for which no better explanation could be developed on the account of these interfacial processes, regardless of the plausibility of the proposed model in the particular circumstances.

The basic reasoning is simple: A "series" bulk resistance and a capacitance arising from interfacial blocking action form a $R-C$ network that has a Debye response (Section IV). There is likely to exist a "distribution" of these $R-C$ networks in the material, forming an interconnecting three-dimensional lattice with a wide distribution of parameters. We are thus left with another version of the familiar DRT argument, equally appealing to the imagination and equally impossible to substantiate in detail. The literature abounds in treatments of this subject, loosely known as the Maxwell–Wagner–Sillars effect, and we merely cite two fairly recent papers where further references may be found (*132, 133*). They are used especially in situations where a strong low-frequency dispersion is to be seen. We argue in Section VI, that many, at least, of these phenomena can be much more convincingly explained within the framework of our universal model with small values of n. The remarkable similarity of the response of so many different materials, some of them classical Maxwell–Wagner systems, e.g., sand and biological samples, represents our principal argument. What remarkable coincidence of $R-C$ network parameters should always result in the same type of response? To satisfy everyone that this is the correct interpretation will be as difficult as with the other DRT "theories, but a wider recognition of the existence of the observed regularity of the low-frequency dispersion may help to awaken a renewed interest in this long-standing if poorly understood problem.

8. Conclusions Regarding the Accepted Interpretations

This brief review of the accepted theoretical approaches to the interpretation of the dielectric response of solids should have convinced the reader that there exists an unbridgeable "credibility gap" between the physical reality and the theory. The well-defined experimental trends in the data relating to a vast range of materials described in Section VI simply defy interpretation by traditional methods, not only for particular materials, but much more dramatically for the range of materials. There is little hope of escaping from this impasse by yet further adjustments and refinements of the existing theories. Their basic principles had been known for many years and they have reached a point of saturation for the simple reason that in the last analysis, these theories attempt to explain the cooperative behavior of many-body systems in terms of distributions of one-particle processes.

This has led to the present unsatisfactory division into, on the one hand, "popular" interpretations, for the most part accepted uncritically by experimentalists as "working models" not requiring serious verification of their physical plausibility and, on the other hand, "respectable" models attracting the interest of theoreticians but relating to the high-frequency quantum limit of response. In the former category, we place most DRTs, the formal descriptions of the complex permittivity, and the equally formal correlation function arguments. The second category does not concern us here, since it does not explain the behavior of solids at "low" frequencies.

The way out is to approach the entire question of the dielectric response of solids from the entirely fresh standpoint of many-body interactions, giving the universal response as a natural primary result, and not through manipulations of Debye responses. It has been known for some time that the Curie–von Schweidler law represents an exact solution to certain many-body processes in theoretical physics, such as X-ray singularities (134) and infrared divergence (135), albeit in a region of times between 10 and 20 decades shorter than those of interest in the dielectrics context. A many-body solution has been proposed for glassy materials (136).

VIII. The Many-Body Interpretation

The treatment of many-body phenomena is notoriously difficult, especially in the time-dependent regime, since "common-sense" or intuitive reasoning does not give reliable results. We know that a system of identical interacting particles may respond completely differently from an individual particle. In this section we propose to outline first a very simple model based on the singular property of the universal response expressed as the constant energy ratio (51) in the frequency domain. We follow this with a brief discussion of the nature of loss peaks in this model. We then mention a recent application of the Ising model to the derivation of α peaks in polymers and glasses and also in ferroelectrics, and finally we outline the most recent rigorous many-body approach to the problem of the universal law, based on a theory by Ngai.

1. THE SCREENED HOPPING MODEL (33–35)

This very simple and to some extent naive model was conceived specifically to account for the strikingly simple physical consequence of the universal law of dielectric response, given by the energy relation (51). The overriding consideration was that the model should have the generality required by the empirical data and that it should therefore be simple and make minimum specific assumptions regarding the material properties.

The basic argument starts from the proposition that if a model can be found that satisfies Eq. (51), then there is no need to solve specifically any particular dynamic differential equations describing the dielectric response, since the universally applicable Kramers–Kronig relations will ensure that the frequency dependence will follow the universal law of Eq. (62) as the only one compatible with the energy criterion. This procedure represents a considerable simplification of the argument and, despite possible objection to it, we are encouraged to present it here by the fact that the end result is confirmed by the more rigorous theory, which in certain respects reflects some of the basic postulates of the simple model. One important point in favor of this model is its inherent simplicity, which may be useful in gaining a deeper insight into the full significance of the more detailed theoretical treatment.

As the basis of our many-body model we have chosen the *Coulomb interaction* between charged species, resulting in their mutual self-consistent *screening*. We recall here that screening is widely accepted in free-particle systems such as gas plasmas and free electron gases in semiconductors and metals (*137*) even though the arguments leading to the derivation of the Debye screening radius $\mathfrak{r}$ in nondegenerate systems are not really rigorous. With a charged-particle density N in a medium of permittivity ϵ the potential of any one particle is given by the *screened Coulomb potential*

$$\Psi(r) = \frac{q}{4\pi\epsilon r} \exp -r/\mathfrak{r} \qquad (69)$$

with

$$\mathfrak{r} = (\epsilon kT/q^2 N)^{1/2} \qquad (70)$$

This means that every charged particle of charge q is surrounded self-consistently by a screening charge $-q$. It is not necessary to figure out exactly how this screening should apply to every particle—this is part of the accepted wisdom—and we propose to apply it directly to the situation arising in the presence of localized charges, which spend most of the time at defined localized sites and only make abrupt transitions between them under thermal excitation. These charges could be electrons or ions or, alternatively, they could be dipoles constrained to assume certain preferred orientations in space.

We suggest that the situation here is similar to that applying to free-charge carriers—mutual interactions will tend to screen charges from one another through repulsion of like and attraction of oppositely charged species. The fundamental difference in comparison with free-charge systems is that the charges are no longer entirely free to adjust their density in a fully self-consistent manner to the local potential. We take account of this fact by the assumption that instead of complete screening by a charge $-q$, we now have incomplete screening by a charge $-(1 - p)q$, where p is a screening parameter that may take values between 0 and 1.

Figure 60 shows the situation of a localized charge q occupying one of two allowed lattice sites i and j and surrounded by the opposite screening charge. If the charge now hops over to the other site in a time that is infinitesimally short in comparison with other time constants of the system, then the screening charge is left behind, since it cannot follow such a rapid movement and this is shown in Fig. 60b. There then follows a relatively slow transfer of the screening charge to the new site, Fig. 60c, until a new equilibrium is established, Fig. 60d. The rate of transfer of the

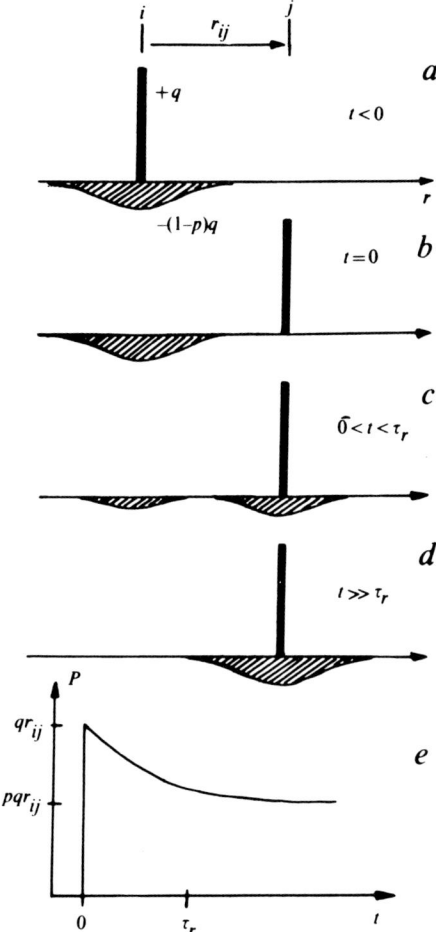

FIG. 60. The screened-hopping model in which a charge is constrained to hop between two localized sites i and j at a distance r_{ij} apart. The charge q is screened by an opposite charge $-(1 - p)q$ (a). When the charge jumps discontinuously to the other site, the screening charge is left behind (b), following relatively slowly (c), until a new steady state is established (d). The time dependence of the polarization resulting from this sequence of events is shown in diagram (e). Jonscher (33).

screening charge is determined by the natural rate of transitions between localized sites for all other charges in the system, since the screening of any one hopping charge is effected by other similar hopping charges. We define for this purpose a characteristic screening time τ_s, which is not to be confused with the Debye relaxation time. We do not imply that the time dependence of the screening transfer is in any sense exponential—the exact law of time dependence is of no consequence to the following argument, the only important point being that the time scale of τ_s is completely different from the abrupt transition of the initial charge.

We now consider the polarization arising from the abrupt transfer of the charge followed by the slower movement of the hopping screen charge. Figure 60e shows that the initial polarization is qr_{ij}, where r_{ij} is the distance between the localized levels, while the final polarization is pqr_{ij} after the arrival of the screening charge. Noting that this entire process takes place in an electric field E, which may be considered constant on the time scale of these events, the work done by the field is Eqr_{ij}, while the energy stored in the system is $Epqr_{ij}$. The energy loss or dissipation arising from this process is due to the fact that the screening charge has to move against the applied field E under the influence of the strong local interactions with the hopping charge. Since the same energy loss and storage are involved in every hopping transition, we obtain the energy ratio

$$\frac{\text{energy lost per transition}}{\text{energy stored}} = \frac{1-p}{p} \tag{71}$$

which must be equal to the ratio given by the energy criterion (51), thus giving the relation between the screening coefficient p and the exponent n in Eq. (62):

$$(1-p)/p = \cot(n\pi/2) \tag{72}$$

The limit $p = 0$ corresponds to complete screening, as in free charge systems; the opposite limit $p = 1$ corresponds to the absence of screening. The functional relationship between the parameters n and p is shown in Fig. 61, which shows how the full range of the exponents n between 0 and 1 can be obtained, in agreement with the empirical observations, by varying the screening parameter p, which here represents in a simplified way the strength of many-body interactions between charges and dipoles in the dielectric system under consideration.

Our screened hopping model involves three characteristic time scales, in addition to the practically instantaneous actual hopping transitions: the natural frequency v with which the charges jump in equilibrium between the sites, the screening adjustment time τ_S, and the frequency ω of the applied electric field. Since the time τ_S is determined by the individual transitions, it is reasonable to suppose that $\tau_S > 1/v$. We may therefore

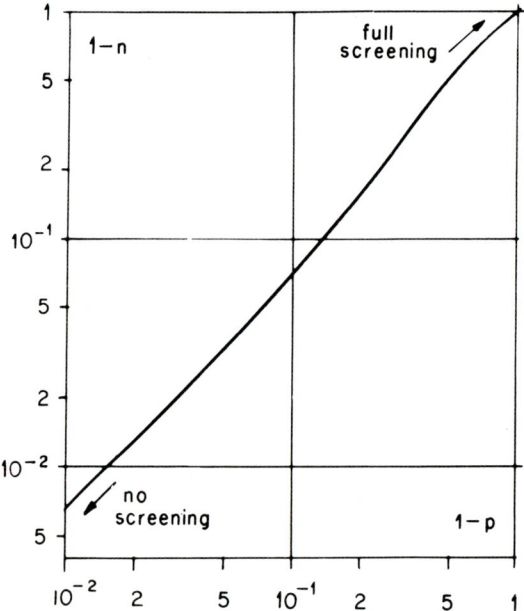

FIG. 61. The relation between the exponent $1 - n$ in the universal relation and the screening parameter $1 - p$ in the model, as given by Eq. (72). Small values of n correspond to "full" screening, values of n close to unity to "weak" screening. From Jonscher (35).

distinguish the following three ranges of behavior depending upon the relative value of the operating frequency ω:

(a) $\omega < 1/\tau_s < \nu$. Particles and screening mobile universal response governed by the energy criterion.

(b) $1/\tau_s < \omega < \nu$. Screening cannot follow the movement of particles and the system behaves in a similar manner to the Debye model below the relaxation frequency.

(c) $1/\tau_s < \nu < \omega$. The system cannot follow the field, the response drops off as in the Debye model beyond the relaxation frequency.

A careful analysis of experimental data for a wide range of materials shown in the present review fails to reveal the presence of any well-developed Debye "tail" up to frequencies in excess of some 10 GHz where quantum and phonon effects take over. We conclude, therefore, that regions (b) and (c) are not visible, which implies that the frequencies $1/\tau_s$ and ν fall above the 10 GHz limit. This seems, in any case, perfectly reasonable in terms of the discussion given in Section VII,2 and relating to the likely orders of magnitude of atomic and molecular frequencies.

The arguments presented above should be valid equally for electronic and ionic charges, and we would expect little change with the variation of the nature of the short-range order in the material. They should be valid regardless of the nature of the chemical bonds involved. They could also be readily transferred to the dipolar situations in which the dipoles have finite lengths—as is the case with all physically realistic dipoles. These dipoles can screen one another in exactly the same manner as do hopping charges, except possibly that the screening may be weaker than in charge carrier systems.

We therefore have an admittedly qualitative model, which nevertheless has the required property of satisfying the energy criterion and which is generally applicable to a wide range of physical and chemical conditions. This model explains the observed universal frequency dependence, without any reference to distributions of parameters. This does not imply that we deny the existence in physical systems of distributions of energies, relaxation frequencies, etc.; it only suggests that these distributions are not necessary as such to explain the universal dielectric response. The variation of the exponent n with temperature and with the type of material may be understood in terms of the variation of the screening parameter p: systems with more "mobile" carriers would have smaller values of n, i.e., stronger dispersion with frequency in view of a more effective screening; systems with weaker screening, e.g., induced as compared with permanent dipoles, would have values of n closer to unity, i.e., a flatter frequency response.

The essential features of our model are restated once again: (1) the existence of screening as an expression of many-body interactions; and (2) the existence of two distinctly separate time scales—the practically instantaneous transition time for the individual hopping movements and the relatively much longer screening adjustment time τ_s. It turns out that both these features are also required in the more rigorous theory to be described and this supports the value of the screened hopping model as a qualitative guide to a better understanding of the polarization processes.

2. The Loss Peak in the Screened Hopping Model

Our screened hopping model does not, in its present form, predict the appearance of a loss peak—the universal relation continues indefinitely toward low frequencies. This is physically inadmissible and we must admit that at sufficiently low frequencies some other behavior, not allowed for in the present model, will take over. We have suggested that while the exposition of the screened hopping model is best made in the frequency domain, the understanding of loss peaks may be easier in the time domain (27). With reference to Section V,4 and to Fig. 12, we recall the simple fact that a loss peak in the frequency domain must correspond in the time domain to a transition from a region of time dependence as t^{-n}, with $n < 1$, to a second region with a logarithmic slope steeper than unity.

We now suggest that this transition corresponds physically to the onset of a second stage of relaxation after step function excitation, following on the "primary" relaxation giving the Curie–von Schweidler law. At the present time we understand the nature of this secondary relaxation giving rise to a loss peak (see Section VIII,5) and we note the fact that the carrier polarization does not appear to exhibit this feature, which is "submerged" in the strong low-frequency dispersion or in direct current conduction. In dipolar systems this secondary mechanism is necessary to give a finite total polarization, but a deeper understanding of this important question becomes possible in the light of the most recent theoretical developments described in Section VIII,5.

3. THE APPLICATION OF THE ISING MODEL

The Ising model represents a theoretical method of analyzing nearest-neighbor interactions in ferromagnetic systems that has been developed recently by Glauber (138). This model has now been adapted to the analysis of one-dimensional dipolar systems, such as polar long-chain polymers and ferroelectrics (139). It appears possible to explain by this means the slightly broadened symmetric peaks and also the asymmetric α peaks with $\lambda < 2.5$, say. The highly significant feature of this analysis is the fact that the onset of many-body interactions between nearest neighbors shifts the initial Debye peak of the noninteracting system to lower frequencies, as well as making the peak broader. This effect is obtained in a system of identical dipoles, in full agreement with the ideas explained above regarding the nature of the response of many-body systems.

At present, we do not understand the reasons why the Ising model of nearest neighbor interactions is applicable only to α relaxations and what the differences in the behavior of the processes leading to β peaks are that make them not suitable for this type of analysis. It is likely that β peaks, like the universal relation for charge carriers, may require the inclusion of more distant interactions, but further study is needed before more detailed answers can be given to this question.

It is likewise too early to say what the relative regions of validity are of the Ising and screening hopping models. At present it would appear that they are complementary, but ultimately they should be subsumed in a complete theory.

4. Ngai's Infrared Divergence Model

We now come to the most recent developments in the interpretation of the dielectric response of solids, which are contained in two complementary papers. The chief difficulty with their exposition lies in our inability to form visually intuitive models based on sense perception of ordinary daily experiences, to correspond to the results of the highly abstract and powerful quantum mechanical analysis being employed. The fundamental reason for this is to be found in the simple fact that many-body interactions transcend the limits of our sense perception just as, for example, four- and more-dimensional phenomena do. This does not mean that, in the course of time, we shall not be able to form some intuitive feeling, but at present our ability to do so is strictly limited. In the present exposition, we avoid all mathematical complications and restrict our attention to an attempt to present a sense of the physical reality behind the mathematical analysis.

The first breakthrough in the formulation of the universal dielectric response in terms of many-body theory is due to Kai Ling Ngai, who noted that the Curie–von Schweidler law of dielectric response, Eq. (54), is formally identical with the theoretically expected time dependence of a wide class of phenomena known under the general name of *infrared divergence* (IRD), which arises in any many-body system that satisfies the following conditions: (1) the system can be excited by the sudden switching of a potential, and (2) there exist lower energy excitations that form a continuous spectrum of constant density in energy and that become excited by this abrupt change of potential, subsequently radiating a "packet" of quanta with energies tending to zero on a time scale tending to infinity.

In the case of the X-ray edge anomaly, which is similar in principle, but involves a completely different range of energies, the original perturbation is the emission of an inner core electron that excites the outer electronic shells, while the slower low-energy excitations correspond to the generation of electron–hole pairs. In the dielectric case, the sudden excitation arises from the abrupt transitions of charged species whether electronic, ionic, or dipolar, as explained in Section VIII,1.

In a recent paper, Ngai, Jonscher, and White argue (*140*) that the second condition is also met in a very wide range of dielectric materials through the presence of ubiquitous *correlated states* arising from manybody interactions between pairs of particles–electrons, ions, dipoles, atoms—and forming a narrow continuous band in energy satisfying the stated requirements. These states are not contained in the ordinary band theory of solids, which is based expressly on a one-electron approximation, in the same way as the plasmon effects, another type of many-body excitation not included in the band model.

One consequence of this statement is that correlations between particles involve energies that are much lower than the energies required to form individual particles, e.g., electrons and holes. This in turn means that near-simultaneous correlated transitions of many particles may be produced with much lower energies than one-particle transitions, and here lies the clue to the remarkably large low-temperature losses in many materials, as described in Section VI.

A second point of fundamental significance for our view of the nature of dielectric polarization is the existence of a very powerful channel for the transfer of energy between the polarizing species—electrons, ions, or dipoles—and the lattice. This is the important missing link in the Debye philosophy, which had never provided any specific framework in which the energy transfer could be envisaged, since the only process that mattered was the change of polarization. We may now state that where the correlated states can be excited by absorbing the polarization energy, the time response is of the universal type and the behavior is necessarily non-Debye. Only in situations in which the correlated states do not exist, for one reason or another, is the Debye response to be expected and the fact that it is hardly ever seen in condensed matter testifies to the importance of these correlated interactions.

This model explains in a very natural manner the observed universality of the dielectric response, regardless of the nature of the interacting polarizing species, of the type of chemical bonding, and of the long-range order. The basis of the response lies in the nature of the time-dependent interactions between "fast" transitions of whatever charged species and the "slowly responding" surrounding matter.

5. Dissado and Hill's Analysis of Two-Level Systems

The same fundamental argument was taken a significant step further in two papers by Dissado and Hill (*141, 142*), who discuss the general properties of the transient response of a many-body system consisting of "fast" and "slow" responders—essentially a refinement of the "jump-and-screen" concepts of the earlier model described in Section VIII,1 and a direct counterpart of the excitations of correlated states by charge jumps envisaged by Ngai *et al.* (*140*). Some further comments on this model, providing a closer physical understanding of the mathematical formulations in these two papers and giving a deeper insight into the rather unfamiliar model of correlated states in its application to dielectric theory, are contained in a recent paper by Jonscher, Dissado, and Hill (*143*).

The many-body model is firmly based on the experimental observation of the universal power law relationship in the frequency domain given by Eq. (52) and in the time domain in the approximate form given by Eq. (53). The essential point is that the two exponents m and n are independent, so that they must reflect the existence of two distinct physical processes (see Section VI,5). We already saw that Ngai *et al.* sought to associate the t^{-n} power law with an infrared divergencelike process, through the intermediary of the correlated states. The Dissado–Hill model is based on the following concepts, which we discuss in turn (*141–143*): (a) interactions, (b) the role of disorder in dielectric polarization, (c) relaxation in a two-level system through "large" and "small" transitions, (d) the time dependence of the relaxation process, (e) the energy relations in dielectric relaxation, and (f) high-frequency response.

a. Interactions. This concept has already been outlined in Section VIII,4 and we need only add that interactions among charges, dipoles, and molecules result in the formation of a new set of correlated states that do not fit into the ordinary band diagram of solids, based as it is on the one-electron approximation, and therefore have to be placed alongside the band diagram.

Unlike the one-particle states, which describe the vibrational motions of single particles, e.g., electrons, the correlated states correspond to the motions of the *centroids of assemblies* of particles. Since typical vibrational motions involve energies in the range 10^{-3}–10^{-2} eV, and since the interaction energies are smaller than these, we expect the range 2ζ of the energies of correlated states to be very small compared to the thermal energies at normal temperatures. Because their energy is much smaller than typical phonon energies, the cooperation of many phonons would be required to provide thermal excitations within the correlated states band. We may say, therefore, that these states do not couple directly to the phonons, and consequently their occupancy cannot be "washed out" by thermal excitations, even at relatively high temperatures. The excitation of these states can occur only through the rapid dipolar or charge carrier transitions, as described earlier.

Although the energy range 2ζ is small, the total number of correlated states is high and their density in energy is correspondingly large. Moreover, from the equivalence of the occupied and unoccupied states we conclude that their total numbers are equal—these states are therefore half-filled, with a sharp dividing energy between them in the absence of excitations. It is expected that the density distribution in energy would be of a parabolic shape, as shown in Fig. 62, so that the transitions involving a fraction of the total energy 2ζ fall in an approximately constant density region in energy, as postulated by Ngai *et al.* as a condition of infrared divergence.

b. The Role of Disorder in Dielectric Polarization (143). Recent years have brought a considerable development of the concept of disorder in solids, especially in the context of the amorphous and strongly disordered materials. The presence of disorder leads to the formation of localized levels in semiconductors and these give rise to electronic transport processes under conditions that would give completely insulating proper-

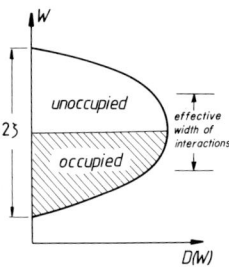

FIG. 62. The density of states and the occupancy of the correlated states band. The total width 2ζ of the band is small but the maximum density $D(W)$ is correspondingly large in view of the large total number of these states. The boundary energy between the occupied and unoccupied states is sharp even at normal temperatures, in view of the thermal inaccessibility of these states. The effective range of interactions is shown to indicate the almost constant density of states over this range of energies. This diagram is not part of the normal one-electron energy band model and should be placed alongside it.

ties in more perfect solids. We believe that disorder plays an equally important role in dielectric polarization, although this point is not normally considered in the discussion of dielectric processes.

We note that orientational polarization, which is the only process of interest in the present review, involves disorder at three different levels. First, dipoles or dipolelike point defects are normally randomly distributed in whatever "medium" they may be embedded in, which means that their mutual interactions are different. Even if these dipoles were themselves distributed perfectly on a regular sublattice, the presence of other defects would modify the local environment of these dipoles in different ways. Finally, randomness is necessarily involved in any system with orientational polarization, which would only become perfectly ordered if all dipoles were identically oriented. We may therefore regard any orientationally polarizable system as containing an inherent degree of disorder and therefore being susceptible of weak but finite modifications of energy resulting from interaction with any movements in the system. The disordered dipolar system may be regarded, therefore, as being "soft," in the sense of being capable of responding to small changes in the disorder of a type that would leave a "hard" perfect system completely unaffected.

This softness of polarizable media is the ultimate reason for their strongly interactive behavior—the same many-body interactions would be much more difficult to observe in more rigid systems.

c. Relaxation in a Two-Level System through "Large" and "Small" Transitions. A system consisting of interacting dipoles jumping between preferred orientations, or the corresponding charge carrier system, may be represented by a double potential well corresponding to the macroscopic dipole being oriented in one or other of the preferred senses. The application of an external field changes the relative positions of the energy minima, thereby causing a change of the distribution of dipoles between them. To that extent the situation is as shown in Fig. 8 for a noninteracting system. The presence of interactions manifests itself in the appearance of the correlated states, which cause a "softness" of the bottoms of the potential wells, as shown in Fig. 63, where the shaded regions represent the same states represented in terms of a density of states distribution in Fig. 62.

The other feature of the presence of interactions is that the energy separating the bottoms of the potential wells is itself a function of the dipole moment M_e of the system:

$$B_{\text{eff}} = B + kT_c M_e \tag{73}$$

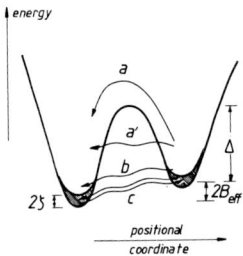

FIG. 63. The potential energy diagram of a many-body two-level system, representing the energy of a large number of individual interacting systems. The potential wells represent the preferred orientations for dipoles or positions for hopping ions or electrons. The shaded regions at the bottom of the potential wells represent the "softness" arising from the presence of the correlated states, whose density distribution in energy is shown in Fig. 62. The energy separation of the mean energy of the minima is given by $2B_{\text{eff}}$, the width of the correlated states band is 2ζ. Four types of transitions are shown by the arrows. The "large" transitions a and a' involve thermal excitation over the energy barrier Δ without and with some contribution of tunneling, respectively, with one particle making the complete transition. The "small" transitions corresponding to configurational tunneling in which small displacements of a large number of particles contribute to the formation of a new configuration without the need to exchange energy with the phonon bath, are denoted by b and c. The former represent a net change of the dipole moment and are known as the flip transitions, while the latter do not change the dipole moment since they involve synchronous transitions in opposite senses at different positions in the system, being referred to as flip–flop transitions.

where B is the average splitting, and T_c a critical temperature. The moment then satisfies

$$M_e = \tanh\left(\frac{B + kT_cM_e}{kT}\right) \tag{74}$$

The relaxation of the system upon a sudden change of the external field may be envisaged in terms of three processes proposed by Dissado and Hill.

One process is the thermal excitation over the barrier Δ separating the potential wells; these transitions involve significant amounts of energy and relatively large spatial displacement of individual particles. We refer to these as "large" transitions and they are denoted in Fig. 63 by the arrow (a). In the present convention these large transitions may involve an element of thermally assisted tunneling of the type indicated in Fig. 63 by the arrow (a').

The other transitions are represented by configurational tunneling between minima—a process involving neither thermal excitation nor actual large displacements of any single particle, but which may be regarded as the creation of a new configuration by a large number of small displacements of individual particles in the disordered system. We shall refer to these tunneling transitions as "small", both because the energies involved are small and because the individual particle displacements are small compared with the large transitions. These configurational tunneling transitions represent an essentially new element in our understanding of the dielectric response of solids, without which we would not be able to explain the universal power law of decay of polarization. Equally significantly, however, the configurational tunneling provides the only explanation of the persistence of dielectric loss down to the lowest temperatures—a feature of the experimental data that defies interpretation in terms of the conventional thermally activated processes.

Among the subset of all "small" transitions we may distinguish two classes: those giving rise to a net change in the dipole moment, which are denoted by the arrow (b) in Fig. 63 and which are referred to as "flip" transitions, and those in which the transition of one particle is accompanied by a synchronous transition of another particle at another place in the opposite direction, leaving the total dipole moment unchanged. The latter are referred to as "flip–flop" transitions and are denoted by the arrow (c). The flip–flop transitions contribute to the exchange of energy between different places and may be regarded as electrical noise on the relaxation process, which takes place through the flip transitions and the large thermal jumps. We may say, therefore, that the flip–flops delay the process of relaxation by preventing the other transitions from taking place.

d. The Time Dependence of the Relaxation Process. Each of the three types of transitions has a specific contribution to make to the time dependence of the relaxation process. The most obvious of these is the thermally assisted relaxation through the "large" transitions, which is exactly equivalent to the classical Debye relaxation and follows the familiar exponential relation given by Eq. (37) for which the depolarization current is given by

$$i(t) \propto \exp(-\omega_p t), \qquad t \simeq 1/\omega_p \qquad (75)$$

The flip processes that dominate the relaxation at short times give the power law relationship expected in many-body systems obeying the criteria of infrared divergence:

$$i(t) \propto t^{-n}, \qquad 0 < n < 1, \qquad t \ll 1/\omega_p \qquad (76)$$

and we may now interpret the exponent n as the degree of correlation of the individual flip transitions, $n = 1$ corresponding to fully correlated transitions, $n = 0$ to uncorrelated transitions.

The flip–flop processes become dominant in the later stages of the relaxation and they do not affect the rate of change of polarization directly, since they are not dipole-changing transitions, but they influence the rearrangement of energy between different places in the material. The time dependence associated with these transitions is the other power law

$$i(t) \propto t^{-1-m}, \qquad 0 < m < 1, \quad t \gg 1/\omega_p \tag{77}$$

Dissado and Hill have developed a detailed theoretical treatment that derives the complete time dependence of the relaxation current in the general form:

$$i(t) \propto \exp(-\omega_p t) t^{-n} {}_1F_1(1 - m; 2 - n; \omega_p t) \tag{78}$$

where ${}_1F_1(\ ;\ ;\ )$ is the confluent hypergeometric function. This general expression reduces to the three forms (75)–(77) in the time intervals indicated.

This theoretical analysis provides a complete and rigorous confirmation of the qualitative picture presented in Fig. 12, where we had pointed out that the empirically derived time domain devolution of the universal relations was based on the sequential unfolding of two separate processes, each being characterized by a specific time dependence of power law form. We are now able to identify these two processes with the flip and the flip–flop transitions, respectively. To that extent, the original intuitive approach proposing the explanation of the loss peaks in the non-Debye relaxation in terms of the sequential devolution of two processes (27) is confirmed as basically sound, but in addition we find the role of the Debye relaxation as the intermediate process given by Eq. (76). This is very gratifying, since the significance and role of the loss peak frequency ω_p is well appreciated in the science of dielectrics and there is ample interpretational material of the significance of these loss peaks in terms of molecular dynamics of the processes in question. In that sense, therefore, the present interpretation does not invalidate the traditional arguments relating to the loss peak frequency, but provides a proper general setting for them as the "pivot" on which the power law "wings" of the universal decay law are "hinged." There is therefore no need to have recourse in the new interpretation to artificial concepts of distributions of relaxation times in order to understand the departures from the Debye response, and it is possible instead to view the complete relaxation process as composed of three different types of transitions.

We may, indeed, reverse the argument by presenting it not in terms of justifying the departures from the Debye response, but by taking as the point of departure the proposition that a purely exponential decay is physically impossible in an interactive system.

e. The Energy Relations in Dielectric Relaxation. While the concept of energy stored in static polarization is very familiar, nothing is said about the process of energy transfer from the dipole system to the "matrix" during the transient depolarization process. In fact, the classical Debye process does not provide any explicit means of energy transfer, except insofar as the time delay τ inherent in this mechanism necessarily involves an energy loss in the case of sinusoidal excitation.

By contrast, our new many-body approach places the energy at the center of the picture of the transient depolarization. The sudden dipolar or charge transitions cause excitations of the correlated states, which are not otherwise susceptible to thermal excitation, and a considerable amount of energy may be stored in these states without being dissipated into the lattice directly.

In particular, every "small" transition of either the flip or the flip–flop type gives rise to the excitation of these correlated states and the energy so transferred is irretrievably lost through the ultimate excitation of energy-dissipating large transitions. We may therefore regard a definite fraction of the energy stored in every dipolar orientation as being irretrievably lost in the process of reorientation after the removal of the polarizing field.

Considering therefore the initial stages of the relaxation process, where the flip transitions are the dominant depolarization mechanism, we see that the energy is being stored primarily in the correlated states and is only being released from these states as the "large" transitions are being excited rather sooner than would have been the case in the purely Debye mechanism. At long times, by contrast, the excitations stored in the correlated states are being gradually released in the form of large transitions under the influence of flip–flop transitions.

Thus, at times $t \ll 1/\omega_p$ and at the corresponding frequencies $\omega \gg \omega_p$, the power law relation (76) corresponds to a constant energy loss per transition, i.e., to a frequency-independent ratio of energy lost per cycle to the energy stored at the peak of the cycle. This reasoning provides a very natural explanation for the empirically observed relation (51) which forms the cornerstone of our "universal" approach and for which we now have a physical interpretation.

We may therefore look upon the power law (76), which is associated in theoretical physics with infrared divergence, as a consequence of a fixed loss of energy per transition. We note that the universal presence of this law in dielectric relaxation provides firm evidence of the constancy of this energy loss per transition—the higher the polarization, the more dipoles or charges have to make the transitions and the higher the amount of energy lost. We also note the significance of the exponent n in the context of energy dissipation, through Eq. (51), which implies that $n = 1$ corresponds to zero energy loss per transition and $n = 0$ gives total loss of energy.

f. High-Frequency Responses. It is estimated that the time required to set up a "synchronous" transition is of the order of 10^{-8}–10^{-10} sec (*144*), so that on a time scale shorter than this limit the flip–flop processes do not come into play. In this limit the exponent m becomes unity, and this becomes evident in the case of those loss peaks falling above the frequency 10^8–10^9 Hz, as is clear from the compilation of Fig. 33, where most of the data on the top edge of the square correspond to this type of response.

Work is currently in progress on the more detailed quantitative interpretation of the specific features of the dielectric response of many materials, with particular emphasis on the significance of the temperature dependence of the exponents m and n, which may be seen experimentally in the representation of Fig. 29. The very richness of the experimental material relating to a wide range of solids promises, in the presence of a physically meaningful theoretical framework, to give valuable insight into the detailed nature of many-body interactions in condensed matter. Here we wish to stress the fact that the present study of the dielectric response of solids has shown clearly that the time and frequency dependence of the dielectric parameters offers a unique tool for the study of many-body interactions in general, with the additional advantage that the apparatus required is relatively simple and readily accessible. The conclusions of the Dissado–Hill analysis are valid in much more general conditions than merely in the dielectric context—they are applicable quite generally to any system that involves interacting two-level systems and they show that such systems must necessarily show nonexponential and therefore non-Debye relaxation

IX. Concluding Comments

There is no doubt that the vast majority of solids show a dielectric response that cannot be explained in terms of any form of one-particle processes of the Debye type, nor by any form of transport process such as diffusion or injection of space charge. The broad similarity between the universal responses of dipoles, electrons, and ions strongly suggests the dominant role of many-body interactions and this is amply confirmed by the recently developed theoretical analyses. The fact that the quantum-mechanical theory based on Ngai's correlated states and the Dissado–Hill theory of time dependence of two-level systems both explain the entire empirically determined range of dielectric response types represents a major achievement of the new approach and is already giving us an entirely new understanding of the dielectric response, which hitherto appeared as a complex and seemingly arbitrary process.

A good deal of theoretical work remains to be done and there is need for further careful experimental studies on a range of selected materials, with a view to providing more reliable data for the theory. In particular, the temperature dependence of the dielectric response promises to offer ample scope for the refinement of the theory. However, it is already clear that the new model has the required generality and potential for development into a fully acceptable and quantitatively valid theory of dielectric response.

Throughout the present treatment we have been stressing that many-body interactions explain the flat and broad universal type of dielectric response without any need for distributions of parameters. This does not imply that distributions do not exist—it is evident that few systems are sufficiently perfect not to involve any distributions at all—the essential point is that any distributions that may be present will involve additional contributions to the overall dielectric response, but all these contributions will be of the universal type. Thus, distributions are admissible, but they are not necessary for the explanation of the dielectric response of solids.

We stress the point that the new many-body approach involves a fundamental break with the hitherto widely accepted Debye philosophy, requiring a profound change in deeply ingrained attitudes. For this reason, we prefer not to use the terms *dielectric relaxation* and *relaxation time,* since they are inseparably connected with the exponential Debye process. The screening adjustment time τ_s should not under any circumstances be confused with the Debye relaxation time.

The new approach also does away with that other traditional and ill-defined concept of *internal field,* which was necessary when there was no better way of accounting for many-body processes, but which becomes superfluous when a direct attack can be made on these.

Finally, we wish to note that that we do not use the often made distinction between *bound* and *free charges,* which we believe to be physically misleading. Although the division appeals to those who like to have a neat division into dipoles and free charge carriers, the physical reality is a good deal more complicated, since hopping charges may act as either dipoles or free charges, and experiment shows that they follow the same laws of time dependence.

Acknowledgments

First, I would like to thank my Colleagues in the Chelsea Dielectrics Group whose experimental work formed the backbone of these ideas. An acknowledgment also is due to Kai Ling Ngai in Washington, who first recognized the possibility of applying to the dielectric response his recently developed theory of many-body phenomena in solids, as well as to Len Dissado and Robert Hill at Chelsea, who have adapted and significantly extended this theory. The present Review is based on experimental and theoretical work of the Chelsea Dielectrics Group, much of which was supported by the Science Research Council.

References

1. A. K. Jonscher and R. M. Hill, in "Physics of Thin Films" (G. Hass, M. H. Francombe, and R. W. Hoffman, eds.), Vol. 8, pp. 169–249. Academic Press, New York, 1975.
2. A. K. Jonscher, *Nature (London)* **267,** 673–679 (1977).
3. H. Wintle, *Solid-State Electron.* **18,** 1039 (1975).
4. S. I. Rudenko, *Sov. Phys.—Semicond. (Engl. Transl.)* **9,** 729 (1975).
5. R. M. Hill, *J. Phys. C* **8,** 2488–2501 (1975).
6. J. van Turnhout, "Thermally Stimulated Discharge of Polymer Electrets." Elsevier, Amsterdam, 1975.
7. M. E. Baird, *Rev. Mod. Phys.* **40,** 219 (1968).
8. R. Chahine and T. K. Bose, *J. Chem. Phys.* **65,** 2211 (1976).
9. B. Guestblom and E. Noreland, *J. Phys. Chem.* **80,** 1631 (1976).
10. A. Erdelyi, "Tables of Integral Transforms," Vol. II, p. 249. McGraw-Hill, New York, 1954.
11. L. D. Landau and E. M. Lifshitz, "Electrodynamics of Continuous Media." Pergamon, Oxford.
12. R. Lovell, *J. Phys. C* **7,** 4378–4384 (1974).
13. A. K. Jonscher, *Thin Solid Films* **36,** 1 (1976).
14. A. K. Jonscher, *Thin Solid Films* **50,** 187 (1978).
15. J. G. Simmons and G. W. Taylor, *Phys. Rev. B* **6,** 4793 (1972).
16. A. K. Jonscher and M. S. Frost, *Thin Solid Films* **37,** 267–273 (1976).
17. K. S. Cole and R. H. Cole, *J. Chem. Phys.* **9,** 341 (1941).
18. R. J. Elliott and A. F. Gibson, "Solid State Physics," p. 201, Macmillan, New York, 1974.
19. P. Debye, "Polar Molecules." Dover, New York, 1945.
20. N. F. Mott and E. A. Davis, "Electronic Processes in Non-crystalline Materials." Oxford Univ. Press, London and New York, 1971.
21. F. Argall and A. K. Jonscher, *Thin Solid Films* **2,** 185 (1968).

22. D. W. Davidson and R. H. Cole, *J. Chem. Phys.* **19**, 1484 (1951).
23. S. Havriliak and S. Negami, *J. Polym. Sci., Part C* **14**, 99 (1966).
24. R. M. Fuoss and J. G. Kirkwood, *J. Am. Chem. Soc.* **63**, 385 (1941).
25. D. J. Denney, *J. Chem. Phys.* **27**, 259 (1957).
26. A. K. Jonscher, *Colloid Polym. Sci.* **253**, 231–250 (1975).
27. A. K. Jonscher, *Nature (London)* **256**, 566–568 (1975).
28. E. von Schweidler, *Ann. Phys. (Leipzig)* [4] **24**, 711 (1907).
29. D. K. Das Gupta and K. Joyner, *J. Phys. D* **9**, 829, 2041 (1976).
30. M. J. C. van Gemert, *Philips Res. Rep.* **28**, 530 (1973).
31. J. Ross Macdonald, *J. Chem. Phys.* **61**, 3977 (1974).
32. J. Ross Macdonald and J. A. Garber, *J. Electrochem. Soc.* **124**, 1022 (1977).
33. A. K. Jonscher, *Nature (London)* **253**, 717–719 (1975).
34. A. K. Jonscher, *Phys. Status Solidi B* **83**, 585 (1977).
35. A. K. Jonscher, *Phys. Status Solidi B* **84**, 159 (1977).
36. N. G. McCrum, B. E. Read, and G. Williams, "Anelastic and Dielectric Effects in Polymeric Solids." Wiley, New York, 1967.
37. S. R. Gough and D. W. Davidson, *J. Chem. Phys.* **52**, 5442 (1970).
38. R. J. Meakins, *Trans. Faraday Soc.* **52**, 320 (1956).
39. C. Pawlaczyk, *Pr. Kom. Mat.-Przyr., Fiz. Dielektr. Radiospektrosk., Poznan. Tow. Przyj. Nauk* **9**, 53.
40. C. Druon and J. M. Wacrenier, *J. Phys. (Paris)* **38**, 47 (1977).
41. H. E. Taylor, *J. Soc. Glass Technol.* **41**, 350T (1957); **43**, 124T (1959).
42. M. Hakim and D. R. Uhlmann, *Phys. Chem. Glasses* **14**, 81 (1973).
43. Y. Ishida and S. Matsuoka, *Polym. Prepr., Am. Chem. Soc., Div. Polym. Chem.* **6**, No. 2, 795–798 (1965).
44. Y. Ishida, M. Matsuo, and K. Yamafuji, *Kolloid-Z. & Z. Polym.* **180**, 108 (1962).
45. Y. Ishida, *J. Polym. Sci., Part A-2* **7**, 1835–1861 (1969).
46. G. P. Johari, *Ann. N.Y. Acad. Sci.* **279**, 117 (1976).
47. J. le G. Gilchrist, *J. Phys. Chem. Solids* **38**, 509–516 (1977).
48. J. le G. Gilchrist, private communication.
49. W. A. Phillips, *Proc. R. Soc. London, Ser. A* **319**, 565–581 (1970).
50. R. A. Carson, *Proc. R. Soc. London, Ser. A* **332**, 255–268 (1973).
51. I. Barsony and A. K. Jonscher, *Solid-State Electron.* **21**, 471–473 (1978).
52. M. Matsuo, Y. Ishida, K. Yamafuji, M. Takayanagi, and F. Ire, *Kolloid-Z. & Z. Polym.* **201**, 89–93 (1965).
53. S. R. Gough, R. E. Hawkins, B. Morris, and D. W. Davidson, *J. Phys. Chem.* **77**, 2969 (1973).
54. F. Meca and A. K. Jonscher, *Thin Solid Films* **59**, 201 (1979).
55. G. Frossati, J. le G. Gilchrist, and J. C. Lasjaunias, private communication.
56. R. M. Hill, *Nature (London)* **275**, 96 (1978).
57. G. Williams, D. C. Watts, S. B. Dev, and A. M. North, *Trans. Farady Soc.* **67**, 1323 (1971).
58. M. Pollak and T. H. Geballe, *Phys. Rev.* **122**, 1745 (1961).
59. R. J. Grant, I. M. Hodge, M. D. Ingram, and A. R. West, *Nature (London)* **266**, 42 (1977).
60. M. Abkovitz, P. G. Le Comber, and W. E. Spear, *Commun. Phys.* **1**, 175 (1976).
61. G. C. Roberts and J. I. Polanco, *Solid State Commun.* **10**, 709 (1972).
62. M. Abkovitz, D. F. Blossey, and A. I. Lakatos, *Phys. Rev. B* **12**, 3400 (1975).
63. D. Hughes and R. Pethig, *IEE Conf. Publ.* **129**, 52 (1975).
64. M. Abkovitz, A. I. Lakatos, and H. Scher, *Phys. Rev. B* **9**, 1813 (1974).

65. L. Murawski and O. Gzowski, *Phys. Status Solidi A* **24**, K115 (1974).
66. M. Sayer, M. Mansingh, J. M. Reeves, and C. J. Rosenblatt, *J. Appl. Phys.* **42**, 2857 (1971).
67. M. S. Frost and A. K. Jonscher, *Thin Solid Films* **29**, No. 1, 7–18 (1975).
68. M. A. Careem, A. K. Jonscher, and F. Taiedy, *Philos. Mag.* [8] **35**, No. 6, 1503–1508 (1977).
69. A. I. Lakatos and M. Abkovitz, *Phys. Rev. B* **3**, 1791 (1971).
70. U. Strom and P. C. Taylor, *in* "Amorphous and Liquid Semiconductors" (J. Stuke and W. Brenig, eds.), p. 375. Taylor & Francis, London, 1974.
71. S. R. Elliott, *Philos. Mag.* [8] **36**, 1291–1304 (1977).
72. R. M. Hill and A. K. Jonscher, *J. Non-Cryst. Solids* **32**, 53 (1979).
73. C. T. Moynihan, R. D. Bressel, and C. A. Angell, *J. Chem. Phys.* **55**, 4414 (1971).
74. J. S. Thorp and R. I. Sharif, *J. Mater. Sci.* **12**, 2274 (1977).
75. H. M. Kizilyalli and P. R. Mason, *Phys. Status Solidi A* **36**, 499 (1976).
76. A. K. Jonscher and D. C. Dube, *Ferroelectrics* **17**, 533 (1978).
77. A. K. Jonscher, *Phys. Status Solidi A* **32**, 665 (1975).
78. A. K. Jonscher, *J. Mater. Sci.* **13**, 553 (1978).
79. A. K. Jonscher, K. L. Deori, J. M. Reau, and J. Moali, *J. Mater. Sci.* **14**, 1308–1320 (1979).
80. A. K. Jonscher and J. M. Reau, *J. Mater. Sci.* **13**, 563 (1978).
81. K. L. Deori and A. K. Jonscher, *J. Phys. C* **12**, L289–L292 (1979).
82. A. K. Jonscher, *Philos. Mag., [Part] B* **38**, 587 (1978).
83. M. A. Careem and A. K. Jonscher, *Philos. Mag.* [8] **35**, 1489 (1977).
84. Unpublished data from the Laboratoire de Génie Eliectrique de Toulouse, by courtesy of the Director.
85. R. A. Thomas and C. N. King, *Appl. Phys. Lett.* **26**, 406 (1975).
86. O. Yano, K. Saiki, S. Tarucha, and Y. Wada, *J. Polym. Sci., Polym. Phys. Ed.* **15**, 43 (1977).
87. M. Shahidi, J. B. Hasted, and A. K. Jonscher, *Nature (London)* **258**, 595 (1975).
88. M. Shahidi, Ph.D. Thesis, University of London, 1977.
89. I. Szundi, private communication.
90. E. Warburg, *Ann. Phys. Chem.* [N.S.] **67**, 493 (1899).
91. E. Warburg, *Ann. Phys. (Leipzig)* [4] **6**, 125 (1901).
92. S. P. Mitoff and R. J. Charles, *J. Appl. Phys.* **43**, 927 (1972).
93. M. Broadhurst, *Natl. Bur. (V.S.), Circ. Stand.* p. 9592 (1970).
94. A. K. Jonscher, *in* "Electrets, Charge Storage and Transport in Dielectrics" (M. M. Perlman, ed.), pp. 269–284. Electrochem. Soc., 1973.
95. E. J. Le Sueur and A. K. Jonscher, *Thin Solid Films* **00**, S9 (1972).
96. A. K. Jonscher and M. A. Careem, *Phys. Lett. A* **55**, 257 (1975).
97. A. K. Jonscher, H. Carchano, and Y. Segui, *Rev. Gen. Electr.* **87**, 866 (1978).
98. A. K. Jonscher, "Principles of Semiconductor Device Operations." Bell, London, 1960.
99. H. J. Wintle, *IEEE Trans. Electr. Insul.* **ei-12**, No. 2, 97–113 (1977).
100. S. Sapieha and H. J. Wintle, *Can. J. Phys.* **55**, 646 (1977).
101. A. K. Jonscher and F. Taiedy, *J. Phys. C* **8**, L107 (1975).
102. A. K. Jonscher and S. Buddhabadana, *Solid-State Electron.* **21**, 991 (1978).
103. A. K. Jonscher, F. Meca, and M. H. Millany, *J. Phys. C* **12**, L293–L296 (1979).
104. A. K. Jonscher, *J. Phys. D* **13**, L89 (1980).
105. A. K. Jonscher, Chelsea Dielectrics Group, Sixth Progress Report, July 1980.
106. A. K. Jonscher, *J. Phys. D* **13**, L137 (1980).

107. C. J. F. Böttcher and P. Bordewijk, "Theory of Electric Polarisation," 2nd ed., Vol. II. Elsevier, Amsterdam, 1978.
108. C. G. Garton, *Trans. Faraday Soc.* **42A**, 56 (1946).
109. H. Fröhlich, "Theory of Dielectrics." Oxford Univ. Press, London and New York, 1949.
110. J. R. Macdonald, *Physica (Utrecht)* **28**, 485 (1962).
111. J. R. Macdonald, *J. Chem. Phys.* **36**, 345 (1962).
112. J. R. Macdonald, *J. Appl. Phys.* **34**, 538 (1963).
113. B. Stoll, W. Pechhold, and S. Blasenbrey, *Kolloid-Z. & Z. Polym.* **250**, 1111–1130 (1972).
114. J. I. Lauritzen, *J. Chem. Phys.* **28**, 118 (1958).
115. M. Pollak, *Philos. Mag.* [8] **23**, 519 (1971).
116. H. Scher and M. Lax, *Phys. Rev. B* **7**, 4491, 4502 (1973).
117. P. N. Butcher and P. Morys, *J. Phys. C* **6**, 2147 (1973).
119. V. Halpern, *Physica (Utrecht)* **79B**, 323, 336 (1975).
120. I. G. Austin and N. F. Mott, *Adv. Phys.* **18**, 41 (1969).
120. I. G. Austin and N. F. Mott, *Adv. Phys.* **18**,41 (1969).
121. J. R. Macdonald, *J. Chem. Phys.* **61**, 3977 (1974).
122. J. R. Macdonald, *in* "Electrode Processes in Solid State Ionics" (M. Kleitz and J. Dupuy, eds.). Reidel, Dordrecht, The Netherlands, 1976.
123. S. H. Glarum, *J. Chem. Phys.* **33**, 639 (1960).
124. S. Bozdemir, Ph.D. Thesis, University of London, 1978.
125. M. Cook, D. C. Watts, and G. Williams, *Trans. Faraday Soc.* **66**, 2503 (1970).
126. S. H. Glarum, *J. Chem. Phys.* **33**, 1371 (1960).
127. R. H. Cole, *J. Chem. Phys.* **42**, 637 (1965).
128. T. W. Nee and R. Zwanzig, *J. Chem. Phys.* **52**, 6353 (1970).
129. G. Williams, *Chem. Rev.* **72**, 55 (1972).
130. R. H. Cole, *Mol. Phys.* **26**, 969 (1973).
131. R. L. Fulton, *J. Chem. Phys.* **62**, 4355 (1975).
132. J. Volger, *Prog. Semicond.* **4**, 205 (1960).
133. F. Haberey and H. P. J. Wijn, *Phys. Status Solidi* **26**, 231 (1968).
134. G. D. Mahan, *Solid State Phys.* **29**, 75 (1974).
135. J. J. Hopfield, *Comments Solid State Phys.* **11**, 40 (1969).
136. M. Pollak and G. E. Pike, *Phys. Rev. Lett.* **28**, 1449 (1972).
137. P. M. Platzman and P. A. Wolff, "Waves and Interactions in Solid State Plasmas." Academic Press, New York, 1973.
138. R. J. Glauber, *J. Math. Phys.* **4**, 294 (1963).
139. S. Bozdemir, Ph.D. Thesis, University of London, 1978.
140. K. L. Ngai, A. K. Jonscher, and C. T. White, *Nature (London)* **277**, 185 (1979).
141. L. A. Dissado and R. M. Hill, *Nature (London)* **279**, 685 (1979).
142. L. A. Dissado and R. M. Hill, *Philos. Mag.* (in press).
143. A. K. Jonscher, L. A. Dissado, and R. M. Hill, *Phys. Status Solidi* (in press).
144. J. Joffrin and A. Levelut, *J. Phys. (Paris)* **36**, 11 (1975).

NOTE ADDED IN PROOF

The latest information regarding the progress of the dielectric studies at Chelsea may be found in the annual *Progress Reports of the Chelsea Dielectrics Group*, which may be obtained from the author.

Author Index

Numbers in parentheses are reference numbers and indicate that an author's work is referred to although the name is not cited in the text. Numbers in italics show the page on which the complete reference is listed.

A

Abkovitz, M., 256(60, 62, 64, 69), 257(60, 62, 64, 69), 260(69), *315, 316*
Abstreiter, G., 45(57), *97*
Albrecht, M. G., 198, *203*
Alfieri, I., 113(51), *200*
Allgaier, R. S., 114(60), 120, 121(72), *201*
Almassy, R. J., 56(93), 57(93), *98*
Amano, O., 254, *255*
Amano, S., 36, (13), *96*
Anderson, J. C., 189(152), *203*
Andrews, A. M., 157(129), 171, *200, 202, 203*
Antcliffe, G. A., 121(73), *201*
Argall, F., 226(21), *314*
Arthur, J. R., 36(1), 42(1, 38), 44(1, 38), 56(1), 57(1), 58(1), 61, 63(111), 65(1), 66(1, 38), 67(1), 73(1), 79(1), 86, 89(1), *96, 97, 99*
Asahi, H., 67(133b), 80(177), *100, 101*
Asch, A. E., 158(131), 164, 165, 191, 192, 193, 194, 195, 196, 197, *203*
Austin, I. G., 290(120), *317*

B

Baars, J., 111(39), *200*
Baba, S., 68, 69, *100*
Bach, B. W., 3(1), 4(1), 5(1), 6(1), 8(1), 14(1), 19(1), *33*
Bachmann, K. M., 47(60), 66(60), 67(60), *97*
Baird, M. E., 211(7), *314*
Balanski, M., 132(94), 133(94), 142(94), *202*
Balon, J. R., 146(116), 149(116), 151(116), *202*
Bandy, S. G., *100*
Bardeen, J., 132, *202*

Barker, A. S., Jr., 84(189, 193), *101, 102*
Barrera, J., 53(88), *98*
Barsony, I., 244(51), 245, *315*
Bassett, D. W., 111(40), 119(40), *200*
Bate, R. T., 169, *203*
Bauser, E., 45(57), *97*
Bean, J.C., 36(29), *97*
Beck, W. A., 155(127), 198, 199, *202, 203*
Bellamy, W. C., 42(47), 89(232), *97, 103*
Bellevance, D. W., 132(90), *201*
Berger, H. H., 88(222), *102*
Berry, J., 78(166), 83(166), *101*
Besson, J. M., 132(94), 133(94), 142(94), *202*
Bis, R. F., 109, 111, 116(66), *200, 201*
Bishop, S. G., 50(73), *98*
Black, G. M., 111(41), 119(71), 132(41), 165(71), 166(71), 186(41), 187, *200, 201*
Blasenbrey, S., 289(113), *317*
Blatt, F. J., 132(92), *202*
Blossey, D. F., 256(62), 257(62), *315*
Board, K., 76(156a), 88(156a), *100*
Bode, D., 106(4), 108(4), *199*
Bordewijk, P., 288, *317*
Bose, T. K., 211(8), 232(8), *314*
Böttcher, C. J. F., 288, *317*
Bozdemir, S., 293(124), 302(139), *317*
Bradford, A., 111(47), *200*
Brailsford, A. D., 184(147), *203*
Brandt, G. E., 10(4), *33*
Brebrick, R. F., 108(22, 23), *200*
Broadhurst, M., 271(93), *316*
Buchner, S. P., 153(126), 155, 199, *202, 203*
Buddhabadana, S., 279(102), 280, *316*
Burrell, G. J., 142(113), 143(113), *202*
Burrus, C. A., 78, *101*
Butcher, P. N., 290(117), *317*
Butler, J. F., 158, *203*
Byer, N. E., 153(125, 126), 155(127), 198(157), 199(127, 157), *202, 203*
Bylander, E. G., 108, *200*

C

Cachet, H., 253, *255*
Calawa, A. R., 36(18, 23), 49(67), 57, 58, 65(95), 67(67), 81(185a), *96, 97, 98, 99, 101,* 128(78, 80), 155(78), *201*
Callender, R. E., 119(67), 131, 168, *201, 203*
Carchano, H., 278(97), *316*
Careem, M. A., 256(68), 257(68), 263(83), 265, 266, 267, 275(96), *316*
Carson, R. A., 241(50), 244(50), *315*
Carter, D. L., 169(142), *203*
Caruso, A. J., 24(18), 25(18), 27(18), 28(18), 29(18), *34*
Casey, H. C., 75(148), 78(148), 79(148), *100*
Casey, H. C., Jr., 51(80, 81), 54(80), 66(121), 79(169, 171), 81(184), 89, *98, 99, 101, 103*
Chahine, R., 211(8), 232(8), *314*
Chan, W. S., 119(68), *201*
Chandra, A., 51(74), 77(163), *98, 101*
Chang, C. A., 67(131), 68, 70(137), 71, 73(137, 145), 74, 79(137), *99, 100*
Chang, L. L., 36, 41, 42(40), 44(40, 54), 47(3), 58(3), 66, 67, 68, 70(137), 71(137), 73(130, 137, 145), 74(137, 145), 79(3, 137), 84, 85(130, 197, 198), 86(204, 205), 87(214), 89(3), *96, 97, 99, 100, 102*
Chaplart, J., 42(43), *97*
Chapman, R. A., 132(89, 90), 146(117), 148(117), 149(117), *201, 202*
Chapman, R. L., 81(185a), *101*
Chaporon, A., 254, *255*
Charles, R. J., 271(92), *316*
Chen, D. R., 76, *100*
Chen, J. M., 153(125, 126), *202*
Chia, P. S., 146(116), 149(116), 151(116), *202*
Cho, A. Y., 36, 42, 44(1), 50, 51(75, 80, 81), 53(89), 54(80, 89), 55(89), 56(1), 57(1, 75), 58(99, 102), 59(75, 102, 104), 61(1), 65(1), 66, 67, 73(1), 75(46, 148), 76, 77(157, 159, 161), 78(148, 168, 172, 173), 79, 80, 81, 82(186), 84(186), 86, 88, 89, 91(233), *96, 97, 98, 99, 100, 101, 102, 103*
Choyke, W. J., 10, *33*
Clarke, J. E., 171(144), *203*
Clawson, A. R., 57(97), 77(164), *99, 101*
Clegg, B., 59(96), 76(96), *99*

Chien, W. Y., 81(184), *101*
Chinn, S. R., 36(18), *96*
Coghill, H. D., 111, *200*
Cole, K. S., 222(17), 227, *314*
Cole, R. H., 222(17), 227, 228, 293(127, 130), *314, 317*
Coleman, J. J., 66(122), *99*
Collins, D. A., 77(164), *101*
Collins, D. M., 36, 63, 64, 67(110), *96, 99, 100*
Comas, J., 66(120), *99*
Constant, E., 254, *255*
Cook, M., 293(125), *317*
Corsi, C., 113, *200*
Covington, D. W., 56(91, 92, 93), 57(93), 77(158, 160), 88, *98, 100, 102*
Cox, J. T., 132(91), 156(91), *202*
Crawley, R. L., 109(29), *200*
Cuff, K. F., 142(110), *202*

D

Dalven, R., 106(8), *199*
Das Gupta, D. K., 230(29), *315*
Das Gupta, S., 254, *255*
Davidson, D. W., 228, 236(37), 249(53), 251(53), 253, *255, 314, 315*
Davies, G., 49(67), 67(67), *98*
Davies, M., 254, *255*
Davis, E. A., 226(20), 256(20), 258, 290(20), *314*
Day, W. B., 77(160), *100*
Debye, P., 223(19), *314*
Deck, R. J., 114(53), *201*
Demske, D. L., 119(71), 165(71), 166(71), *201*
Denney, D. J., 229(25), *315*
Deori, K. L., 261(79, 81), 263(79, 81), 270(79), 272, 273(79), *316*
Dernier, P. D., 86(215), *102*
Deutsch, T. F., 198, *203*
Dev, S. B., 256(57), 257(57), *315*
DeVaux, L. H., 146, 149(116), 151, *202*
Deveaud, B., 49(71), *98*
Devlin, J., 76(156), 88(156, 225), *100, 102*
Devoldre, P., 42(43), 47(65), 48(65, 66), 86(217, 218), *97, 98, 102*

DiLorenzo, J. V., 42(46), 50(46), 75(46), 76(46, 155), 88, 91(233), *97, 100, 102, 103*
Dimmock, J. O., 106(5), 132, 143, *199, 202*
Dingle, R., 57(94), 79(176), 84, 85, 86(207), *99, 100, 101, 102*
DiSimone, D., 51(79), 54(79), 57(98), 60(98), 62(98), 76(154), *98, 99, 100*
Dissado, L. A., 305, 306(143), *317*
Dixon, J. R., 111(36), *200*
Dixon, R. W., 79(169), *101*
Dohler, G. H., 59, 67(105), 82, 83(187), *99, 101*
Donnelly, J. P., 106(7), 115(7), 130(85, 86), 131(7, 85), 171(143), *199, 201, 203*
Dove, D. B., 42, 44, *97*
Druon, C., 236(40), 244, *315*
Dryden, J. S., 253, 254, *255*
Dube, D. C., 261(76), 262, *316*
Duh, K., 111(38), 114(38), *200*
Dunn, C. N., 77(161), *100*

E

Eastman, L. F., 51(74, 79), 54(79, 90), 56(90), 57(98), 60(98), 62(98), 66(128), 70(140a), 76(154, 156, 156a), 77(162, 163), 78(166), 83(188a), 85(162), 88(156, 156a, 225), *98, 99, 100, 101, 102*
Easton, B. C., 81(185), 82(185), *101*
Eddols, D. V., 166(120), *202*
Egerton, R. F., 114(54), 153(123), *201, 202*
Ehrlick, D. J., 198(159), *203*
Elleman, A. J., 107(9), *200*
Elliott, R. J., 223, *314*
Elliott, S. R., 258(71), *316*
Ellis, B., 142(113), 143(113), *202*
Emtage, P. R., 142, 143(109), 146, 147, 148, *202*
Erdelyi, A., 214(10), *314*
Esaki, L., 36(3), 41(3, 34), 44(53), 45(53, 61), 47(53, 61), 47(3), 58(3), 67(130, 131), 68(131, 137), 70(137), 71(137), 73(130, 137, 145), 74(137, 145), 79(3, 137), 81(185a), 85(130, 185b, 197, 198), 86(202, 203, 205), 87(214), 89(3), *96, 97, 99, 100, 101, 102*

Etienne, P., 36(9), 41(35), 42(43), 45(55), 47(65), 48(65), 86(216, 218), *96, 97, 98, 102*

F

Fahrinre, T. O., 115, *201*
Fan, J. C., 81(185a), *101*
Farabaugh, E. N., 116(66), *201*
Farrow, R. C., 36, 42(6), 48(32), 49(68), 66(125, 126), 67(125), 79(6), 88, *96, 97, 98, 99*
Favenec, P. N., 79(71), *98*
Finn, D., 66(122), *99*
Fischer, A. C., 42(45), 44(49), 45(45, 49, 56, 59), 59(49), 83(188), 89(230, 231), *97, 101, 103*
Flannery, R. E., 168(141), *203*
Foster, L. M., 73(146, 146a), *100*
Foxon, C. T., 36, 64(115), 66(124), 70, 89(4), *96, 97, 99*
Foy, P. W., 75(148), 78(148), 79(148), 86(210), *100, 102*
Foyt, A. G., 106(7), 115(7), 130(85), 131(7, 85), 171(143), *199, 201, 203*
Francombe, M. H., 171(145), 183(145), *203*
Franks, A., 21(14), *33*
Freller, H., 69(139), *100*
Fröhlich, H., 288(109), *317*
Frossati, G., 249(55), 252, *315*
Frost, M. S., 222, 256(67), 257(67), 270, 271 *314, 316*
Fujü, T., 50(72), 70(72), *98*
Fujü, Y., 67(130a), *99*
Fujuki, T., 70(141), 71(141), 72(141), *100*
Fukai, F., 92(238), *103*
Fulton, R. L., 293(131), *317*
Fuoss, R. M., 229(24), *315*
Furuichi, J., 253, *255*

G.

Gale, B., 21(14), *33*
Garber, M., 132(91), 156(91), *202*
Garrett, J. P., 65(119), *99*
Garber, J. A., 230(32), *315*
Garton, C. G., 288(108), *317*

Geballe, T. H.,256(58), 257(58), *315*
Gertner, E. R., 107(18), 171(18, 144), *200, 203*
Gesi, K., *255*
Ghoshtagore, R. N., 171(145), 183(145), *203*
Gibbs, P., *255*
Gibson, A. F., 223, *314*
Gilchrist, J. le G., 241(47, 48), 243, 249(55), 252(55), 253, 254, *255, 315*
Glarum, S. H., 293(123, 126), *317*
Glauber, R. J., 302, *317*
Goebloed, J. J., 135(98), *202*
Goldstein, M., 254, *255*
Golovashkin, A. I., 137(101), *202*
Gonda, S., 42(44), 45(58), 67(130a, 133), 73(123, 144), *97, 99, 100*
Goodwin, C. A., 36(26), *97*
Gorski, D. A., 130, 158(131), 164(137), 192, 196(81), *201, 203*
Gossard, A. C., 53(89), 54(89), 55(89), 78(165), 79(174, 176), 81(178, 179), 84(189, 192, 193), 85(165, 195, 196, 199, 200, 201), 86(179, 207), *98, 101, 102*
Gough, S. R., 236(37), 249(53), 251(53), 253, *255, 315*
Grange, J. D., 68, *100*
Grant, R. J., 256(59), 257(59), *315*
Grant, R. W., 157(129), *202*
Green, M., 198, *203*
Groves, S. H., 36(18, 23), *96, 97*, 107(19), *200*
Groves, W. O., 66(122), *99*
Guestblom, B., 211(9), *314*
Guidi, R. L., 132(89, 90), *201*
Gzowski, O., 256(65), 257(65), *316*

H

Haas, L. D., 119(70), 130(87), 131(70), *201*
Haberey, F., 294(133), *317*
Hakim, M., 238(42), *315*
Hall, L. H., 132(92), *202*
Halpern, V., 290(119), *317*
Hansen, E. E., 108(24), 112(24), *200*
Harman, T. C., 36(18), *96*, 106(6, 7), 115(7), 128(78, 80), 130(85), 131(7, 85), 142, 146, 155(78), 166(6), *199, 201, 202*

Harris, J. J., 51(77), 57(77), 58(77), 59(77), 60(77), 65(77), 77(77), 88(77), *98*
Hartman, R. L., 79(169), *101*
Harvey, J. A., 66(124), *99*
Hass, G., 12(5), 14(10), 15(10), 16(10), *33*
Hasted, J. B., 253, 270(87), *255, 316*
Havriliak, S., 229(23), *315*
Hawkins, R. E., 249(53), 251(53), *315*
Hayashi, I., 58(99), 59(104), *99*
Heavens, O. S., 136(100), *202*
Heinrich, H., 130(87), 145, *201, 202*
Henry, C. H., 84(190), *101*
Herkert, R., 112(46), *200*
Herrmann, K. H., 147, *202*
Hewitt, B. S., 42(46), 50(46), 75(46), 76(46), *97*
Hicklin, W. H., 77(158), *100*
Hiehaus, W. C., 88(223), *102*
Higgins, J. A., 107(18), 171(18), *200*
Hikichi, K., 253, *255*
Hill, G. N., 77(160), *100*
Hill, R. M., 206(1), 207(1), 209(5), 251, 253, 256, 258, 259, 278(5), 290(72), 291, 305, 306(143), *314, 315, 316, 317*
Hintenberger, H., 106(1), *199*
Hirofuji, Y., 67(133), *100*
Hirose, M., 89(231), *103*
Hisatsugu, T., 70(141), 72(141), 71(141), *100*
Hiyakawa, H., 67(130a), *99*
Hiyamizu, S., 50(72), 70, 71(141), 72, *98, 100*
Hoai, T. X., 147, *202*
Hodge, I. M., 256(59), 257(59), *315*
Hoenig, R. E., 61(108), *99*
Hohnke, D. K., 108(25), 109(25, 29), 115(25, 65), 119(25), 120(25), 121(76), 163(136), 164(65, 137), 166(140), *201, 203*
Holden, W. S., 78(168), *101*
Holloway, H., 107(12, 13, 14), 108(17), 109, 114(58), 115, 119(64), 120(31, 64), 121, 128(13), 130(14, 82, 86), 131(13, 82), 135(13, 99), 142, 151(122), 155(13), 158(131), 163(136), 164(137), 166(63, 139, 140), 173, 178, 180, 184(147), 188, 189(148, 149), 190(64, 153), 191(149), 192(154), *200, 201, 202, 203*
Holnyak, N., 66(122), *99*
Honke, D. K., 36(21), 41(21), *96*
Hood, D., 81(185), 82(185), *101*
Hopfield, J. J., 295(135), *317*

AUTHOR INDEX

Horita, H., 68(138), 69(138), *100*
Hornung, J., 111(39, 40), 119(40), *200*
Houston, B. B., 114(60), *201*
Howard, W. E., 36(3), 41(3), 47(3), 58(3), 79(3), 85(198), 89(3), *96, 102*
Huber, A. M., 49, 50(70), *98*
Huber, W., 142(111), 144, 145(115), *202*
Hudock, P., 111, *200*
Hughes, D., 256(63), 257(63), *315*
Hunter, W. R., 5(2), 7(2), 13(6, 7), 14(2, 9, 10), 15(10), 16, 17(11, 12), 19(2, 11), 22(2), 23(16, 17), 24(17, 18), 25(17, 18), 26(17), 27(17, 18), 28(18), 29(18, 19), 30(2, 19), 31(2, 20), 32(11), *33, 34*
Hurley, M. D., 36(21, 22), 41(21), *96, 97,* 115(65), 121(76), 164(65), 166(140), *201, 203*
Hutley, M. C., 9(3), 10(3), 19(3), 21, 23(17), 24(17), 25(17), 26(17), 27(17), *33, 34*

I

Ikeda, M., 67(133b), *100*
Ilegems, M., 53(89), 54(89), 55(89), 57(94), 65, 66(117, 120, 121), 79(171), 86(208, 209, 210, 210a, 211, 212), 91, *98, 99, 101, 102, 103*
Ingram, M. D., 256(59), 257(59), *315*
Ishida, Y., 238(43), 239, 240, 247, *255, 315*
Itoh, A., 67(130a), *99*

J

Jantsch, W., 119(70), 130(87), 131(70), *201*
Jasper, M., 132(91), 156(91), 157(128), *202*
Jensen, J. D., 111(41, 42), 113(52), 119(42, 71), 132(41), 135(52), 153(124), 165(71), 166, 186, 187, *200, 201, 202*
Jesion, G., 189(149), 191(149), *203*
Joffrin, J., 312(144), *317*
Johari, G. P., 240(46), 246, 253, 254, *255, 315*
Johnson, F. S., 33, *34*
Johnson, M. R., 146, 148, 149(117), *202*
Johnson, W. J., 130(83), 131(83), 190(153), *201, 203*
Jones, S. J., 253, *255*
Jonscher, A. K., 206(1), 207(1, 2), 208(2), 217(13, 14), 221(16), 222, 226(21), 230(26, 27), 232(33, 34, 35), 234, 236, 238, 239, 240, 241, 244, 245, 247, 248, 249, 250, 252, 256(2, 67, 68), 257, 258, 259, 261(76, 77, 78, 79, 80, 81, 82), 262, 263, 265, 266, 267, 270, 271, 272, 273, 274(94, 95, 96), 275(54), 276, 277, 278(97, 98), 279, 280, 281(54, 103), 282, 283, 286, 290(72), 291, 292(77, 78), 296(33, 34, 35), 298, 300, 304, 305, 306(143), 310(27), *314, 315, 316, 317*
Joosten, J., 135(98), *202*
Joyce, B. A., 36, 41(36), 42(36, 39, 42), 45, 48(36), 59(36), 60(36), 61, 64(115), 66(124), 70, 89(4), *96, 97, 99*
Joyner, K., 230(29), *315*
Judaprawira, S., 57(98), 60(98), 62(98), 76(154), 83(188a), *99, 100, 101*
Juhasz, C., 153(123), *202*

K

Kaiser, S. W., 32(22), *97,* 108(25), 109(25), 115(25), 119(25), 120(25), *200*
Kamei, K., 77(162), 85(162), *101*
Kamimura, K., 76(149), *100*
Kanedo, M., 253, *255*
Kaplan, D. R., 168(141), *203*
Kasai, I., 111(39, 40), 119(40), *200*
Kasper, E., 36(30), *97*
Katama, Y., 36(28), *97*
Kawai, N. J., 86(205), *102*
Kawamura, Y., 67(133b), *100*
Keune, D. L., 66(122), *99*
Kibbel, H., 36(30), *97*
Kicinski, F., 106(3), 108(3), *199*
Kimata, M., 67(133), 68(134), *100*
Kimmerling, L. C., 52(83), *98*
Kimura, H., 146(116), 149(116), 151(116), *202*
Kinbara, A., 68(138), 69(138), *100*
King, C. N., 268(85), 269, *316*
King, R. M., 66(127), 68(135, 136), *99, 100*
Kirchner, P. D., 54(90), 56(90), *98*
Kirkwood, J. G., 229(24), *315*
Kizilyalli, H. M., 260(75), 261, *316*
Kobayashi, K. L. I., 36(28), *97*
Koller, L. R., 111, *200*

Koma, A., 44(54), 48(64), 84(194), 97, 98, 102
Komatsubara, K. F., 36(28), 97
Konig, U., 36(30), 97
Koschel, W. H., 50(73), 98
Kramer, D. A., 61(108), 99
Kramer, G., 189(150), 203
Krikorian, E., 113, 200
Kroemer, H., 81(184), 101
Kruse, P. W., 123(77), 181(77), 201
Kunzel, H., 83(188), 101
Kuvas, R. L., 77(161), 100

L

Lakatos, A. I., 256(62, 64, 69), 257(62, 64, 69), 315, 316
Lambert, V. L., 121(75), 201
Landau, L. D., 215(11), 314
Lang, D. V., 51(81), 52(82, 83), 53(89), 54, 55, 89(227), 98, 103
Lanir, M., 160, 203
Lauritzen, J. I., 289(114), 317
Lax, M., 290(116), 317
Lebrun, A., 254, 255
LeComber, P. G., 256(60), 257(60), 315
Lee, T. P., 78(168), 79(175), 80, 101
Lestrade, J. C., 253, 255
LeSueur, E. J., 274(95), 316
Levelut, A., 312(144), 317
Levine, M. A., 189(150), 203
Lifshitz, E. M., 215(11), 314
Lile, D. L., 77(164), 101
Lindley, W. T., 106(7), 115(7), 131(7), 171(143), 199, 203
Link, N. T., 36(9), 38(31), 41(35), 42(43), 45(55), 47(65), 48(65, 66), 49(71), 86(216, 217, 218), 96, 97, 98, 102
Linke, R. A., 77(159), 100
Lischka, K., 119(70), 130(87), 131(70), 142(111), 144, 145(115), 201, 202
Lisle, D., 189(152), 203
Litton, C. W., 56(93), 57(93), 58(101), 98, 99
Lloyd, J. M., 162, 203
Lockwood, A. H., 146(116), 149(116), 151(116), 202
Loewen, E. G., 14(8), 33
Logan, R. A., 52(82, 83), 79(174, 176), 81(178, 182), 86(207), 98, 101, 102

Logothetis, E. M., 107(12, 13, 14), 115(12, 61), 119(61), 121, 128(13), 130(14, 83), 131(13), 135(13), 155(13), 166(63, 139), 200, 201, 203
Longo, J. T., 107(18), 157(129), 171(18, 144), 200, 202, 203
Longshore, R., 157(128), 202
Lopez-Ottero, A., 111, 112, 114(37, 59), 119, 121, 130(87), 131(70), 145(115), 200, 201, 202
LoVecchio, P., 132(91), 156(91), 157(128), 202
Lovell, R., 215(12), 314
Lowney, J. R., 111(36), 200
Ludeke, R., 36(3, 15), 41(3), 42, 43, 44(40, 53), 45(41, 53, 61), 47(3), 48(63, 64), 58(3), 67(41, 131), 68, 70(137), 71(137), 73(137, 145), 74, 79(3, 137), 85(197, 198), 87, 89(3), 96, 97, 98, 99, 100, 102
Lum, W. Y., 50(73), 57(97), 98, 99
Luscher, P. E., 36, 96

M

McCammon, R. D., 253, 255
McCombe, B. D., 50(73), 98
McCoy, G. L., 56(93), 57(93), 98
McCrum, N. G., 235(36), 238(36), 315
Macdonald, J. R., 288(110, 111, 112), 292(121, 122), 317
McFee, J. H., 45(48), 47, 58(103), 66(60), 67, 80(103), 97, 99
McGlauchlin, L. D., 123(77), 181(77), 201
McLevige, W. V., 66(120), 99
McMahon, T. J., 119(69), 201
McQuistan, R. D., 123(77), 181(77), 201
Maekawa, S., 36(13, 14), 50(72), 70(72, 141), 71(141), 72(141), 96, 98, 100
Mahan, G. D., 295(134), 317
Mahoney, G. E., 91(233), 103
Makita, V., 73(123), 99
Makita, Y., 36(13, 14), 96
Malitson, I. H., 135(96), 202
Manasevit, H. M., 108(20), 121(20), 200
Mansingh, M., 256(66), 257(66), 316
Martin, G. M., 52(86), 56, 98
Martin, R. J., 58(103), 80(103), 99
Maruyama, S., 44(51), 74, 97, 100
Mason, P. R., 253, 260(75), 261, 255, 316

Massies, J., 36, 41(35), 42(43), 45(55), 47, 48, 86, *96*, *97*, *98*, *102*
Mathur, D. P., 131(88), 168, *201*
Matsunaga, N., 64(114), *99*
Matsuo, M., 239(44), 247(44), 254, *255*, *315*
Matsuoka, S., 238(43), 254, *255*, *315*
Matsushima, Y., 42(44), 45(58), 67, 68(134), 73, *97*, *99*, *100*
Maystre, D., 21(15), 29(19), 30(19), *34*
Meakins, R. J., 236(38), 253, 254, *255*, *315*
Meca, F., 249(54), 252, 277, 275(54), 281(54, 103), 282(103), 283(103), *315*, *316*
Meeks, E. L., 56(91, 92), 77, 88, *98*, *100*, *102*
Meggett, B. T., 66(127), 68, *99*, *100*
Melngailes, I., 106(5, 6), 142, 146, 166(6), *199*, *202*
Mertz, J. L., 79(172, 174), 81(172, 178, 179, 182), 84(189, 193), 86(179), *101*, *102*
Messick, L., 77(164), *101*
Metze, G. M., 78(166), 83(166), *99*, *101*
Meyers, J. H., 114(53), *201*
Michels, D. J., 17(11), 18, 19(11, 13), 32(11), *33*
Mikes, T. L., 14(10), 15(10), 16(10), 17(11), 18(11), 19(11), 32(11), *33*
Mikulyak, R. M., 81(183), 86(210, 210a), *101*, *102*
Millany, M. H., 281(103), 282(103), 283(103), *316*
Miller, B I., 45(48), 47, 58(103), 66(60), 67, 78(167), 80(103), *97*, *99*, *101*
Miller, R. C., 79(176), 86(207), *101*, *102*
Mircea, A., 52(84, 86), 56(86), *98*
Misawa, S., 67(130a), *99*
Mitoff, S. P., 271(92), *316*
Mitsui, S., 51(76, 78), 52(78), 60(76), 76(78, 152), *98*, *100*
Mitsui, Y., 76(152), *100*
Mittonneau, A., 52(84, 86), 56(86), *98*
Moali, J., 261(79), 263(79), 270(79), 273(79), *316*
Moore, E. J., 290(118), *317*
Moore, L., 253, *255*
Morillot, G., 49(71), *98*
Morkoc, H., 51(75), 57(75), 59(75), *98*
Morris, B., 249(53), 251(53), *315*
Morris, H. B., 132(89, 90), *201*
Morriss, R. H., 114(53), *201*
Morys, P., 290(117), *317*

Moss, T. S., 142(113), 143(113), *202*
Mott, N. F., 226(20), 256(20), 258, 290(20, 120), *314*, *317*
Mukai, S., 67(133), 73(123), *99*, *100*
Munir, Z. A., 108(24), 112(24), *200*
Murawski, L., 256(65), 257(65), *316*
Murotani, T., 51(76, 78), 52, 60, 76, *98*
Muruyama, S., 73(146), *100*
Muth, E. P., 116(66), *201*
Myoshi, Y., 36(14), *96*

N

Naganuma, M., 44(50), 64, 65(113), 70(140), 73(140), 76, *97*, *99*, *100*
Nagata, S., 91(237), 92(237, 238), *103*
Nakatawi, M., 76(152), *100*
Namioka, T., 23(16), 24(18), 25(18), 28(18), 29(18), *34*
Nanbu, K., 50(72), 70(72, 141), 71(141), 72(141), *98*, *100*
Neave, J. H., 42(39, 42), *97*
Nee, T. W., 293(128), *317*
Negami, S., 229(23), *315*
Nelson, R. J., Jr., 66(122), *99*
Neviere, M., 14(8, 9), 29, 30(19), *33*, *34*
Ngai, K. L., 285(140), 304, 305, *317*
Nicollian, E. H., 51(80, 81), 54(80), 89(227), *98*, *103*
Niehaus, W. C., 42(46), 50(46), 75(46), 76(46, 155), *97*, *100*
Nill, K. W., 107(19), 128, 129, 155, 175, *200*, *201*
Nishimoto, C. K., *100*
Nogami, M., 68(134), *100*
Nordland, W. A., 79(176), *101*
Nordland, W. A., Jr., 86(207), *102*
Noreland, E., 211(9), *314*
Norris, M. T., 67(133a), *100*
North, A. M., 256(57), 257(57), *315*
Northrop, D. A., 108, *200*
Noreika, A. J., 171, 183, *203*
Noyce, R. N., 140(103), *202*

O

Ogawa, S., 44(51), 73(146), 74(147), *97*, *100*
Ohno, H., 49(67), 67(67), 70, *98*, *100*

Okamoto, H., 67(133b), 80(177), *100, 101*
Osgood, R. M., 198(159), *203*
Ota, Y., 36, *97*

P

Paclaczyk, C., 236(39), 237, *315*
Pagel, B. R., 150(121), *202*
Palmer, E. W., 21, *33*
Palmetshofer, L., 130(87), *201*
Pandy, R. K., 112, *200*
Pannish, M B., 66(121), *99*
Parker, E. H. C., 36(24, 25), 66(127), 68(135, 136), *97, 98, 100*
Parnieux, J. P., 254, *255*
Partlow, W. D., 10(4), *33*
Pasko, J. G., 107(18), 157(129), 171(18), *200, 202*
Pechhold, W., 289(113), *317*
Pethig, R., 256(63), 257(63), *315*
Petritz, R. L., 150(121), *202*
Petrocco, G., 113(51), *200*
Petroff, P. M., 85(195, 196), *102*
Pfeiffer, H., 112(46), *200*
Phillips, W. A., 241(49), *315*
Piccioli, N., 132(94), 133, 142(94, 114), *202*
Pickhardt, V. Y., 36(11, 12, 19, 20), 59(11), *96*, 109, 116(30), 119(30), *200*
Pike, G. E., 295(136), *317*
Platzman, P. M., 297(137), *317*
Plew, L., 66(120), 82, *99, 101*
Ploog, K., 42(45), 44(49), 45(45, 49, 56, 59), 59, 67(105), 82, 83(187, 188), 89(230, 231), *97, 99, 101, 103*
Polanco, J. I., 256(61), 257(61), *315*
Pollak, M., 256(58), 257(58), 290(115), 295(136), *315, 317*
Porter, R. F., 108(21), *200*
Preier, H., 111(38), 112(46), 114(38), 146, 147, 148, 149, *200, 202*
Prinz, D. K., 17(12), 18, *33*

R

Raisch, F., 45(59), *97*
Rathbun, L., 70(140a), *100*
Read, B. E., 235(36), 238(36), *315*
Read, W. T., 142, *202*

Reau, J. M., 261(79, 80), 263(79), 270(79), 273(79), *316*
Reeves, J. M., 256(66), 257(66), *316*
Renda, F. J., 146(116), 149(116), 151(116), *202*
Reinhart, F. K., 77(157), 79(173), 80, 81, *100, 101*
Reynolds, D. C., 56(93), 57(93), *98*
Rideout, V. L., 87(213), *102*
Roberts, G. C., 256(61), 257(61), *315*
Roberts, J. S., 65(118), 66(118), 70(142), *99, 100*
Rolls, W. H., 166(120), *202*
Rosenblatt, C. J., 256(66), 257(66), *316*
Ross Macdonald, J., 230(31, 32), *315*
Rudenko, S. I., 209(4), 211(4), *314*
Rudolph, D., 23(17), 24(17), 25(17), 26(17), 27(17), *34*
Rupp, L W., 57(94), *99*

S

Saba, R. G., 253, *255*
Sah, C. T., 140(103), *202*
Sai-Halasz, G. A., 68(137), 70(137), 71(137), 73(137), 74(137), 79(137), 86(202, 205), *100, 102*
Saiki, K., 268(86), 269(86), *316*
Saito, K., 253, *255*
Sakaki, H., 68(137), 70, 71, 73(137), 74, 79(137), *100*
Saki, Y., 76(149), *100*
Sapieha, S., 278(100), *316*
Savage, A., 85(195, 196), *102*
Sayer, M., 256(66), 257(66), *316*
Scanlon, W. W., 120, 121(72), *201*
Scher, H., 256(64), 257(64), 290(116), *315, 317*
Schlosser, W. O., 42(46), 50(46), 75(46), 76(46), *97*
Schmahl, G., 23(17), 24(17), 25(17), 26(17), 27(17), *34*
Schneider, M. V., 77(159), *100*
Schoolar, R. B., 107, 108(10), 111(41), 113(52), 114(11, 55, 56, 57), 119(71), 128, 131(56), 132, 135(52), 153(124), 160, 165, 166, 186, 187, *200, 201, 202, 203*

Schroeder, W. E., 77(161), *100*
Schul, G., 36(3), 41(3), 47(3), 58(3), 79(3), 89(3), *96*
Scott, G. B., 70, *100*
Segmuller, A., 67(130), 71, 73(130), 85(130), *99, 100*
Segui, Y., 278(97), *316*
Sergent, A. M., 81(182), *101*
Shahidi, M., 270(87, 88), 272, *316*
Sharif, R. I., 260, *316*
Shaw, D. W., 76(153), *100*
Sheets, M J., 146(116), 149(116), 151(116), *202*
Shimano, T., 51(76, 78), 52(78), 60(76), 76(78, 152), *98, 100*
Shimanoe, T., 58(100), 66(100), *99*
Shiraki, Y., 36, *97*
Shockley, W., 140(102, 103), 142, 143, *202*
Simmons, J. G., 218(15), *314*
Simpson, O., 106(2), 108(2), *199*
Simpson, W. I., 108(20), 121(20), *200*
Skalski, J. F., 189(151), *203*
Smith, D. L., 36, 59, 63, *96, 99,* 109, 116(30), 119(30), *200*
Smith, R. A., 135(97), *202*
Snow, E. H., *255*
Somekh, S., 66(121), 79(171), *99, 101*
Spear, W. E., 256(60), 257(60), *315*
Sosnowski, L., 106(2), 108(2), *199*
Stall, R. A., 51(79), 54(79, 90), 56(90), 76(156, 156a), 88(156, 156a, 225), *98, 99, 100, 102*
Stanley, C. R., 67(133a), *100*
Starkiewicz, J., 106(2), 108(2), *199*
Stern, F., 142(112), *202*
Stillman, G. E., 61(107), *99,* 121(74), *201*
Stoll, B., 289(113), *317*
Stormer, H. L., 78(165), 85(165, 199, 200, 201), *101, 102*
Strauss, A. J., 106(5), 107(19), 108(22, 23), *199, 200*
Streete, J. L., 162(134), *203*
Streetman, B., 66(120), *99*
Strom, U., 256(70), 257(70), *316*
Stuart, P. R., 23(17), 24(17), 25(17), 26(17), 27(17), *34*
Summer, B., 157(128), *202*
Sun, T. S., 153, 155(127), 198(157), 199(127), *202, 203*

Sunier, J. W., 130(84), 131(84), *201*
Supertzi, E. P., 10(4), *33*
Susuki, T., 64(114), *99*
Swain, J., 254, *255*
Sze, S. M., 141(104), *202*
Szundi, I., 271(89), 274, *316*

T

Taiedy, F., 256(68), 257(68), 279(101), *316*
Takada, S., 67(130a), *99*
Takahashi, K., 44(50), 64(112, 113, 114), 65(113), 70(140), 73(140), 76(149), *97, 99, 100*
Takayanagi, M., 254, *255, 315*
Takei, W. J., 171(145), 183(145), *203*
Tanaka, T., 91(237), 92(237, 238), *103*
Tao, T. F., 113, 130(84), 131(84), *200, 201*
Tarucha, S., 268(86), 269(86), *316*
Tateishi, K., 44(50), 70, 73(140), *97, 100*
Taylor, G. W., 218(15), *314*
Taylor, H. E., 238(41), 242, *315*
Taylor, P. C., 256(70), 257(70), *316*
Thomas, R. A., 268(85), 269, *316*
Thorp, J. S., 260, *316*
Tien, D. K., 58(103), 80(103), *99*
Timothy, J. G., 24(18), 25(18), 27(18), 28(18), 29(18), *34*
Toulouse, B., 49(71), *98*
Tousey, R., 13(7), 33(21), *33, 34*
Trommer, R., 89(230), *103*
Tsang, W. T., 67(129), 79, 88, 89, 91, *99, 101, 102, 103*
Tsu, R., 81(185a), 82(185b), 86(202), *101, 102*
Turner, G. W., 81(185a), *101*
Turney, A., 253, *255*

U

Ueda, R. M., 36(16), *96*
Uhlmann, D. R., 238(42), *315*

V

Vaidyanathan, K. V., 66(120), *99*
Vanderwyck, A. H. B., 160(133), *203*

Van der Ziel, J. P., 79(176), 81(183), 86(208, 209, 210, 210a, 211, 212), *101, 102*
van Gemert, M. J. C., 230(30), *315*
van Roosbroeck, W., 142(107), 143, *202*
Varga, A. J., 107(13), 128(13), 130(83), 131(13, 83), 135(13), 155(13), 158(131), 190(153), *200, 201, 203*
Venskytis, F. J., 10(4), *33*
Verrill, J. F., 21(14), *33*
Vilms, J., 65(119), *99*
Volger, J., 294(132), *317*
von Schweidler, E., 230, *315*

W

Wacrenier, J. M., 236(40), 244, *315*
Wada, Y., 268(86), 269(86), *316*
Wadsley, A. D., *255*
Wagenhuber, M., 130(87), *201*
Waho, T., 44(51), 73(146), 74(147), *97, 100*
Walpole, J. N., 36, *96, 97,* 107, 108, 109, 128, 129, 155(78), 175, *200, 201*
Wang, C. C., 113, 130(84), 131(84), 160(133), *200, 201, 203*
Warburg, E., 271, 292(90, 91), *316*
Washwell, E. R., 142(110), *202*
Watanabe, M., 254, *255*
Wataze, M., 76(152), *100*
Watts, D. C., 256(57), 257(57), 293(125), *315, 317*
Weber, W. H., 107(14), 130(14), *200*
Wentworth, E., 111(47), *200*
Wentz, J. L., 171(145), 183(145), *203*
West, A. R., 256(59), 257(59), *315*
White, C. T., 304, 305(140), *317*
Wieder, H. H., 50(73), 57(97), 89(226), *98, 99, 102*
Wiegmann, W., 53(89), 54(89), 55(89), 78(165), 79(174, 176), 81(178, 179), 84(190, 192, 193, 193a), 85(165, 195, 196, 199, 200, 201), 86(179, 207), *98, 101, 102*
Wijn, H. P. J., 294(133), *317*
Wilkes, E., 107(12, 13), 109(29), 115(12), 121(12), 128(13), 131(13), 135(13), 155(13), *200*

Williams, D., 36, *96, 97*
Williams, G., 235(36), 238(36), 254, *255,* 256(57), 257(57), 293(125, 129), *315, 317*
Williams, R. E., 76(153), *100*
Wilman, H., 107(9), *200*
Wintle, H. J., 209(3), 278(3, 99, 100), *314, 316*
Wolfe, C. M., 61(107), *99,* 121(74), *201*
Wolff, P. A., 297(137), *317*
Wood, C. E. C., 41(36), 42(36), 45, 48(36), 49(67, 69), 50, 51(74, 77, 79), 52(85, 87), 54(79, 90), 56(90), 57(77, 98), 58(77), 59, 60, 61, 62, 63(106), 65(77, 118), 66(87, 118), 70(140a), 76(96, 154, 156, 156a), 77, 78(166), 81(185), 82, 83(166, 188a), 88(69, 77, 156, 156a, 220, 225), *97, 98, 99, 100, 101, 102*
Woodard, D. W., 51(74), *98*
Woodcock, J., 51(77), 57(77), 58(77), 59(77), 60(77), 65(77), 77(77), 88(77), *98*
Work, R. N., 253, *255*
Wright, P. D., 66(122), *99*
Wrobel, J. S., 121(73), 146(117), 148(117), 149(117), 169(142), *201, 202, 203*

Y

Yamafuji, K., 239(44), 247(44), 254, *255, 315*
Yano, M., 68, *100*
Yano, O., 268(86), 269, *316*
Yao, T., 36(13, 14), *96*
Yariv, A., 81(180), 86(180), *101, 102*
Yeh, P., 81(180), 86, *101, 102*
Yeung, K. F., 151(122), 158(131), 164(137), 166(140), *202, 203*
Yoshida, S., 67(130a), *99*

Z

Zamel, J. N., 107, 108(10), 109, 113, 114(15), 115, 135(52), *200, 201*
Zappella, P. I., 192(154, 156), *203*
Zwanzig, R., 293(128), *317*

Subject Index

A

AC conduction, theory of, 206
AC conductivity
  vs. DC conductivity, 258
  in dielectric response, 217
  frequency dependence in, 290
  for various materials, 257
AES, *see* Auger electron spectroscopy
Alkali halides, IV–VI layers in, 113–115
Alumina, electrical conductivity for, 261
Aluminum evaporation, on arsenic-rich GaAs surfaces, 87
Aluminum-surfaced diffraction gratings, coating for, 13
Antimony content, in films for MBE fluxes, 73–74
Arrhenius plots, of Pb barrier PbTe photodiodes, 160–161
Auger electron spectroscopy, 44–45

B

Background noise limited detection, 124–126, 158
Band filling, quantum efficiency and, 139
Barium fluoride substrates, IV–VI semiconductor layers and, 115–122, 198
Beryllium, in MBE doping, 65–66
Blazed concave holographic grating, efficiency maps of, 24
BNL detection, *see* Background noise limited detection
Bound carriers, 226
Bound charges, vs. free, 313–314

C

Carbon, as MBE impurity, 57–58
Chalcogen deficiency, n-type conductivity and, 107
Charge carriers, DC conductivity of, 231–232
Charge hopping, frequency dependence in, 291
  see also Hopping electronic systems
Charge injection, in solids, 278–279
Charging currents, logarithmic representation of, 276
Chlorobenzene-cis-Decalin system, loss peak for, 246
Circular collectors, efficiency of, 181–182
Coatings, reflectance, *see* Reflectance coatings
Cole–Cole diagram, 222, 226–228
Complex dielectric susceptibility, formal expression for, 288
Complex impedance diagrams, 218–220, 244
Complex impedance plot, representation of, 222
Complex permittivity, frequency dependence of, 220
Complex permittivity diagrams, for triglycine sulfate, 237
Correlated states, in many-body interactions, 304
Correlation junctions, in dielectric response, 293–294
Curie law, 225
Curie–von Schweidler law, 230, 232, 275, 295, 302

D

Dark discharge, boundary and initial conditions applicable to, 281
DC conductivity
  vs. AC conductivity, 258
  charge carriers and, 231–232
  frequency dependence in, 271
  theory of, 206
Debye mechanisms, 223

Debye peaks, distribution of, 288–289
Debye relaxation time, 224, 299
Debye response
  broadened, 236–237
  dipole relaxation and, 293
  ideal, 219, 234
  loss peak frequency and, 236
  polarization decay and, 224
  pure, 227
Debye system, ideal, 225
Delta function, in dielectric response, 210
Detector (D*)
  BNL (background noise limited), 124–128, 148–149, 158–159, 172–173
  JNL (Johnson noise limited), 124–128, 148–149, 158–159, 172–173
Detector noise, in thin-film IV–VI semiconductor photodiodes, 149–152
DHLs, see Double heterojunction lasers
Dielectric data, methods of presentation of, 217–223
Dielectric induction, defined, 210
Dielectric lattice, 232–233, 259
Dielectric lattice response, 267–270
Dielectric loss
  frequency dependence on, 269
  for hollandite, 263
  for polyethylene, 268–269
  for stearic acid films, 265–267
  temperature dependence in, 268–269
  for triglycine sulfate, 262
Dielectric loss characteristics, empirical classification of, 226–235
Dielectric loss peaks, 229
  for ceramics, 260
  general analysis of, 251–256
  in polydian carbonate, 238
  for polyethylene, 243
  for polyethylene terephthalate, 247
  for polymeric solids, 250
  for polymethyl methacrylate, 248
  for poly-n-butyl methacrylate, 240
  for polyvinyl acetate, 239
Dielectric loss spectrum, for silica glass, 252
Dielectric matrix, 232–233
Dielectric permittivity
  defined, 213
  of free space, 210
  high-frequency, 216

Dielectric polarization, disorder in, 306–307
Dielectric properties, of various materials, 253–255
Dielectric relaxation
  energy relations in, 311–312
  time-dependence in, 309–311
Dielectric response
  charge carriers and, 285
  correlation function approach to, 293–294
  defined, 206–207, 209–210
  diffusive models in, 292–293
  frequency dependence of, 207, 245
  Ising model of, 302–303
  many-body interpretation in, 296–312
  Maxwell–Wagner effects in, 294–295
  in optical region, 207–208
  Ngai's infrared divergence model in, 303–304
  screened hopping model in, 296–301
  superposition principle in, 211
  temperature dependence in, 233–235
  time dependence in, 206–207
  "true," 219
  two-level systems in, 305–312
  types of, 284–286
  universal, see Universal dielectric response
  universal law of, 229–230
Dielectric response function, 210
Dielectrics, time domain response and, 272–277
Dielectric susceptibility
  defined, 213
  formal expression for, 288
  glancing incidence reflection electron type, 42–44
Diffraction gratings
  aluminum coatings for, 3, 13
  aluminum–gold interdiffusion in, 14–17
  blazed concave holographic, 23
  characteristics of, 13–14
  in conical diffraction mountings, 28–30
  echelle, 3
  efficiency maps for, 18–25
  efficiency vs. wavelength in, 25–29
  grating anomalies in, 19–21
  groove efficiency curve for, 26

holographic, 9–11, 23
intermetallic diffusion in, 14–17
master conventional, 11
monopartite ruling for, 19
nonuniform surface of, 19–20
overcoating of, 12
reflectance coatings for, 1–33
replica conventional, 11–12
roughness in, 8
ruling of, 2–3, 19
ruling errors in, 6
stray light measurements for, 30–33
target pattern in, 21
tripartite ruling for, 19
wavelength vs. efficiency in, 25–29
Diffraction grating efficiency
in conical diffraction mountings, 28–30
improvement in, 13–14
measurement of, 17–33
uniformity of, 17
wavelength and, 25–29
Dipolar β peaks, 246–251
Dipolar materials, dielectric properties of, 253–255
Discharging currents
logarithmic representation of, 276
time domain response of, 230–231
Dissado–Hill analysis, of two-level dielectric response systems, 305–312
Distribution of relaxation times, dielectric susceptibility and, 288
Dopants, N-type, 57–63
Doping, of MBEs, 52–66
Double heterojunction lasers
GaAlAs heterojunctions in, 78–79
minority traps and, 53
DRT, see Distribution of relaxation times

E

Echelle gratings, coating thickness for, 3
see also Diffraction gratings
Efficiency maps
of blazed concave holographic gratings, 24
for other types of diffraction gratings, 18–25

Electron traps
DLTS spectra of, 54–55
in MBE, 52–56
Epitaxy, molecular beam, see Molecular beam epitaxy

F

FETs, see Field effect transistors
Field effect transistors
doping profile in, 75–76
high-power, 75
low-noise, 75–76
MBE substrate and, 49–51
power layers of, 76
Flash evaporation, of IV–VI pseudobinary alloys, 112–113
Focused stray light
in diffraction gratings, 31–32
in vacuum ultraviolet, 32
Free carriers, 225–226
Free charges, vs. bound, 313–314
Frequency, lateral shift of, 233–235
Frequency dependency, dielectric response to, 207–208, 269
Frequency dispersion, 270–272
Frequency spectrum, dielectric response and, 214
FSL, see Focused stray light

G

Gallium–aluminum–arsenide alloys, in MBE, 70–72
Gallium arsenide, in MBE doping, 52–69
Gallium arsenide antimony alloys, in MBE, 72–73
Gallium arsenide field effect transistors, 75
Gallium arsenide phosphorus alloys, in MBE, 72–73
Germanium, as dopant in MBE, 58–59, 65
Glancing incidence reflection electron diffraction, 42–44
Gold coatings
groove density and, 5
holes or voids in, 5–7
Gold film, for reflectance gratings, 3–4

Grating anomalies, in diffraction gratings, 21
  see also Diffraction gratings
Groove efficiency curve, for diffraction gratings, 26
Groups III–V binaries, in MBE doping, 66–69
Group IV–VI compounds, surface effects in, 153–157
Group IV–VI layers
  in alkali halides, 113–115
  in insulating substrates, 113–122
  vacuum deposition on, 108–113
Group IV–VI semiconductor photodiodes, 127–157, see also Thin-film IV–VI semiconductor diodes
  maximum blackbody photon flux of, 127
Group IV–VI semiconductors
  on BaF_2 and SrF_2 substrates, 115–122
  epitaxial growth of, 108
  growth of on alkali halide substrates, 107
  infrared detectors and, 106
  quantum efficiencies of, 132–140
  static dielectric constants of, 168
  surface effects in, 153–157
  thin films of, 105–128
Gunn devices, millimeter-wave, 77

H

Hall coefficient, temperature dependence of, 122
Hall mobility, temperature dependence of, 122
HEED, see High-energy electron diffraction
Heterojunction structures, MBE in, 84–86
High-energy electron diffraction, 42
High vacuum, coating deposition in, 2
Hollandites
  dielectric loss for, 261–264
  strong dispersion in, 270
Holographic gratings, 9–11
  blazed concave, 23
  energy distribution in, 23
  supporting blank and, 10
Hopping-charge carriers, 259
Hopping electronic systems, 290–291

Hopping model, screened, 296–302
Hopping transition, energy loss and storage in, 299
Hot-well vacuum deposition technique, 111–112
Hydrogen-background pressure, in MBE, 58

I

Ideal Debye response, 234, see also Debye response
Ideal Debye system, 275
Ideal photodiode, quantum efficiency of, 125
IMPATT diode devices, 77
Infrared photodiodes, properties of, 122
Infrared radiation, IV–VI compound semiconductors in detection of, 106–107
Insulator–semiconductor systems, MBE and, 89
Integral interference filter, in thin-film photodiodes, 188–189
Integrated optical systems, MBE and, 80
Intermetallic diffusion, prevention of, 16
Internal field, concept of, 313
Ionic conductors, 292
Ionic polarization, 223
Ising model, in dielectric response, 302–303
Isoelectronic doping, in MBE, 66

J

JFET, see Junction field effect transistors
Johnson noise, in thin-film IV–VI semiconductor photodiodes, 150–151, 165
Johnson noise limited detector, 124–128, 148–149, 158–159, 172–173
Junction capacitance, reduction of, 168–171
Junction field effect transistor, 190

K

Kramers–Kronig relations, 214, 232, 245, 267, 270

L

Lateral collection photodiodes, 177–185
  in bulk crystals, 183–184
  collector geometries for, 178
  with integral interference filter, 188
  potential barrier in, 178
Lead, as dopant in MBE, 60–61
Lead barrier thin-film PbTe devices, 130–132
  earliest work with, 135–136
Lead chalcogenides, 62, 116
Lead telluride
  Auger lifetimes of, 143, 146
  detectivity for, 146
  diffusion-limited behavior in, 158–159
  on mica substrates, 153
  parlodian replica of layer of, 117–120
  RLL quantum efficiency for, 137–138
Lead telluride devices
  Arrhenius slopes of, 160–161
  heat treatment for, 155
Lead telluride photodiode array, configuration of, 195–196
Lead telluride phototransistors, 189–191
Lead telluride thin-film detectors, 130
LEDs, see Light-emitting diodes
LEED, see Low-energy electron diffraction
Light-emitting diodes
  high-intensity, 78
  minority traps in, 53
Liquid-phase epitaxy, in PbTe substrates, 171
Localized carriers, 226
Low-energy electron diffraction, 45–46

M

Magnesium, in MBE doping, 64–65
Majority carrier devices, 75–77
Manganese, in MBE doping, 65
Many-body phenomena, in dielectric responses, 296–312
Master conventional gratings, 11–12, see also Diffraction gratings
Maxwell–Wagner mechanism, 272
Maxwell–Wagner–Sillars effect, 294
Maxwell–Wagner systems, 294
MBE, see Molecular beam epitaxy
Mean free time between collisions, defined, 226
Medium-energy electron diffraction, 42
Metal-barrier devices, surface effects and nature of, 153–157
Metal deficiency, p-type conductivity and, 107
Metal semiconductor systems, MBE in, 86–88
Millimeter-wave Gunn devices, 77
Minority carriers
  applications of, 78
  lateral collection of, 179–180
Minority traps, MBE and, 53
MISFETs, 77–78
Molecular beam epitaxy, 75–86
  advantages of, 95–96
  beryllium as dopant in, 65–66
  buffer layers in, 49–52
  carbon as impurity in, 57–58
  computer-controlled, 41
  diagram of typical system in, 37
  doping of, 52–66
  electron traps in, 52–56
  FETs and, 49–51
  in future majority carrier device applications, 77–78
  gallium arsenide dopants in, 52–69
  germanium as atmospheric dopant in, 58–59, 65
  Group III–V binaries other than GaAs in, 66–69
  Group IV elements as dopants in, 57–61
  Group VI elements as dopants in, 61–63
  growth stages in, 48–52
  heterojunction structures and, 84–86
  hydrogen background pressure in, 58
  impurities in, 95
  in situ assessment techniques in, 40–48
  in insulator–semiconductor systems, 89
  isoelectronic doping in, 66
  intentional doping control in, 53
  lead as dopant in, 60–61
  liquid nitrogen-cooled shroud for, 38
  magnesium as dopant in, 64–65
  majority carrier devices in, 75–77, 109
  manganese dopant in, 65

in metal semiconductor systems, 86–88
minority carrier devices and, 78–81
in multistage structures, 82–84
N-type dopants in, 57–63
periodic structures in, 86
problems and applications in, 35–96
projections for, 92–95
P-type dopants in, 63–66
quadrupole mass spectrometry in, 41
quaternary III–V layer growth in, 74–75
reflection electron diffraction and, 42–44
research in, 93
in selected-area epitaxy, 89–92
silicon as dopant in, 58
source assemblies in, 38
substrate preparation in, 48–49
surface analytical facilities for, 40
system design in, 36–40
technological developments in, 86–92
ternary III–V compound alloys in, 69–75
tin as dopant in, 59–60
unintentional impurities in, 95
and vacuum deposition on IV–VI layers, 108–110
Molecular beam evaporation, of IV–VI semiconductors, 110
Monopartite ruling, for diffraction gratings, 19
MOSFETs, high-dielectric-constant films for, 89
Mother Nature, in time domain, 212, 274
Multilayer structures, MBE in, 82–84

N

Ngai infrared divergence model, 303–304
Non-Debye loss peak, 288
Non-Debye response, 234–235
N-type dopants, in MBE, 57–63

O

Ohmic contacts, preparation of, 88
Oxidized aluminum, reflecting properties of, 13

P

Photoconductors, types of, 106
Photocurrent, amplification of, 189
Photodiodes, see also Thin-film IV–VI semiconductor photodiodes
  high-performance Pb barrier PbTe type, 130
  ideal, 125
  junction capacitance of, 168–172
  lateral-collection, 177–185
  pinched-off, 172–177
  self-filtered narrow-band, 186–189
  theoretical D^* of, 124
  time response to, 279–281
Photorelaxation, boundary and initial conditions applicable to, 281
Photoresist, for holographic gratings, 9–10
Phototransistors, 189–191
Pinched-off photodiodes, 172–177
p–n junction, dielectric loss in, 244
Polarization
  basic mechanisms of, 223–226
  contribution of charge carriers to, 281–283
  dielectric, 306–307
Polarization increment, 215
Polydian carbonate, dielectric loss peaks in, 238
Polyvinyl acetate, dielectric loss peaks for, 239
Pseudobinary alloys, band crossings in, 106

Q

Quadrupole mass spectrometry, 41
Quantum efficiency
  band filling and, 139
  of IV–VI semiconductors, 132–140

R

RED, see Reflection electron diffraction
Reflectance coatings, see also Dopants; Gold coatings
  adherence of, 9

aluminum in, 3
aluminum–gold interdiffusion in, 14–17
cleaning of, 9–12
for diffraction gratings, 1–33
gold as, 3
grating performance and, 12–14
intermetallic diffusion in, 14–17
overcoating and, 12
techniques for, 9–17
thickness of, 3
Reflection electron diffraction
applications of, 42
glancing incidence, 42–44
patterns of, 43
scanning high energy, 44–48
Reflection loss limit, 135–137
Reflection loss limit quantum efficiency
for lead telluride, 137–138
for strip p–n junctions, 178–179
Relaxation process, time dependence of, 309–311
Relaxation times, distribution of, 288–289
Replica conventional gratings, 11–12
RLL, see Reflection loss limit
Rock-salt-structured IV–VI semiconductors, on fluorite-structured substrates, 121

S

Scanning high-energy reflection electron diffraction, 44–48
Schottky barrier, series capacitor and, 219
Schottky barrier hyperabrupt varactors, 77
Screened hopping model
on dielectric surface, 296–302
loss peak in, 301–302
Secondary ion mass spectrometry, 45, 49
Selected-area epitaxy
MBE and, 89–92
shadow technique for, 91
Self-filtered narrow-band photodiodes, 186–188
SHEED, see Scanning high-energy reflection electron diffraction
Shockley–Read recombination mechanisms, 147
Silicon, as dopant in MBE, 58

SIMS, see Secondary ion mass spectrometry
Solar heat collectors, intermaterial diffusion in, 16
Solids, charge injection into, 278–279
Stearic acid films, capacitance and dielectric loss in, 265–267
Stray light measurements, for diffraction gratings, 30–33
Stripe collectors, collection efficiency of, 180
Strong low-frequency dispersion, in ionic conductors, 270–272
Strontium fluoride, IV–VI semiconductor layers on, 115–122
Submicron structures, growth techniques for, 94
Surface analytic features, in molecular beam epitaxy, 40

T

Ternary III–V alloys, in MBE, 69–75
Thin-film devices, unconventional, 168–199
Thin-film IV–VI photodiodes, 128–157, see also Photodiodes
first report of, 128
response of, 163
temperature-dependent properties of, 158
noise in, 150
Thin-film PbTe detectors, 130
Thin-film PbTe photodiodes
configuration of, 195–197
temperature-dependent noise properties of, 151
thermal image from, 197
Thin-film IV–VI semiconductor photodiodes
detector noise in, 149–153
for 8–12 μm operation, 166–168
flip-chip mounting for, 192
with integral interference filter, 188
properties of at 80 K, 167
resistance estimates for, 140–149
surface effects in, 153–157
for 3–5 μm operation, 157–165

Thin film lateral-collection photodiodes, for intermediate-temperature operation, 181–182
Thin-film photodiode arrays, 191–199, *see also* Thin-film IV–VI devices
  configuration of, 194–197
  schematic diagram of, 193
Three-dimensional waveguide structure, preparation of, 81
Time domain
  Mother Nature in, 212, 274
  response of dielectrics in, 230–231, 272–277
Tin, as dopant in MBE, 59–60
Traps, electron, 52–56
Trap-tunneling mechanism, in MBE, 53
Triglycine sulfate
  complex permittivity diagrams for, 237
  dielectric loss for, 262
Tripartite ruling, for diffraction gratings, 19
Two-level system, Dissado–Hill model of, 305–312

U

Unintentional impurities, in MBE, 95
Universal dielectric response, 205–314
  current interpretations of, 287–295
  defined, 229–230
  experimental evidence for, 283–286
  historical background of, 287–288
  local field in, 287
  without loss peaks in carrier systems, 256–257
Universal power law, 232

V

Vacuum deposition, hot-wall technique of, 111–112
Vacuum ultraviolet
  focused stray light in, 32
  reflectance coatings for diffraction gratings in, 1–33
  spectroscopy in, 17
Vacuum ultraviolet monochromator, 17
Varactors, Schottky barrier hyperabrupt, 77
Valence polarization, 223
VUV, *see* Vacuum ultraviolet

W

Warburg mechanism, 271, 292
Wavelength
  diffraction grating efficiency and, 25–29
  maximum groove efficiency and, 26